Modern Computational Finance

Modern Computational Finance

AAD and Parallel Simulations

with professional implementation in C++

ANTOINE SAVINE

Preface by Leif Andersen

WILEY

Published by John Wiley & Sons, Inc., Hoboken, New Jersey.

Published simultaneously in Canada.

Library of Congress Cataloging-in-Publication Data

Names: Savine, Antoine, 1970- author.
Title: Modern computational finance : AAD and parallel simulations / Antoine Savine.
Description: Hoboken, New Jersey : John Wiley & Sons, Inc., [2019] | Includes bibliographical references and index. |
Identifiers: LCCN 2018041510 (print) | LCCN 2018042197 (ebook) | ISBN 9781119539544 (Adobe PDF) | ISBN 9781119539520 (ePub) | ISBN 9781119539452 (hardcover)
Subjects: LCSH: Finance—Mathematical models. | Finance—Computer simulation. | Automatic differentiation.
Classification: LCC HG106 (ebook) | LCC HG106 .S28 2019 (print) | DDC 332.01/5195—dc23
LC record available at https://lccn.loc.gov/2018041510

Cover Design: Wiley
Cover Image: ©kentarcajuan/E+/Getty Images

Printed in the United States of America

V10005101_101118

To my wife Audrey, who taught me to love.

To my children, Simon, Sarah, and Anna, who taught me to care.

To Bruno Dupire, who taught me to think.

To Jesper Andreasen, who believed in me when nobody else would.

And to my parents, Svetlana and Andre Savine,
who I wish were with us to read my first book.

—Antoine Savine

Contents

Modern Computational Finance

Computational concerns, the ability to calculate values and risks of derivatives portfolios practically and in reasonable time, have always been a major part of quantitative finance. With the rise of bank-wide regulatory simulations like CVA and capital requirements, it became a matter of survival. Modern computational finance makes the difference between calculating CVA risk overnight in large data centers and praying that they complete by morning, or in real-time, within minutes on a workstation.

KØBENHAVNS
UNIVERSITET

Computational finance became a key skill, now expected from all quantitative analysts, developers, risk professionals, and anyone involved with financial derivatives. It is increasingly taught in masters programs in finance, such as the Copenhagen University's MSc Mathematics - Economics, where this publication is the curriculum in numerical finance.

Danske Bank

Danske Bank's quantitative research built its front office and regulatory systems combining technologies such as model hierarchies, scripting of transactions, parallel Monte-Carlo, a special application of regression proxies, and Automatic Adjoint Differentiation (AAD).

In 2015, Danske Bank demonstrated the computation of a sizeable CVA on a laptop in seconds, and its full market risk in minutes, without loss of accuracy, and won the In-House System of the Year Risk award.

Wiley's Computational Finance series, written by some of the very people who wrote Danske Bank's systems, offers a unique insight into the modern implementation of financial models. The volumes combine financial modeling, mathematics, and programming to resolve real-life financial problems and produce effective derivatives software.

The scientific, financial, and programming notions are developed in a pedagogical, self-contained manner. The publications are inseparable from the professional source code in C++ that comes with them. The books build the libraries step by step and the code demonstrates the practical application of the concepts discussed in the publications.

This is an essential reading for developers and analysts, risk managers, and all professionals involved with financial derivatives, as well as students and teachers in Masters and PhD programs in finance.

ALGORITHMIC ADJOINT DIFFERENTIATION

This volume is written by Antoine Savine, who co-wrote Danske Bank's parallel simulation and AAD engines, and teaches volatility and computational finance in Copenhagen University's MSc Mathematics - Economics.

Arguably the strongest addition to numerical finance of the past decade, Algorithmic Adjoint Differentiation (AAD) is the technology implemented in modern financial software to produce thousands of accurate risk sensitivities within seconds on light hardware. AAD is one of the greatest algorithms of the 20th century. It is also notoriously hard to learn.

This book offers a one-stop learning and reference resource for AAD, its practical implementation in C++, and its application in finance. AAD is explained step by step across chapters that gently lead readers from the theoretical foundations to the most delicate areas of an efficient implementation, such as memory management, parallel implementation, and acceleration with expression templates.

The publication comes with a self-contained, complete, general-purpose implementation of AAD in standard modern C++. The AAD library builds on the latest advances in AAD research to achieve remarkable speed. The code is incrementally built throughout the publication, where all the implementation details are explained.

The publication also covers the application of AAD to financial derivatives and the design of generic, parallel simulation libraries. Readers with working knowledge of derivatives and C++ will benefit most, although the book does cover modern and parallel C++.

The book comes with a professional parallel simulation library in C++, connected to AAD. Some of the most delicate applications of AAD to finance, such as the differentiation through calibration, are also explained in words, mathematics, and code.

Preface by Leif Andersen

It is now 2018, and the global quant community is realizing that size does matter: big data, big models, big computing grids, big computations – and a big regulatory rulebook to go along with it all. Not to speak of the *big* headaches that all this has induced across Wall Street.

The era of "big finance" has been creeping up on the banking industry gradually since the late 1990s, and got a boost when the Financial Crisis of 2007–2009 exposed a variety of deep complexities in the workings of financial markets, especially in periods of stress. Not only did this lead to more complicated models and richer market data with an explosion of basis adjustments, it also emphasized the need for more sophisticated governance as well as quoting and risk managements practices. One poster child for these developments is the new market practice of incorporating portfolio-level funding, margin, liquidity, capital, and credit effects (collectively known as "XVAs") into the prices of even the simplest of options, turning the previously trivial exercise of pricing, say, a plain-vanilla swap into a cross-asset high-dimensional modeling problem that requires PhD-level expertise in computing and model building. Regulators have contributed to the trend as well, with credit capital calculation requirements under Basel 3 rules that are at the same level of complexity as XVA calculations, and with the transparency requirements of MiFID II requiring banks to collect and disclose vast amounts of trade data.

To get a quick sense of the computational effort involved in a basic XVA calculation, consider that such a computation typically involves path-wise Monte Carlo simulation of option trade prices through time, from today's date to the final maturity of the trades. Let us say that 10,000 simulations are used, running on a monthly grid for 10 years. As a good-sized bank probably has in the neighborhood of 1,000,000 options on its books, calculating a single XVA adjustment on the bank's derivatives holding will involve in the order of $10^3 \cdot 10 \cdot 12 \cdot 10^6 \approx 10^{11}$ option re-pricings, on top of the often highly significant effort of executing the Monte Carlo simulation of market data required for pricing in the first place. Making matters significantly worse is then the fact that the traders and risk managers looking after the XVA positions will always require that sensitivities (i.e., partial derivatives) with respect to key risk factors in the market data are returned along with the XVA number itself. For complex portfolios, the number of sensitivities that one needs to compute can easily be in the order of 10^3; if these are computed naively (e.g., by finite difference approximations), the number of option re-pricings needed will then grow to a truly unmanageable order of 10^{14}.

There are many interesting ways of chipping away at the practical problems of XVA calculations, but let us focus on the burdens associated with the computation of sensitivities, for several reasons. First, sensitivities constitute a perennial problem in the quant world: whenever one computes some quantity, odds are that somebody in a trading or governance function will want to see sensitivities of said quantity to the inputs that are used in the computation, for limit monitoring, hedging, allocation, sanity checking, and so forth. Second, having input sensitivities available can be very powerful in an optimization setting. One rapidly growing area of "big finance" where optimization problems are especially pervasive is in the machine learning space, an area that is subject to enormous interest at the moment. And third, it just happens that there exists a very powerful technique to reduce the computational burden of sensitivity calculations to almost magically low levels.

To expand on the last point above, let us note the following quote by Phil Wolfe ([1]):

> *There is a common misconception that calculating a function of n variables and its gradient is about $n + 1$ times as expensive as just calculating the function. This will only be true if the gradient is evaluated by differencing function values or by some other emergency procedure. If care is taken in handling quantities, which are common to the function and its derivatives, the ratio is usually 1.5, not $n + 1$, whether the quantities are defined explicitly or implicitly, for example, the solutions of differential equations...*

The "care" in "handling quantities" that Wolfe somewhat cryptically refers to is now known as *Algorithmic Adjoint Differentiation* (AAD), also known as *reverse automatic differentiation* or, in machine learning circles, as *backward propagation* (or simply *backprop*). Translated into our XVA example, the promise of the "cheap gradient" principle underpinning AAD is that computation of all sensitivities to the XVA metric – no matter how many thousands of sensitivities this might be – may be computed at a cost that is order $\mathcal{O}(1)$ times the cost of computing the basic XVA metric itself. It can be shown (see [2]) that the constant in the $\mathcal{O}(1)$ term is bounded from above by 5. To paraphrase [3], this remarkable result can be seen as somewhat of a "holy grail" of sensitivity computation.

The history of AAD is an interesting one, marked by numerous discoveries and re-discoveries of the same basic idea which, despite its profoundness,[1]

[1]Nick Trefethen [4] classifies AAD as one of the 30 greatest numerical algorithms of the 20th century.

has had a tendency of sliding into oblivion; see [3] for an entertaining and illuminating account. The first descriptions of AAD date back to the 1960s, if not earlier, but did not take firm hold in the computer science community before the late 1980s. In Finance, the first published account took another 20 years to arrive, in the form of the award-winning paper [5].

As one starts reading the literature, it soon becomes clear why AAD originally had a hard time getting a foothold: the technique is hard to comprehend; is often hidden behind thick computer science lingo or is buried inside applications that have little general interest.[2] Besides, even if one manages to understand the ideas behind the method, there are often formidable challenges in actually implementing AAD in code, especially with management of memory or retro-fitting AAD into an existing code library.

The book you hold in your hands addresses the above challenges of AAD head-on. Written by a long-time derivatives quant, Antoine Savine, the exposition is done at a level, and in an applications setting, that is ideal for a Finance audience. The conceptual, mathematical, and computational ideas behind AAD are patiently developed in a step-by-step manner, where the many brain-twisting aspects of AAD are demystified. For real-life application projects, the book is loaded with modern C++ code and battle-tested advice on how to get AAD to run *for real*.

Select topics include: parallel C++ programming, operator overloading, tapes, check-pointing, model calibration, and much more. For both newcomers and those quaint exotics quants among us who need an upgrade to our coding skills and to our understanding of AAD, my advice is this: start reading!

[2]Some of the early expositions of AAD took place in the frameworks of chemical engineering, electronic circuits, weather forecasting, and compiler optimization.

Acknowledgments

The Computational Finance series is co-authored by Jesper Andreasen, Brian Huge, and Antoine Savine, who worked together in Danske Bank's quantitative research, and wrote its award winning systems, together with Ove Scavenius, Hans-Jorgen Terp Flyger, Jakob Nielsen, Niels Sonderby, Marco Haller Schultz and the rest of the department. Brian Fuglsbjerg conducted a last-minute review of the text and suggested meaningful improvements in the interest of clarity and correctness.

We extend thanks to Danske Bank for providing a work environment that encouraged and nurtured the development of the advanced technologies described in these publications. For avoidance of doubt, the code provided with the books is not the one implemented in the bank's systems and was developed independently for the purpose of the publications.

Irrespective of its intrinsic quality, quantitative research is only as effective as its relationship to business. We are extending special thanks to Danske Bank's exotics and xVA trading desks, who, under the particularly enlightened leadership of Peter Honore, Martin Linderstrom, and Nicki Rasmussen, offered valuable feedback, effective partnership, and unabated support during the development of the bank's systems. We could not have done it without them.

The publications are also based on our articles and talks, as well as our lectures at Copenhagen University. We extend special thanks to Professor Rolf Poulsen, head of the MSc Mathematics–Economics, for his continuous support, valuable feedback, and the multiple discussions that helped us tremendously throughout our work. Rolf also kindly reviewed and helped improve the more theoretical Chapters 4 and 5.

Before publication, the books were submitted for review to Leif Andersen, global head of quantitative research at BAML, and co-author, with Vladimir Piterbarg, of the three volumes of Interest Rate Models – in our opinion, the clearest, most comprehensive and useful reference in financial models [6]. Leif and his teams thoroughly reviewed our publication and found hundreds of language, grammar, style, mathematical, and programming mistakes. We are extending a million thanks to Leif and his researchers who performed this considerable work and saved us some serious embarrassment. The mistakes that remain are of course our own responsibility.

Finally, we extend thanks to our multiple friends and colleagues from other institutions who contributed through lively discussions and debates. All these brainstorms sharpened our understanding of the field and made our work a particularly enjoyable one. Specifically, we are extending very special thanks to our dear friends, Bruno Dupire from Bloomberg, Guillaume Blacher from BAML, and Jerome Lebuchoux from Goldman Sachs, for all the hours spent in such meaningful, rich, and pleasant discussions, of which they will certainly find traces in this publication.

Introduction

In the aftermath of the global financial crisis of 2008, massive regulations were imposed on investment banks, forcing them to conduct frequent, heavy regulatory calculations. While these regulations made it harder for banks to conduct derivatives businesses, they also contributed to a new golden age of computational finance.

A typical example of regulatory calculation is the CVA (Counterparty Value Adjustment),[1] an estimation of the loss subsequent to a future default of a counterparty when the value of the sum of all transactions against that counterparty (called netting set) is positive, and, therefore, lost. The CVA is actually the value of a real option a bank gives away whenever it trades with a defaultable counterparty. This option is a put on the netting set, contingent on default. It is an exotic option, and a particularly complex one, since the underlying asset is the netting set, consisting itself in thousands of transactions, some of which may be themselves optional or exotic. In addition, the netting set typically includes derivatives transactions in different currencies on various underlying assets belonging to different asset classes. Options on a set of heterogeneous underlying assets are known to the derivatives industry and called hybrid options. Investment banks' quantitative research departments actively developed hybrid models and related numerical implementations in the decade 1998–2008 for the risk management of very profitable transactions like Callable Reverse Power Duals (CRPD) in Japan.

The specification, calibration, and simulation of hybrid models are essentially well known; see, for example, [7], [8], and [9]. What is unprecedented is the vast number of cash flows and the high dimension of the simulation. With a naive implementation, the CVA on a large netting set can take minutes or even hours to calculate.

In addition, the market and credit risk of the CVA must be hedged. The CVA is a cost that impacts the revenue and balance sheet of the bank, and its value may change by hundreds of millions when financial and credit markets move. In order to hedge the CVA, it is not enough to compute it.

[1]We have a lot of similar regulatory calculations, collectively known as xVA. The capital charge for derivatives businesses and the cost of that capital are also part of these calculations.

All its sensitivities to market variables must also be produced. And a CVA is typically sensitive to thousands of market variables: all the underlying assets that affect the netting set, all the rate and spread curves and volatility surfaces for all the currencies involved, as well as all the foreign exchange rates and their volatility surfaces, and, of course, all the credit curves. In order to compute all these risks with traditional finite differences, the valuation of the CVA must be repeated thousands of times with inputs bumped one by one.

This is of course not viable, so investment banks first implemented crude approximations, at the expense of accuracy, and distributed calculations over large data centres, incurring massive hardware, development, and maintenance costs.

Calculation speed became a question of survival, and banks had to find new methodologies and paradigms, at the junction of mathematics, numerics, and computer science, in order to conduct CVA and other regulatory calculations accurately, practically, and quickly on light hardware.

That search for speed produced new, superior mathematical modeling of CVA (see, for instance, [10]), a renewed interest in the central technology of scripting derivatives cash flows (see our dedicated volume [11]), the systematic incorporation of parallel computing (Part I of this volume), and exciting new technologies such as Algorithmic Adjoint Differentiation (AAD, Part III of this volume) that computes thousands of derivatives sensitivities in constant time.

In the early 2010s, head of Danske Markets Jens Peter Neergaard was having lunch in New York with quantitative research legend Leif Andersen. As Leif was complaining about the administration and cost of data centres, JP replied:

We calculate our CVA on an iPad mini.

In the years that followed, the quantitative research department of Danske Bank, under Jesper Andreasen's leadership, turned that provocative statement into reality, by leveraging cutting-edge technologies in a software efficient enough to conduct CVA risk on light hardware without loss of accuracy.

In 2015, at a public Global Derivatives event in Amsterdam, we demonstrated the computation of various xVA and capital charge over a sizeable netting set, together with all risk sensitivities, within a minute on an Apple laptop. That same year, Danske Bank won the In-House System of the Year Risk award.

We have also been actively educating the market with frequent talks, workshops, and lectures. These publications are the sum of that work.

Modern quantitative researchers must venture beyond mathematics and numerical methods. They are also expert C++ programmers, able to leverage modern hardware to produce highly efficient software, and master key technologies like algorithmic adjoint differentiation and scripting. It is the purpose of our publications to teach these skills and technologies.

It follows that this book, like the other two volumes in the series, is a new kind of publication in quantitative finance. It constantly combines financial modeling, mathematics, and programming, and correspondences between the three, to resolve real-life financial problems and improve the accuracy and performance of financial derivatives software. These publications are inseparable from the professional source code in C++ that comes with them. The publications build the libraries step by step and the code demonstrates the practical application of the concepts discussed in the books.

Another unique aspect of these publications is that they are not about models. The definitive reference on models was already written by Andersen and Piterbarg [6]. The technologies described in our publications: parallel simulations, algorithmic adjoint differentiation, scripting of cash-flows, regression proxies, model hierarchies, and how to bring them all together to better risk manage derivatives and xVA, are all model agnostic: they are designed to work with all models. We develop a version of Dupire's popular model [12] for demonstration, but models have little screen time. The stars of the show are the general methodologies that allow the model, any model, to compute and differentiate on an iPad mini.

This volume, written by Antoine Savine, focuses on algorithmic adjoint differentiation (AAD) (Part III), parallel programming in C++ (Part I), and parallel simulation libraries (Part II). It is intended as a one-stop learning and reference resource for AAD and parallel simulations, and is complete with a professional implementation in C++, freely available to readers in our source repository.

AAD is a ground-breaking programming technique that allows one to produce derivative sensitivities to calculation code, automatically, analytically, and most importantly *in constant time*. AAD is applied in many scientific fields, including, but not limited to, machine learning (where it is known under the name "backpropagation") or meteorology (the powerful improvement of expression templates, covered in Chapter 15, was first suggested by a professor of meteorology, Robin Hogan). While a recent addition, AAD has quickly become an essential part of quantitative finance and an indispensable centrepiece of modern financial libraries. It is, however, still misunderstood to a large extent by a majority of finance professionals.

This publication offers a complete coverage of AAD, from its theoretical foundations to the most elaborate constructs of its efficient implementation. This book follows a pedagogical logic, progressively building intuition and taking the time to explain concepts and techniques in depth. It is also a complete reference for AAD and its application in the context of (serial and parallel) financial algorithms. With the exception of Chapters 12 and

13,[2] Part III covers AAD in itself, without reference to financial applications. Part III and the bundled AAD library in C++ can be applied in various contexts, including machine learning, although it was tested for maximum performance in the context of parallel financial simulations.

A second volume [11], co-authored by Jesper Andreasen and Antoine Savine, is dedicated to the scripting of derivatives cash flows. This central technology is covered in detail, beyond its typical usage as a convenience for the structuring of exotics. Scripts are introduced as a practical, transparent, and effective means to represent and manipulate transactions and cash flows in modern derivatives systems. The publication covers the development of scripting in C++, and its application in finance, to its full potential. It is also complete with a professional implementation in C++. Some advanced extensions are covered, such as the automation of fuzzy logic to stabilize risks, and the aggregation, compression, and decoration of cash flows for the purpose of xVA.[3]

A third (upcoming) volume, written by Jesper Andreasen, Brian Huge, and Antoine Savine, explores effective algorithms for the computation and differentiation of xVA, and covers the details of the efficient implementation, applications, and differentiation of the LSM[4] algorithm.

---

[2]As well as part of 14.

[3]We briefly introduced CVA. The other value adjustments banks calculate for funding, liquidity, cost of capital, etc. are collectively known as xVA and the computation techniques detailed for CVA are applicable with some adjustments.

[4]Least Squares Method, sometimes also called the Longstaff-Schwartz Model, invented in the late 1990s by Carriere [13] and Longstaff-Schwartz [14], and briefly introduced in Section 5.1.

About the Companion C++ Code

This book comes with a professional implementation in C++ freely available to readers on:

www.wiley.com/go/computationalfinance

In this repository, readers will find:

1. All the source code listed or referenced in this publication.
2. The files AAD*.*[1] constitute a self contained, general-purpose AAD library. The code builds on the advanced techniques exposed in this publication, in particular those of Chapters 10, 14, and 15, to produce a particularly fast differentiation library applicable to many contexts. The code is progressively built and explained in Part III.
3. The files mc*.*[2] form a generic, parallel, financial simulation library. The code and its theoretical foundations are described in Part II.
4. Various files with support code for memory management, interpolation, or concurrent data structures, such as threadPool.h, which is developed in Part I and used throughout the book to execute tasks in parallel.
5. A file main.h that lists all the higher level functions that provide an entry point into the combined library.
6. A Visual Studio 2017 project wrapping all the source files, with project settings correctly set for maximum optimization. The code uses some C++ 17 constructs, so the project setting "C++ Language Standard" on the project property "C/C++ / Language / C++ Language Standard" must be set to "ISO C++ 17 standard." This setting is correctly set on the project file xlComp.vcxproj, but readers who compile the files by other means must be aware of this.
7. A number of xl*.* files that contain utilities and wrappers to export the main functions to Excel, as a particularly convenient front end for the library. The project file xlComp.vcxproj is set to build an xll, a file that is opened from Excel and makes the exported library functions callable from Excel like its standard functions. We wrote a tutorial that explains

[1]With a dependency on gaussians.h for the analytic differentiation of Gaussian functions, and blocklist.h for memory management, both included.
[2]With dependency on various utility files, all included in the project.

how to export C++ code to Excel. The tutorial ExportingCpp2xl.pdf is available in the folder xlCpp along with the necessary source files. The wrapper xlExport.cpp file in our project precisely follows the directives of the tutorial and readers can inspect it to better understand these techniques.

8. Finally, we provide a pre-built xlComp.xll[3] and a spreadsheet xlTest.xlsx that demonstrates the main functions of the library. All the figures and numerical results in this publication were obtained with this spreadsheet and this xll, so readers can reproduce them immediately. The computation times were measured on an iMac Pro (Xeon W 2140B, 8 cores, 3.20 GHz, 4.20 max) running Windows 10. We also carefully checked that we have *consistent* calculation times on a recent quad core laptop (Surface Book 2, i7-8650U, 4 cores, 1.90 GHz, 4.20 max), that is, (virtually) identical time in single threaded mode, twice the time in multi-threaded mode.

The code is entirely written in standard C++, and compiles on Visual Studio 2017 out of the box, without any dependency to on a third-party library.

[3]To run xlComp.xll, readers may need to install Visual Studio redistributables VC_redist.x86.exe and VC_redist.x64.exe, also included in the repository.

PART I

Modern Parallel Programming

Introduction

This part is a self-contained tutorial and reference on high-performance programming in C++, with a special focus on parallel and concurrent programming. The second part applies the notions and constructs developed here to build a generic parallel financial simulation library.

A working knowledge of C++ is expected from readers. We cover modern and parallel C++11, and we illustrate the application of many STL (standard template library) data structures and algorithms, but we don't review the basic syntax, constructs, and idioms of C++. Readers unfamiliar with basic C++ should read an introductory textbook before proceeding. Mark Joshi's website http://www.markjoshi.com/RecommendedBooks.html#C contains a list of recommended C++ textbooks.

Readers already familiar with advanced C++ and concurrent programming, or those using different material, such as Anthony Williams' [15], may skip this part and move on to Parts II and III. They will need the *thread pool*, which we build in Section 3.18 and use in the rest of the publication. The code for the thread pool is in ThreadPool.h, ThreadPool.cpp, and ConcurrentQueue.h in our repository.

PARALLEL ALGORITHMS

Parallel programming may allow calculations to complete faster by computing various parts simultaneously on multiple processing units. The improvement in speed is at best linear in the number of units. To hit a linear improvement (or any improvement at all), parallel algorithms must be carefully designed and implemented. One simple, yet fundamental example is a *transformation*, implemented in the *transform*() function of the C++ standard library, where some function f is applied to all the elements $\{x_i\}$ in a source collection S to produce a destination collection $D = \{y_i = f(x_i)\}$. Such transformation is easily transposed in parallel with the "divide-and-conquer" idiom: partition S into a number of subsets $\{S_j\}$ and process the transformation of the different subsets, simultaneously, on different units.

The chart below illustrates the idiom for the transformation of 8 elements on 2 units.

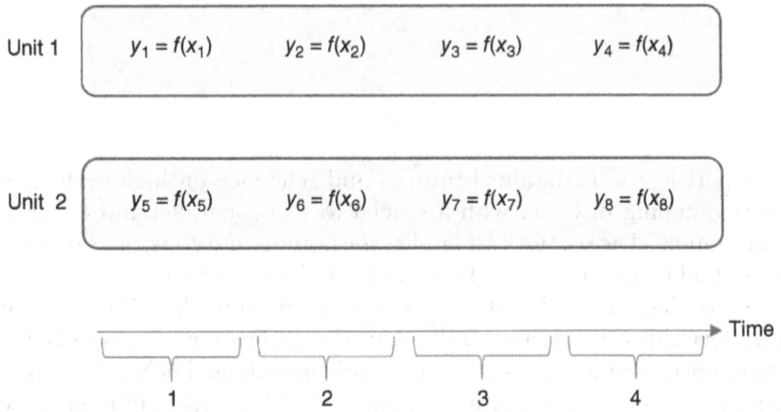

The 8 transformations are processed in the time of 4. Parallel processing cuts time by the number of units, a *linear* speed-up. With n transformations on p units, each unit evaluates f n/p times, and the units work simultaneously, so that the entire transformation completes in the time of n/p evaluations, as opposed to n evaluations in the serial version. In addition, this linear improvement is easily achieved in practice, assuming f computes an output y out of an input x without any side effect (like logging into a console window or a file, or updating a static cache).[1] Such problems that are trivially transposed and implemented in parallel are called "embarrassingly parallel."

Another classical family of fundamental algorithms is the *reduction*, implemented in the standard C++ *accumulate*() function, which computes an aggregate value (sum, product, average, variance...) from a collection of elements. One simple example is the sum $z = \sum_{i=1}^{n} x_i$. In order to compute z in parallel, we can compute partial sums over subsets, which leaves us with a lower number of partial sums to sum-up. A parallel reduction is therefore recursive in nature: we compute partial sums, then partial sums of partial

[1] We call such a function "thread safe." It is a fundamental principle of functional programming that functions should not be allowed to have side effects. This is enforced in pure functional languages like Haskell, where all functions are thread safe and suitable for parallel application. C++ allows side effects, so thread safe design is the developer's responsibility. In particular, it should be made very clear which functions have side effects and which don't. Thread safe design is closely related to const correctness, as we will see in the next chapters.

sums, and so forth, until we get a final sum. The chart below illustrates the sum of 8 terms on 2 units:

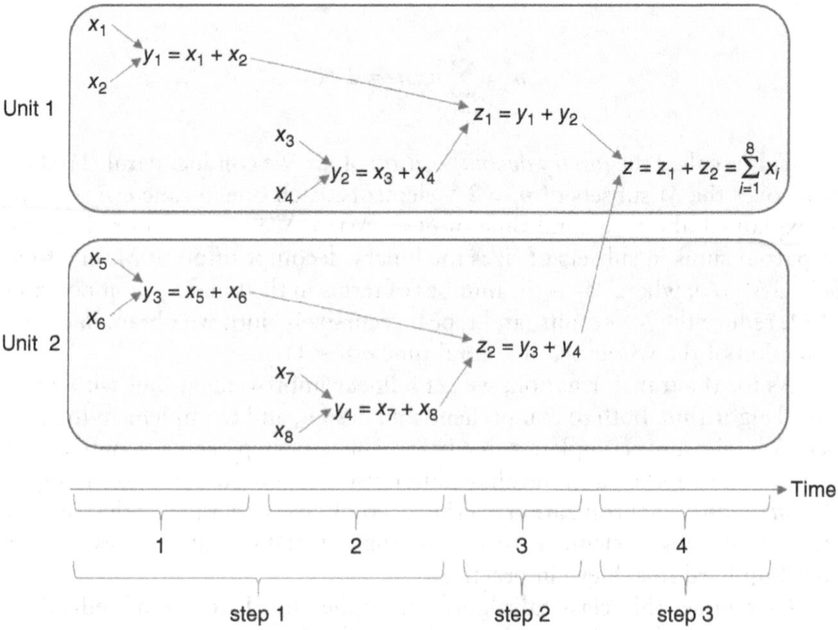

A reduction of 8 elements is, again, completed in the time of 4. A parallel reduction also produces a linear improvement. But the algorithm is no longer trivial; it involves 3 steps in a strict sequential order, and logic additional to the serial version. In general, to reduce n elements in parallel on p units, assuming n is even, we partition $S = \{x_i, 1 \leq i \leq n\}$ into $n/2$ pairs and sum the 2 elements of the $n/2$ pairs in parallel. To process $n/2$ adds in parallel on p units takes the time of $n/2p$ serial adds. Denoting δ the time of a serial add, we reduced S into $n/2$ partial sums in $(n/2p)\delta$ time. Assuming again that $n/2$ is even, we repeat and reduce the $n/2$ partial sums into $n/4$ pairwise sums in a time $(n/4p)\delta$. By recursion, and assuming that n is a power of 2: $n = 2^m$, the entire reduction completes in $m = log_2 n$ steps, where the step number i consists in $n/2^i$ pairwise adds on p units and takes $(n/2^i p)\delta$ time. The total time is therefore:

$$\sum_{i=1}^{m} \frac{n}{2^i p}\delta = \frac{n-1}{p}\delta$$

as opposed to $(n-1)\delta$ for the serial reduction, hence an improvement by p.

In a more general case where n is not necessarily a power of 2, we still achieve a linear improvement. To demonstrate this, we note that n can be uniquely decomposed into a sum of powers of 2:

$$n = \sum_{i=1}^{M} n_i, n_i = 2^{m_i}$$

This is called the *binary decomposition* of n.[2] We conduct parallel reductions over the M subsets of $n_i = 2^{m_i}$ elements, each one in time $\delta(n_i - 1)/p$, as explained above, a total time of $\delta(n - M)/p$. We repeat and reduce the M partial sums in subsets of sizes the binary decomposition of M, in a time $\delta(M - M_2)/p$, where M_2 is the number of terms in the binary decomposition of M, reduce the M_2 results, and repeat recursively until we obtain one final reduction of the whole set, in a total time $\delta(n - 1)/p$.

As for the transformation, we get a linear improvement, but with a less trivial algorithm, both to comprehend and design, and to implement in practice. All units must complete a step before the next step begins, so units must wait on one another, and the algorithm must incorporate some form of *synchronization*. Synchronization overhead, combined with the fact that the last steps involve fewer elements to sum-up than available units, makes a linear speed-up hard to achieve in practice.

Of course, this classical algorithm applies to all forms of reduction, not only sums, but also products, dot products, variance and covariance estimates, etc. Because it applies to dot products, it also applies to matrix products, where every cell in the result matrix is a dot product.

We have introduced two fundamental examples. One is embarrassingly parallel and fully benefits from parallelism without difficulty. The other one takes work and rarely achieves full parallel efficiency. In finance, derivatives risk management mainly involves two families of valuation algorithms: Monte-Carlo (MC) simulations and finite difference methods (FDM). MC simulations compute the average outcome over many simulated evolutions of the world. FDM computes the same result as the solution of a partial differential equation, numerically, over a grid. MC simulations are slow, heavy, and embarrassingly parallel. They are also applicable in a vast number of contexts. MC is therefore, by far, the most widely used method in modern finance; it is easy to implement in parallel, and the linear speed improvement makes a major difference due to the slow performance of the serial version. Part II is dedicated to the implementation of parallel MC. In contrast, FDM is light and fast, but applicable in a limited number of situations. FDM is

[2]To implement a binary decomposition is a trivial exercise, and an unnecessary one, since the chip's native representation of (unsigned) integers is binary.

less trivial to implement in parallel, and the resulting acceleration may not make a significant difference, FDM performing remarkably fast in its serial form. FDM and its parallel implementation are not covered in this text.

THE MANY FLAVORS OF PARALLEL PROGRAMMING

Distributed and concurrent programming

Parallel programming traditionally comes in two flavors: distributed and concurrent. The distributed flavor divides work between multiple *processes*. Processes run in parallel in separate memory spaces, the operating system (OS) *scheduling* their execution over the available processing units. Distributed computing is safe by design because processes cannot interfere with one another. It is also scalable to a large number of processing units, because different processes may run on the same computer or on multiple computers on a network. Since processes cannot communicate through memory, their synchronization is achieved through *messaging*. Distributed computing is very flexible. The cooperation of different nodes through messaging can accommodate many parallel designs.

Distributed programming is not implemented in standard C++ and requires specialized hardware and software.[3] The creation and the management of processes takes substantial overhead. Processes cannot share memory, so the entire context must be copied on every node. Distributed programming is best suited for *high-level* parallelism, like the distribution of a derivatives book, by transaction, over multiple machines in a data center. A lighter form of parallel computing, called *shared memory parallelism* or *concurrent programming*, is best suited for the implementation of parallelism in the lower level valuation and risk algorithms themselves.

Concurrent programming divides work between multiple *threads* that run on the same process and share common memory space. Threads are light forms of processes. Their execution is also scheduled by the OS over the available processing units, but limited to a single computer. The overhead for their creation and management is orders of magnitude lower. They share context and communicate directly through memory. The management of threads in C++ is standard and portable since C++11. All modern compilers, including Visual Studio since 2010, incorporate the Standard Threading Library out of the box. The threading library provides all the necessary tools and constructs for the manipulation of threads. It is also well documented, in particular in the book [15] written by one of its developers.

[3] A number of frameworks exist to facilitate distributed programming, the best known being the Message Passing Interface (MPI).

Concurrent programming is generally the preferred framework for the implementation of parallel algorithms. This text focuses on concurrent programming and does not cover distributed programming. Concurrent computing is not without constraints: it is limited to the processing units available on one computer, and, contrary to the distributed memory model, it is not safe by design. Threads may interfere with one another in shared memory, causing severe problems like crashes, incorrect results, slowdowns, or deadlocks, and it is the developer's responsibility to implement thread safe design and correct synchronization to prevent these. The Standard Threading Library provides a framework and multiple tools to facilitate correct concurrent programming. This library and its application to the development of safe, efficient parallel algorithms, is the main subject of this part.

CPU and GPU programming

A particular form of concurrent computing is with GPU (Graphic Processing Unit) parallelism. Modern GPUs provide a massive number of processing units for limited hardware costs. GPUs were initially designed to process the computation of two- and three-dimensional graphics, and later evolved to offer general-purpose computing capabilities. Nvidia, in particular, offers a freely available C++-like language, CUDA (Compute Unified Device Architecture), for programming their chips. GPU programming is implemented today in many scientific and commercial applications besides graphic design and video games. In particular, GPU accelerates many machine learning and financial risk management softwares. Reliable sources in the financial industry report speed improvements of order 50× for production Monte Carlo code; see for instance [16].

GPU programming is evidently not standard C++. In addition to specialized hardware and software, effective GPU programming requires algorithms to be written in a specific manner, down to low level, to accommodate the design and constraints of GPUs. GPU parallelism may come with unbeatable hardware cost; it is also subject to prohibitive development cost. For this reason, GPUs somewhat fell out of favor recently in the financial industry; see for instance [17].

CPU (Central Processing Unit) manufacturers like Intel have been systematically increasing the number of cores on their chips in recent years, so that CPUs may compute with GPUs, without the need for specialized GPU programming. High-end modern workstations now include up to 48 cores (up to 18 on Apple's slim iMac Pro). CPU cores are truly separate units, so different cores may conduct independent, unrelated work concurrently. GPU cores are not quite that flexible. CPU cores are also typically substantially faster than their GPU counterparts. More importantly, CPU parallelism is programmed in standard C++.

In addition, it is very challenging to effectively run memory-intensive algorithms, like AAD, on GPUs. Algorithmic adjoint differentiation (AAD) produces risk sensitivities in constant time (see our dedicated Part III), something no amount of parallelism can achieve. AAD is the most effective acceleration implemented in the financial industry in the past decade.[4] It is not particularly difficult to implement AAD on parallel CPU. We do that in Part III. On the other hand, to our knowledge, AAD has not yet been convincingly implemented on GPU, despite recent encouraging results by Uwe Naumann from RWTH Aachen University.

For those reasons, we don't explore GPU parallelism further and focus on CPU concurrency. Readers interested in learning GPU programming are referred to one of the many available CUDA textbooks.

Multi-threading and SIMD programming

Finally, concurrent CPU programming itself splits into two distinct, mutually compatible forms of parallelism: multi-threading and SIMD.

Multi-threading (MT) consists in the concurrent execution of different parts of some calculation on different threads running on different cores. The cores are effectively independent CPUs, so they may run unrelated calculations concurrently without loss of performance.[5] For example, the transformation of a collection can be multi-threaded by transforming different subsets over different threads. In this case, the different cores will execute the same sequence of instructions, but over different data. In a simple video game, the management of the player's star ship and the management of the enemies can be processed on two different threads. In that case, the two cores that execute the two threads conduct different, independent work. The game still runs twice as fast, providing a more satisfactory experience.

Multi-threading is a particularly flexible, efficient form of concurrent CPU programming, and the subject of Chapter 3. In Part II, it is applied to accelerate financial simulations.

By contrast, SIMD (Same Instruction Multiple Data) refers to the ability of every core in a modern CPU (or GPU) to apply the same instruction to multiple data, simultaneously. In a transformation, for instance, one single core could apply the function f to multiple xs simultaneously. SIMD applies CPU instructions to a vector of data at a time, as opposed to a single data. For this reason, it is also called "vectorization." Modern mainstream CPUs (implementing AVX2) can process four doubles (32 bytes, 256 bits) at a time.

---

[4]In our opinion, AAD is the most effective algorithm implemented in finance since FDM.

[5]Although they may interfere through shared memory and cache as we will see later in this part.

Higher-end CPUs (AVX512) can process eight doubles (64 bytes, 512 bits) in a single instruction.

SIMD may be combined with multi-threading, in a process sometimes called "GPU on CPU" because GPUs implement a similar technology. For the transformation of a collection, multiple subsets can be processed on different threads running on different cores while every core applies SIMD to process multiple elements of the subset in a single instruction.

The theoretical acceleration from combining MT with SIMD is the product of the number of cores by the SIMD vector width. In a standard 8-core workstation with AVX2, this is 32. On an 18-core high-end iMac Pro (AVX512), this is a vertiginous 144, well above the acceleration reportedly obtained on GPU. But this is theoretical of course. MT usually achieves linear acceleration as long as the code is correctly designed, as we will see in the rest of this book. SIMD acceleration is a different story.

SIMD is very restrictive: only simple functions applied to simple types may be vectorized. In addition, the data must be coalescent and properly aligned in memory. For example, a simple transformation $x \to \cos x$ of a set $S = \{x_i, 1 \leq i \leq n\}$ can be vectorized, and Visual Studio would indeed *automatically* vectorize a loop implementing this transformation. With a different function, like $x \to \log x$ if $x > 0$ or $-\log -x$ otherwise, the transformation cannot (easily) be vectorized: the effective CPU instruction that applies to each data depends on whether the number is positive or negative; we are no longer in a *same* instruction multiple data context. Visual Studio would not vectorize the transformation.

SIMD is low level, close to the metal parallelism. It is not implemented in standard C++ or supported in standard libraries. SIMD is implemented in the compiler. Contrary to multi-threading, it is not scheduled at run time. It is at compile time, when the C++ code is turned into CPU instructions, that vectorization occurs or not. Standard compilers like Visual Studio offer little support for SIMD. Visual Studio automatically vectorizes simple loops, and may be set to reports which loops are vectorized and which are not. Other compilers, like Intel's, offer much more extensive support for vectorization in the form of settings and pragmas.[6]

Intel also offers its Integrated Performance Primitives (IPP) and Math Kernel Library (MKL) as a free download. These libraries include a vast number of vectorized mathematical functions and algorithms. With direct relevance for Monte-Carlo simulations, IPP offers a vectorized implementation of the inverse Normal distribution *ippsErfcInv_64f_A26()*.

---

[6]Special instructions inserted in the body of the code that are not part of the C++ code to be compiled, but some directives for the compiler about how code is to be compiled. Pragmas are therefore compiler specific.

Hence, and even though Visual Studio supports SIMD for the occasional auto-vectorization of loops, an effective, systematic implementation of vectorization takes a specialized compiler and specialized third-party libraries.

In addition, in order to systematically benefit from SIMD, code must be written in a special way, so that mathematical operations are applied simultaneously to multiple data at low level. SIMD acceleration is *intrusive*. We will see in Part II that this is not the case for MT. We implement parallel simulations without modification of the model or the product code (as long as that code is *thread safe*). To vectorize simulations, we would need to rewrite the simulation code in models and products entirely.

A full SIMD acceleration is only achieved when the exact same CPU instructions are applied to different data. Whenever we have control flow in the code (if, while and friends), SIMD may be partially applied at best. Partial parallelism is a known pitfall in parallel programming. When perfect parallelism (acceleration by the number of units p) is achieved in a proportion μ of some algorithm, a proportion $1 - \mu$ remains sequential, resulting in a global acceleration by a factor $p/[\mu + (1 - \mu)p]$,[7] very different from the perhaps intuitive but completely wrong $p\mu$. The resulting acceleration is typically counterintuitively weak. For instance, the perfect parallel implementation of 75% of an algorithm over 8 units results in an acceleration by a factor 2.9, terribly disappointing[8] and far from a perhaps mistakenly expected factor 6. Even when 90% of the algorithm is parallel with efficiency 100%, the resulting global acceleration is only 4.7, a parallel efficiency below 50%. Naive parallel implementation often causes bad surprises, and partial parallelism, in particular, generally produces disappointing results. Therefore, to achieve a linear acceleration with SIMD over complex, structured code is almost impossible. Besides, that theoretical limit is only 4× on most current CPUs. By contrast, a linear acceleration of up to 24× is easily achieved with MT. As long as the multi-threaded code is thread safe and lock-free, high-level multi-threading is, by construction, immune to partial parallelism. Of course, MT code is also vulnerable to sometimes challenging flaws and inefficiencies, which we will encounter, and resolve, in the chapters of this book. We do achieve a linear acceleration for our simulation code, including differentiation with AAD.

Finally, and most importantly, it is extremely challenging to combine SIMD with AAD. Vectorized AAD may be the subject of a future paper, but

---

[7]This result is known as Amdahl's law. It is immediately visible that, when the serial calculation time is Δ, the parallel calculation time is $\Delta[\mu/p + (1 - mu)]$, and the result follows.

[8]The ratio of the speed-up over the number of parallel units is called "parallel efficiency." Parallel efficiency is always between 0 (excluded) and 1 (included). The purpose of parallel algorithms is to achieve an efficiency as close as possible to 1.

it is way out of the scope of this book. In contrast, MT naturally combines with AAD, as demonstrated in detail in Chapter 12, where we apply AAD over multi-threaded simulations.

We see that the combination of MT with SIMD offers similar performance to GPU but suffers from the same constraints. This is not a coincidence. GPUs are essentially collections of a large number of slower cores with wide SIMD capabilities. Systematic vectorization requires rewriting calculation code, down to a low level, with a specialized compiler and specialized libraries. Of course, all of this also requires specialized skills. We will not cover SIMD in detail, and we will not attempt to write specialized code for SIMD acceleration. On the contrary, we will study MT in detail in Chapter 3, and implement it throughout the rest of this text.

We will also apply *casual* vectorization, whereby we write loops in a way to encourage the compiler to auto-vectorize them, for instance, by using the general algorithms of the standard C++ library, like *transform*() or *accumulate*(), in place of hand-crafted loops, whenever possible. Those algorithms are optimized, including for vectorization when possible, and typically produce faster code. In addition, it makes our code more expressive and easier to read. It is therefore a general principle to apply STL (Standard Template Library) algorithms whenever possible. C++ programmers must know those algorithms and how to use them. We use a number of STL algorithms in the rest of the publication, and refer to [18] for a complete reference.

Readers interested in further learning SIMD are referred to Intel's documentation. For an overview and applications to finance, we refer to [19]. To our knowledge, there exists no textbook on SIMD programming. MT programming is covered in detail in [15], and, in a more condensed form, in our Chapter 3.

Multi-threading programming frameworks

There exist many different frameworks for concurrent MT programming. The responsibility for the management of threads and their *scheduling* on hardware cores belongs to the operating system, and all major OS provide specific APIs for the creation and the management of threads. Those APIs, however, are generally not particularly user friendly, and obviously result in non-portable code. Vendors like Intel or Microsoft released multiple libraries that facilitate concurrent programming, but those products, while excellent, remain hardware or OS dependent.

Prior to C++11, two standardization initiatives were implemented to provide portable concurrent programming frameworks: OpenMP and Boost.Thread. OpenMP offers compile time concurrency and is supported by all modern compilers on all major platforms, including Visual Studio on

Windows. Its consists of a number of compiler directives, called pragmas, that instruct the compiler to automatically generate parallel code for loops. OpenMP is particularly convenient for the parallelization of simple loops, and we provide an example in Chapter 1. OpenMP is available out of the box with most modern compilers, and therefore results in portable code. Using OpenMP is also amazingly simple, as demonstrated in our example.

For multi-threading complex, structured code, however, we need the full flexibility of a threading library. Boost.Thread provides such a library, with run-time parallelism, in the form of functions and objects that client code uses to create and manage threads and send tasks for concurrent execution.[9] Boost is available on all major OSs, and offers a consistent API for the management of threads and parallel tasks while encapsulating OS and hardware-specific logic. Many developers consider Boost as "almost standard," and applications written with Boost.Thread are often considered portable. However, compilers don't incorporate Boost out of the box. Applications using Boost.Thread must include Boost headers and link to Boost libraries. This is not a particularly simple, user friendly exercise, especially on Windows. It is also hard work to update Boost libraries in a vast project, for instance, when upgrading to a new version of Visual Studio.

Since C++11, C++ comes with a Standard Threading Library, which is essentially a port of Boost.Thread, better integrated within the C++ language, and consistent with other C++11 innovations and design patterns. Contrary to Boost.Thread, the Standard Threading Library is provided out of the box with all modern compilers, including Visual Studio, without the need to install or link against a third-party library. Applications using the Standard Threading Library are fully portable across compilers, OS, and hardware. In addition, as part of C++11 standard, this library is well documented, and the subject of numerous dedicated books, the best in our opinion being [15]. The Standard Threading Library is our preferred way of implementing parallelism. With very few exceptions, all the algorithms and developments that follow use this library.

WRITING MULTI-THREADED PROGRAMS

So, there exists many flavors of parallel computing and a wide choice of hardware and software environments for a practical implementation of each. Distributed, concurrent, and SIMD processing are not mutually exclusive. They form a hierarchy from the highest to the lowest level of parallelism. In Chapter 1, we implement a matrix product by multi-threading over the

[9] Before Boost.Thread, the pthread C library offered similar functionality in plain C.

outermost loop and vectorizing the innermost loop. In a financial context, the risk of a bank's trading books can be split into multiple portfolios distributed over many computers. Each computer that calculates a portfolio may distribute multiple tasks concurrently across its cores. The calculation code that executes on every core may be written and compiled in a way to enable SIMD. Similarly, banks using GPUs distribute their trading book across several machines, and further split the sub-books across the multiple cores on a computer, each core controlling one GPU that implements the risk algorithm. Maximum performance is obtained with the combination of multiple levels of parallelism. For the purpose of this publication, we made the choice, motivated by the many reasons enumerated above, to cover multi-threading with the Standard Threading Library.

As Moore's law has been showing signs of exhaustion over the past decade with CPU speeds limiting around 4GHz, chip manufacturers have been multiplying the number of cores in a chip as an alternative means of improving performance. However, whereas a program automatically runs faster on a faster chip, a single-threaded code does not automatically parallelize over the available cores. Algorithms other than embarrassingly parallel must be rethought, code must be rewritten with parallelism in mind, extreme care must be given to interference between threads when sharing memory, and, more generally, developers must learn new skills and rewrite their software to benefit from the multiple cores. It is often said that the exhaustion of Moore's law terminated a free option for developers. Modern hardware comes with better parallelism, but to benefit from that takes work. This part teaches readers the necessary skills to effectively program parallel algorithms in C++.

PARALLEL COMPUTING IN FINANCE

Given such progress in the development of parallel hardware and software, programmers from many fields have been writing parallel code for years. However, this trend only reached finance in the very recent years. Danske Bank, an institution generally well known for its cutting-edge technology, only implemented concurrent Monte-Carlo simulations in their production systems in 2013. This is all the more surprising as the main algorithm in finance, Monte-Carlo simulations, is embarrassingly parallel. The reason is there was little motivation for parallel computing in investment banks prior to 2008–2011. Quantitative analysts almost exclusively worked on the valuation and risk management of exotics. Traders held thousands of exotic transactions in their books. Risk sensitivities were computed by "bumping" market variables one by one, then recomputing values. Values and risk sensitivities were additive across transactions. Hence, to produce the value and risk for a portfolio, the valuation of the transactions was repeated for every transaction and in every "bumped" market scenario. A large number of small valuations were conducted, which encouraged high-level parallelism, where valuations were distributed across transactions and scenarios but the internal algorithms that produce a value in a given context remained sequential.

That changed with regulatory calculations like CVA (Counterparty Value Adjustment) that estimates the loss incurred from a counterparty defaulting when the net present value (PV) of all transactions against this counterparty is positive. CVA is not additive due to netting effects across different transactions, and must be computed for all transactions at once, in one (particularly heavy) simulation. A CVA is also typically sensitive to thousands of market variables, and risk sensitivities cannot be produced by bumping in reasonable time. The response to this particular challenge, and arguably the strongest addition to computational finance over the past decade, is AAD, an alternative to bumping that computes all derivatives together with the value in constant time (explained in detail in Part III). So we no longer conduct many light computations but a single, extremely heavy one. The time taken by that computation is of crucial importance to financial institutions: slow computations result in substantial hardware costs and the inability to compute risk sensitivities in time. The computations must be conducted in parallel to take advantage of modern hardware. Hence, parallel computing only recently became a key skill for quantitative analysts and a central feature of modern financial libraries. It is the purpose of this part to teach these skills.

We cover concurrent programming under the Standard Threading Library, discuss the challenges involved, and explain how to develop effective parallel algorithms in modern C++. C++11 is a major modernization of C++ and includes a plethora of new features, constructs, and patterns. The Standard Threading Library is probably the most significant one, and it was designed consistently with the rest of C++11. For this reason, we review the main innovations of C++11 before we delve into its threading library.

Chapter 1 discusses high-performance programming in general terms, shows some concrete examples of the notions we just introduced, and delivers some important considerations for the development of fast applications in C++ on modern hardware. Chapter 2 discusses the most useful innovations in C++11, and, in particular, explains some important modern C++ idioms and patterns applied, among other places, in the Standard Threading Library. These patterns are also useful on their own right. Chapter 3 explores many aspects of concurrent programming with the Standard Threading Library, shows how to manage threads and send tasks for parallel execution, and illustrates its purpose with simple parallel algorithms. We also build a *thread pool*, which is applied in Parts II and III to situations of practical relevance in finance.

CHAPTER 1

Effective C++

It is often said that quantitative analysts and developers should focus on algorithms and produce a readable, modular code and leave optimization to the compiler. It is a fact that substantial progress was made recently in the domain of compiler optimization, as demonstrated by the massive difference in speed for code compiled in release mode with optimizations turned on, compared to debug mode without the optimizations. It is also obviously true that within a constantly changing financial and regulatory environment, quantitative libraries must be written with clear, generic, loosely coupled, reusable code that is easy to read, debug, extend, and maintain. Finally, better code may produce a *linear* performance improvement while better algorithms increase speed by orders of magnitude. It is a classic result that 1D finite differences converge in ΔT^2 and ΔX^2 while Monte Carlo simulations converge in $\sqrt{N}$; hence FDM is preferable whenever possible. We will also demonstrate in Part III that AAD can produce thousands of derivative sensitivities for a given computation *in constant time*. No amount of code magic will ever match such performance. Even in Python, which is quite literally *hundreds of times* slower than C++, a good algorithm would beat a bad algorithm written in C++.

However, speed is so critical in finance that we cannot afford to overlook the low-level phenomena that affect the execution time of our algorithms. Those low-level details, including memory cache, vector lanes, and multiple cores, do not affect algorithmic complexity or theoretical speed, but their impact on real-world performance may be very substantial.

A typical example is memory allocation. It is well known that allocations are expensive. We will repeatedly recall that as we progress through the publication and strive to preallocate at high level the memory required by the lower level algorithms in our code. This is not an optimization the compiler can conduct on our behalf. We must do that ourselves, and it is not always easy to do so and maintain a clean code. We will demonstrate some techniques when we deal with AAD in Part III. AAD records every mathematical operation, so with naive code, every single addition, subtraction, multiplication, or division would require an allocation. We will use custom *memory pools* to eliminate that overhead while preserving the clarity of the code.

Another expensive operation is *locking*. We lock *unsafe* parts of the code so they cannot be executed concurrently on different threads. We call *thread safe* such code that may be executed concurrently without trouble. All code is not always thread safe. The unsafe pieces are called *critical regions* and they may be *locked* (using primitives that we explore later in this part) so that only one thread can execute them at a time. But locking is expensive. Code should be thread safe *by design* and locks should be encapsulated in such a way that they don't produce unnecessary overhead. It is not only explicit locks we must worry about, but also hidden locks. For example, memory allocations involve locks. Therefore, all allocations, including the construction and copy of containers, must be banned from code meant for concurrent execution.[1] We will show some examples in Chapters 7 and 12 when we multi-thread our simulation library and preallocate all necessary memory beforehand.

Another important example is memory caches. The limited amount of memory located in CPU caches is orders of magnitude faster than RAM. Interestingly perhaps, this limitation is not technical, but economical. We *could* produce RAM as fast as cache memory, but it would be too expensive for the PC and workstation markets. We may envision a future where this ultra-fast memory may be produced for a reasonable cost, and CPU caches would no longer be necessary. In the meantime, we must remember caches when we code. CPU caches are a hardware optimization based on a locality assumption, whereby when data is accessed in memory, the same data, or some data stored nearby in memory, is likely to be accessed next. So, every access in memory causes a duplication of the nearby memory in the cache for faster subsequent access. For this reason, code that operates on data stored nearby in memory – or *coalescent* – runs substantially faster. In the context of AAD, this translates into a better performance with a large number of small tapes than a small number of large tapes,[2] despite "administrative" costs per tape. This realization leads us to differentiate simulations pathwise, and, more generally, systematically rely on *checkpointing*, a technique that differentiates algorithms one small piece at a time over short tapes. This is all explained in detail in Part III.

[1] Intel, among others, offers a concurrent lock-free allocator with its freely available Threading Building Blocks (TBB) library. Without recourse to third-party libraries, we must structure our code to avoid concurrent allocations altogether.

[2] The *tape* is the data structure that records all mathematical operations.

For now, we make our point more concrete with the extended example of an elementary matrix product. We need a simplistic matrix class, which we develop as a wrapper over an STL vector. Matrices are most often implemented this way, for example, in Numerical Recipes [20].

```
template <class T>
class matrix
{
        //  Dimensions
        size_t      myRows;
        size_t      myCols;

        //  Data
        vector<T>   myVector;

public:

        using value_type = T;

        //  Constructors
        matrix() : myRows(0), myCols(0) {}
        matrix(const size_t rows, const size_t cols)
            : myRows(rows), myCols(cols), myVector(rows*cols) {}

        //  Access
        size_t rows() const { return myRows; }
        size_t cols() const { return myCols; }
        //  So we can call matrix [i][j]
        T* operator[] (const size_t row)
            { return &myVector[row*myCols]; }
        const T* operator[] (const size_t row) const
            { return &myVector[row*myCols]; }
};
```

We test a naive matrix product code that sequentially computes the result cells as the dot product of each row vector on the left matrix with the corresponding column vector on the right matrix. Such code is a direct translation of matrix algebra and we saw it implemented in a vast number of financial libraries.

```
inline void matrixProductNaive(
const matrix<double>& a,
const matrix<double>& b,
matrix<double>& c)
{
    const size_t rows = a.rows(), cols = b.cols(), n = a.cols();

    //  Outermost loop on result rows
    for (size_t i = 0; i < rows; ++i)
    {
        const auto ai = a[i];
        auto ci = c[i];

        //  Loop on result columns
```

```
15         for (size_t j = 0; j < cols; ++j)
16         {
17             //  Innermost loop for dot product
18             double res = 0.0;
19             for (size_t k = 0; k < n; ++k)
20             {
21                 res += ai[k] * b[k][j];
22             }
23             //  Set result
24             c[i][j] = res;
25         }
26     }
27 }
```

This code is compiled on Visual Studio 2017 in release 64 bits mode, with all optimizations on. Note that the following settings must be set on the project's properties page, tab "C/C++":

- "Code Generation / Enable Enhanced Instruction Set" must be set to "Advanced Vector Extensions 2" to produce AVX2 code.
- "Language / OpenMP Support" must be set to "yes" so we can use OpenMP pragmas.

For two random 1,000 × 1,000 matrices, it takes around 1.25 seconds to complete the computation on our iMac Pro. Looking into the innermost loop, we locate the code on line 21, executed 1 billion times:

```
res += ai[k] * b[k][j];
```

One apparent bottleneck is that $b[k][j]$ resolves into $(\&b.myVector[k * b.myCols])[j]$. The multiplication $k * b.myCols$, conducted a billion times, is unnecessary and may be replaced by an order of magnitude faster addition, at the cost of a somewhat ugly code, replacing the lines 17–22 by:

```
//  Dot product
double res = 0.0;
const double* bkj = &b[0][j];
size_t r = b.rows();
for (size_t k = 0; k < n; ++k)
{
    res += ai[k] * *bkj;
    bkj += r;
}
```

And the result is *still* 1.25 second! The compiler was already making that optimization and we polluted the code unnecessarily. So far, the theory that optimization is best left to compilers holds. We revert the unnecessary modification. But let's see what is going on with memory in that innermost loop.

The loop iterates on k and each iteration reads $ai[k]$ and $b[k][j]$ (for a fixed j). Data storage on the *matrix* class is row major, so successive $ai[k]$ are localized in memory next to each other. But the successive $b[k][j]$ are distant by 1,000 doubles (8,000 bytes). As mentioned earlier, CPU caches are based on locality: every time memory is accessed that is not already duplicated in the cache, that memory *and the cache line around it, generally 64 bytes, or 8 doubles*, are transferred into the cache. Therefore, the access to a is cache efficient, but the access to b is not. For every $b[k][j]$ read in memory, the line around it is unnecessarily transferred into the cache. On the next iteration, $b[k+1][j]$, localized 8,000 bytes away, is read. It is obviously not in the cache; hence, it is transferred along with its line again. Such unnecessary transfer may even erase from the cache some data needed for forthcoming calculations, like parts of a. So the code is not efficient, not because the number of mathematical operations is too large, but because it uses the cache inefficiently.

To remedy that, we modify the order of the loops so that the innermost loop iterates over coalescent memory for both matrices:

```
inline void matrixProductSmartNoVec(
const matrix<double>& a,
const matrix<double>& b,
matrix<double>& c)
{
    const size_t rows = a.rows(), cols = b.cols(), n = a.cols();

    //  zero result first
    for (size_t i = 0; i < rows; ++i)
    {
        auto ci = c[i];
        for (size_t j = 0; j < cols; ++j)
        {
            ci[j] = 0;
        }
    }

    //  Loop on result rows as before
    for (size_t i = 0; i < rows; ++i)
    {
        const auto ai = a[i];
        auto ci = c[i];

        //  Then loop not on result columns but on dot product
        for (size_t k = 0; k < n; ++k)
        {
            const auto bk = b[k];
            //  We still jump when reading memory,
            //          but not in the innermost loop
            const auto aik = ai[k];

            //  And finally loop over columns in innermost loop
            //          without vectorization to isolate impact of cache alone
```

```
34            #pragma loop(no_vector)
35            for (size_t j = 0; j < cols; ++j)
36            {
37                //  No more jumping through memory
38                ci[j] += aik * bk[j];
39            }
40        }
41    }
42 }
```

The pragma on line 34 will be explained ahead.

This code produces the exact same result as before, in 550 milliseconds, more than twice as fast! And we conducted just the same amount of operations. The only difference is cache efficiency. To modify the order of the loops is an operation too complex for the compiler to make for us. It is something we must do ourselves.

It is remarkable and maybe surprising how much cache efficiency matters. We increased the speed more than twice just changing the order of the loops. Modern CPUs operate a lot faster than RAM so our software is *memory bound*, meaning CPUs spend most of their time waiting on memory, unless the useful memory is cached in the limited amount of ultra-fast memory that sits on the CPU. When we understand this and structure our code accordingly, our calculations complete substantially faster.

And we are not quite done there yet.

What does this "#pragma loop(no_vector)" on line 34 stand for? We introduced SIMD (Single Instruction Multiple Data) in the Introduction. SIMD only works when the exact same instructions are applied to multiple data stored side by side in memory. The naive matrix product code could not apply SIMD because the data for *b* was not coalescent. This was corrected in the smart code, so the innermost loop may now be vectorized. We wanted to measure the impact of cache efficiency alone, so we disabled SIMD with the pragma "#pragma loop(no_vector)" over the innermost loop.

Visual Studio, like other modern compilers, *auto-vectorizes* (innermost[3]) loops whenever it believes it may do so safely and efficiently. If we remove the pragma but leave the code otherwise unchanged, the compiler should auto-vectorize the innermost loop[4]. Effectively, removing the pragma

[3]Evidently, given SIMD constraints, innermost loops are the only candidates for vectorization.

[4]Provided the setting "C/C++ / Code Generation / Floating Point Model" is manually set to Fast in the project properties page for the release configuration on the relevant platform, presumably x64. When this is not the case, Visual Studio does not vectorize reductions. Whether the reduction was vectorized or not can be checked by writing "/Qvec-report:2" in the "Additional Options" box in the "C/C++ / Command Line" setting.

further accelerates calculation by 60%, down to 350 milliseconds. The SIMD improvement is very significant, if somewhat short of the theoretical acceleration. We note that the innermost loop is a *reduction*, having explained in the Introduction how parallel reductions work and why they struggle to achieve parallel efficiency. In addition, data must be *aligned* in memory in a special way to fully benefit from AVX2 vectorization, something that we did not implement, this being specialized code, outside of the scope of this text.

What this teaches us is that we must be SIMD aware. SIMD is applied in Visual Studio outside of our control, but we can check which loops the compiler effectively vectorized by adding "/Qvec-report:2" (without the quotes) in the "Configuration Properties/ C/C++ / Command Line/ Additional Options" box of the project's properties. We should strive to code innermost loops in such a way as to encourage the compiler to vectorize them and then check that it is effectively the case at compile time.[5] To fail to do so may produce code that runs at half of its potential speed or less.

Altogether, to change the order of the loops accelerated the computation by a factor 3.5, from 1250 to 350 milliseconds. Over half is due to cache efficiency, and the rest is due to vectorization.

Always try to structure calculation code so that innermost loops sequentially access coalescent data. Do not hesitate to modify the order of the loops to make that happen. This simple manipulation accelerated our matrix product by a factor close to 4, similar to multi-threading over a quad core CPU.

Finally, we may easily distribute the *outermost* loop over the available CPU cores with another simple pragma above line 19:

```
// OpenMp directive: execute loop in parallel
#pragma omp parallel for
for (int i = 0; i < rows; ++i)
{
    const auto ai = a[i];
    auto ci = c[i];

    for (size_t k = 0; k < n; ++k)
    {
        const auto bk = b[k];
        const auto aik = ai[k];
```

[5]Visual Studio is somewhat parsimonious in its auto-vectorization and frequently declines to vectorize perfectly vectorizable loops (although it would never vectorize a loop that should not be vectorized). Therefore we must check that our loops are effectively vectorized and, if not, rewrite them until such time the compiler finally accepts to apply SIMD. This may be a frustrating process. STL algorithms are generally easier auto-vectorized than hand-crafted loops, yet another reason to prefer these systematically.

```
            for (size_t j = 0; j < cols; ++j)
            {
                ci[j] += aik * bk[j];
            }
        }
    }
```

This pragma is an OpenMP directive that instructs the compiler to multi-thread the loop below it over the available cores on the machine running the program. Note that we changed the type of the outermost counter *i* from *size_t* to *int* so OpenMP would accept to multi-thread the loop. OpenMP's auto-parallelizer is somewhat peculiar this way, not unlike Visual Studio's auto-vectorizer.

This code produces the exact same result in just 40 milliseconds, approximately 8.125 times faster, more than our number (8) of cores! This is due to a so-called "hyper-threading" technology developed by Intel for their recent chips, consisting of two *hardware threads* per core that allow each core to switch between threads at hardware level while waiting for memory access. The OS effectively "sees" 16 hardware threads, as may be checked on the Windows Task Manager, and their scheduling over the 8 physical cores is handled on chip. Depending on the context, hyper-threading may increase calculation speed by up to 20%. In other cases, it may *decrease* performance due to excessive *context switches* on a physical core when execution switches between two threads working with separate regions of memory.

Contrary to SIMD, multi-threading is not automatic; it is controlled by the developer. In very simple cases, it can be done with simple pragmas over loops as demonstrated here. But even in these cases, this is not fully satisfactory. The code is multi-threaded at compile time, which means that it always runs concurrently. But users may want to control that behavior. For instance, when the matrix product is part of a program that is itself multi-threaded at a higher level, it is unnecessary, and indeed decreases performance, to run it concurrently. Concurrency is best controlled at run time. Besides, to multi-thread complex, structured code like Monte-Carlo simulations, we need more control than OpenMP offers. In very simple contexts, however, OpenMP provides a particularly light, easy, and effective solution.

Altogether, with successive modifications of our naive matrix product code, but without any change to the algorithm, its complexity, or the mathematical operations involved, we increased the execution speed by a factor of 30, from 1,250 to 40 milliseconds. Obviously, the results are unchanged. We achieved this very remarkable speed-up by tweaking our code to take full advantage of our modern hardware, including on-chip cache, SIMD, and multiple cores. It is our responsibility to know these things and to develop code that leverages them to their full potential. The compiler will not do that for us on its own.

CHAPTER 2

Modern C++

C++ was thoroughly modernized in 2011 with the addition of a plethora of features and constructs borrowed from more recent programming languages. As a result, C++ kept its identity as the language of choice for programming close to the metal, high-performance software in demanding contexts, and at the same time, adapted to the demands of modern hardware, adopted modern programming idioms borrowed from the field of functional programming, and incorporated useful constructs into its syntax and standard library that were previously available only from third party libraries.

The major innovation is the new Standard Threading Library, which is explored in the next chapter. Since we are using new C++11 constructs in the rest of the text, this chapter selectively introduces some particularly useful innovations. A more complete picture can be found online or in up-to-date C++ textbooks. Readers familiar with C++11 may easily skip this chapter.

2.1 LAMBDA EXPRESSIONS

One of the most useful features in C++11, borrowed from the field of *functional programming*, is the ability to define anonymous function objects on the fly with the *lambda* syntax like in:

```
auto myLambda = [] (const double r, const double t)
{
    double temp = r * t; return exp( -temp);
};
```

The new *auto* keyword provides automatic type deduction, which is particularly useful with lambdas that produce a different compiler-generated type for every lambda declaration. A lambda declaration starts with a *capture* clause [], followed by the list of its arguments (a lambda is after all a function, or more exactly a *callable object*), and its *body*, the sequence of instructions that are executed when the lambda is called, just like the body of functions and methods. The difference is that lambdas are declared *within* functions. Once declared, they may be called like any other function or function object, for example:

```
cout << myLambda( 0.01, 10) << endl;
```

One powerful feature of lambdas is their ability to *capture* variables from their environment *on declaration.*

- [] means no capture.
- [=] means capture *by value*, that is, by copy, of all variables in scope *used in the lambda's body.*
- [&] means capture all variables *by reference.*

We may also capture variables selectively with the syntax:

- [x] means capture only x, by value with this syntax or by reference with [&x].
- [=, &x, &y] means capture x and y by reference, and all others by value. Obviously [&, x, y] means capture x and y by value and all others by reference.
- [x, &y] means capture x by value, y by reference and nothing else.

For instance,

```
double mat = 10;
auto myLambdaRef = [&mat] (const double r)
{
    return exp( -mat * r);
};
auto myLambdaCopy = [mat] (const double r)
{
    return exp( -mat * r);
};

cout << myLambdaRef( 0.01) << endl;
//  exp( -10 * 0.01)
cout << myLambdaCopy( 0.01) << endl;
//  exp( -10 * 0.01)

mat = 20;
cout << myLambdaRef( 0.01) << endl;
//  exp( -20 * 0.01)
cout << myLambdaCopy( 0.01) << endl;
//  exp( -10 * 0.01)
```

Behind the scenes, the compiler creates a *function object* when we declare a lambda, that is, an object that defines the operator () (with the arguments of the lambda) and therefore is *callable* (like a function). The captured variables are implicitly declared as data members with a value type when captured by value and a reference type when captured by reference, initialized with the captured data on declaration. Hence, the syntax:

```
void main()
{
    double a, x;
    // ...
    auto l = [=, &x] (const double y) { return a*x*y; }
    // ...
    double z = l(y);
}
```

is equivalent to the (much heavier):

```
class Lambda
{
    const double  myA;
    const double& myX;

public:

    Lambda(const double a, const double& x) : myA(a), myX(x) {}

    operator() (const double y) const { return myA * myX * y; }
};
void main()
{
    double a, x;
    // ...
    Lambda l(a, x);
    // ...
    double z = l(y);
}
```

As a function object, a lambda can be passed as an argument or returned from functions. Functions that manipulate functions are called *higher-order functions*, and the standard <algorithm> library provides a vast number of these.

Lambdas are also incredibly useful as *adapters* and resolve a constant annoyance C++ developers face when calling functions with signatures inconsistent with their data. The Standard Template Library (STL), for instance, includes a wealth of useful generic algorithms. But to use these algorithms we must respect their functions' signatures. Say we hold a vector of times from today:

```
vector<double> times;
```

and we want to compute an annuity given a constant rate *r*. We could write a hand-crafted loop, of course:

```
double ann = 0.0;
for(size_t i = 0; i < times.size(); ++i)
{
    ann += exp(-r*times[i]);
}
```

but it is considered best professional practice to apply generic algorithms instead.[1] The computation we just conducted is a *reduction*, where a collection is traversed sequentially and an *accumulator* is updated for each element. The STL algorithm for reductions is *accumulate*(), located in the <numeric> header. The version of interest to us has the following signature:

```
template< class InputIt, class T, class BinaryOperation >
T accumulate( InputIt first, InputIt last, T init,
              BinaryOperation op );
```

The type T of the accumulator in our case is double, as is *InputIt, so the function *op* that updates the accumulator *acc* for each element x must be consistent with the form:

```
double op(const double& acc, const double& x);
```

but our instruction for the update of the accumulator is:

```
acc += exp(-r*x);
```

and prior to C++11, it would have been such an annoyance to squeeze that line of code into the required signature that we would probably have ended up with the hand-crafted loop. With the lambda syntax, it takes a line to do that right:

```
double ann = accumulate(times.begin(), times.end(), 0.0,
              [r] (const double& acc, const double& x)
              {
                 return acc + exp(-r*x);
              });
```

There are of course many other uses of lambdas, and we will discuss a few later, but their ability to seamlessly adapt data to signatures is the reason why we use them every day.

C++11 also provides dedicated adapter functions *bind*() and *mem_fn*() in the <functional> header (the latter turning member functions *class.func*() into free functions *func*(*class*)), although lambdas can also do this in more convenient manner. The syntax for *bind*() in particular is rather peculiar and it is easier to achieve the exact same behavior with lambdas.

We will be working with lambdas throughout the book.

2.2 FUNCTIONAL PROGRAMMING IN C++

The introduction of lambdas is part of an effort to modernize C++ with idioms borrowed from the growing and fashionable field of functional programming. Although C++ does not, and never will, support functional programming idioms the way a language like Haskell does, C++ does

[1] For reasons explained for instance in Scott Meyers [18].

support some key elements of functional programming, in particular *value semantics for functions* and *higher-order functions.*

Value semantics means that functions may be manipulated just like other types and in particular they can be assigned to variables and passed as arguments or returned as results by higher-order functions. Note that lambdas are literals for functions, which means that the instruction:

```
auto f = [] (const double x) { return 2*x; };
```

assigns a *function literal* to *f* in the same way we assign number or string literals in:

```
double x = 2.0;
string s = "C++11";
```

C++11 defines the *function* template class in the <functional> header as a unique class for holding functions and *anything callable.* That means that a concrete type like

```
function<double(const double)>
```

can hold anything that may be called with a double to return a double: a C style function pointer, a function object, including a lambda, or a member function bound to an object. An object of that type is itself callable of course, and it has value semantics, in the sense that it can be assigned or passed as an argument, or returned as a result from a higher-order function.

It looks peculiar and at first sight impossible in C++ to define a type based on the behavior rather than the nature of the objects it holds.[2] *function* is implemented with an interesting, advanced design pattern called *type erasure.* Unfortunately, this versatility comes with a cost. Type erasure necessarily involves the storage of the underlying objects on the heap. Hence, to initialize, assigning or copying a *function* object involves an allocation.[3] For this reason, we refrain from using this class despite its convenience, and manipulate functions as template types instead.[4]

Composition

As a first example, we consider the composition of functions, and write a (higher-order) function that takes two functions as arguments and returns the *function* resulting from their composition.

---

[2]This is sometimes called "duck-typing": if it walks like a duck and quacks like a duck, then it is a duck.

[3]Visual Studio implements a small object optimization (SMO) whereby a small buffer is allocated for every object on the stack to minimize heap allocations.

[4]The implementation of type erasure is outside of our subject; interested readers can find information online, in particular on Stack Overflow.

```
template<class F1, class F2=F1>
auto compose(const F1& f, const F2& g)
{
    return [=](const auto& x) { return f(g(x)); };
}
```

We use the auto keyword so that types are deduced at compile time. Note that it is a function, not a number, that is returned. For instance, the following code creates a function by composing an exponential with a square root:

```
int main()
{
    auto f = compose([](const double x) { return exp(x); },
        [](const double x) { return sqrt(x); });
    cout << f(0.5) << endl;
}
```

Lambdas are obviously unnecessary here; they wrap the functions *exp*() and *sqrt*() without adapting anything. However, the following does not compile on Visual Studio:

```
int main()
{
    auto f = compose(exp, sqrt);
    cout << f(0.5) << endl;
}
```

Standard mathematical functions are overloaded so they work with many different types, and the compiler doesn't know which overload to pick to instantiate the templates. For this reason, we must explicitly state the function types when we compose standard functions, as follows:

```
auto f = compose<double(double)>(exp, sqrt);
```

We are not limited to numerical functions. Any function that takes an argument of type T_1 and returns a result of type T_2 (which we denote $f : T_1 -> T_2$) may be composed with any function $g : T_2 -> T_3$ to create a function $h : T_1 -> T_3$. We can imagine a function that creates a vector $1..n$ out of an unsigned integer:

```
vector<unsigned> generateVec(const unsigned n)
{
    vector<unsigned> result(n);
    generate(result.begin(), result.end(),
        [counter = 0]() mutable { return ++counter; });
    return result;
}
```

where *generate*() is an STL algorithm from the header <algorithm> that fills a sequence by repeated calls to a function, and the lambda is marked mutable because its execution modifies its internal data *counter*. We can code a function that sums up the values in a vector:

```
template <class T>
T accumulateVec(const vector<T>& v)
{
    return accumulate(v.begin(), v.end(), T());
}
```

where the STL *accumulate*() algorithm was discussed earlier. We could define a (particularly inefficient) way to compute the sum of the first *n* numbers by composition:

```
int main()
{
    auto h = compose(accumulateVec<unsigned>, generateVec);

    cout << h(100) << endl;
}
```

We could even design ways to compose functions of *multiple* arguments, either by binding or currying. We have to stop here and refer interested readers to a specialized publication like [21].

Lifting

Another useful idiom borrowed from functional programming is *lifting*. To lift a function means to turn it into one that operates on compound types. For instance, we may implement a lift that turns a scalar function into a vector function that applies the original function to all the elements of a vector:

```
template <class F>
auto lift(const F& f)
{
    return [f](const vector<double>& v)
    {
        vector<double> result(v.size());
        transform(v.begin(), v.end(), result.begin(), f);
        return result;
    };
}
```

transform() is a generic STL algorithm from header <algorithm> that applies a unary function to all the elements in a collection. What is returned from *lift*() is not a vector but a function of a vector that returns a vector. It can be used as follows (we lift the *exp*() function into a *vExp*() that computes a vector of exponentials from a vector of numbers):

```
int main()
{
    auto vExp = lift<double(double)>(exp);

    vector<double> v = { 1., 2., 3., 4., 5. };
    vector<double> r = vExp(v);

    for_each(r.begin(), r.end(),
        [](const double& x) {cout << x << endl; });
}
```

for_each() is another generic algorithm from the <algorithm> header that sequentially applies an action to all the elements in a collection. We use it to display the entries in the result vector *r*.

As a (slightly) more advanced example, suppose we have a function that implements the Black and Scholes formula from [22]:

```
double blackScholes(const double spot,
    const double strike,
    const double expiry,
    const double vol);
```

We can lift it into a function that computes a vector of option prices from a vector of spots, but we must first turn it into a function of the spot alone by binding the other arguments. That could be done with a lambda, or with the *bind*() function from the header <functional>:

```
#include <functional>
using namespace std;
using namespace placeholders;

int main()
{
    //  Create unary pricing function out of spot alone
    //      by binding the other arguments
    auto BSfromS = bind(
        blackScholes,   //  Function to bind
        _1,             //  spot = 1st arg of bound function
        100.,           //  strike = 100
        1.,             //  maturity = 1
        .10);           //  vol = 10

    //  Lift the unary function into a vector function
    auto vBlackScholes = lift(BSfromS);

    //  Apply the lifted function to a vector of spots
    vector<double> spots = { 50., 75., 100., 125., 150. };
    vector<double> calls = vBlackScholes(spots);

    //  Display results
    for_each(calls.begin(), calls.end(),
        [](const double& x) {cout << x << endl; });
}
```

More information about *bind*() can be found online. It takes a function, followed by its arguments in order, and returns a new function. When we pass a value for an argument, the argument is bound to this value. When we pass a placeholder *_n*, the argument is bound to the *n*th argument of the resulting function. In our example, we created a new function out of *blackScholes*(), by binding its first argument (the spot) to the first argument of the new function, and all other arguments to fixed values.

Alternatively, we could bind the spot and create a function that values a call out of volatility alone, and then lift it so it returns a vector of calls out of a vector of volatilities:

```
int main()
{
    // Create unary pricing function out of vol alone
    auto BSfromSigma = bind(
        blackScholes,     // Function to bind
        100.,             // spot = 100
        120.,             // strike = 120
        1.,               // maturity = 1
        _1);              // vol = (first) argument to created function

    // Lift the unary into a vector function
    auto vBlackScholes = lift(BSfromSigma);

    // Apply the lifted function to a vector of vols
    vector<double> vols = { .05, .10, .15, .20, .25, .50 };
    vector<double> calls = vBlackScholes(vols);

    // Display results
    for_each(calls.begin(), calls.end(),
        [](const double& x) {cout << x << endl; });
}
```

Note that our lifting function is specialized for functions *double*$->$ *double*, lifting into a function *vector* $<$ *double* $>$ $->$ *vector* $<$ *double* $>$. It is possible, with template magic, to produce a generic lifting function for functions $T_1 -> T_2$, lifting into $C < T_1 > -> C < T_2 >$ where C is an arbitrary collection, not necessarily a vector. This exercise is out of scope here, and we refer to specialized publications.

Functional programming idioms are exciting and fashionable. For an excellent introduction to functional programming in its natural habitat Haskell, we refer to [23].

We barely scratched the surface of functional programming in C++11, but hopefully gave a sense of how functions may be created and manipulated like any other type of data. It would take a dedicated publication to cover that subject in full, and, indeed, one such publication exists, [21], where interested readers will find a much more complete discussion of the implementation of functional programming idioms in C++.

2.3 MOVE SEMANTICS

Moving onto a different topic, the following pattern is valid but inefficient in traditional C++:

```
vector<double> f(/*...*/)
{
    vector<double> inner;
    //  ...
    return result;
}

int main()
{
    vector<double> outer = f(/*...*/);
}
```

The *inner* vector is destroyed when *f* returns, but before that, the *outer* vector is allocated and the contents of *inner* are copied. This is of course very inefficient: memory is allocated twice, and an unnecessary duplication of data is conducted from a container that is destroyed immediately afterwards. This inefficiency *might* be caught by the compiler's RVO (Return Value Optimization), whereby the compiler would directly instantiate *outer* inside *f*. RVO is not guaranteed[5] so programmers settled for a less natural syntax where result vectors are passed by reference as arguments rather than returned from functions.

C++11 *move semantics* permanently resolved this situation.

Conventional C++ allows class developers to implement their own *copy constructors* and *copy assignment operators*:

```
struct MyClass
{
    MyClass() { cout << "ctor" << endl; }
    ~MyClass() { cout << "dtor" << endl; }

    MyClass(const MyClass& rhs)
    {
        cout << "copy ctor" << endl;
    }

    MyClass& operator=( const MyClass& rhs)
    {
        if( this == &rhs) return *this;
        cout << "copy =" << endl;
        return *this;
    }
};
```

The code in the body of the copy constructor and assignment is automatically executed whenever an object of that type is initialized or assigned

[5]Although C++ rules are complex and in flux in that area, see for instance [24].

from another object of the same type. The code doesn't have to conduct a copy (our example does not) but it is expected that this is the case, and that such code should result in the duplication of the right-hand side (*rhs*) into the left-hand side (*lhs*).

When the developer does not supply a copy constructor or a copy assignment, the compiler provides default ones that perform copies of all data members (by calling their own copy constructors or assignment operators when these members are themselves classes).

C++11 introduced additional *move* constructors and assignments, with the perhaps unusual "&&" syntax:

```
MyClass(MyClass&& rhs)
{
    cout << "move ctor" << endl;
}

MyClass& operator=(MyClass&& rhs)
{
    if( this == &rhs) return *this;
    cout << "move =" << endl;
    return *this;
}
```

These are automatically invoked whenever the *rhs* is a *temporary object*, like a result returned from a function, as opposed to a *named* object. They can also be explicitly invoked with the *move*() keyword (which is actually a function):

```
MyClass someFunc(const MyClass& x)
{
    MyClass temp;
    //...
    return temp;
}

void myFunc()
{
    MyClass x, y, r;

    x = y;
    //  copy assign

    myClass z(y);
    //  copy construct

    myClass t(move(z));
    //  move construct

    r = move(t);
    //  move assign

    MyClass s = someFunc(r);
    //  temporary detected: auto-move!
}
```

We can (and, in the example, did) code whatever we want in the move constructor and assignment. What is expected is a quick transfer of the ownership of the *rhs* object's resources to the *lhs* object, without modification of the managed resources themselves, leaving the *rhs* empty.

Let us discuss a relevant example. If we wanted to code our own vector class wrapping a dumb pointer, we could proceed as follows (we simplify to the extreme and only show code for the core functionality). We start with the skeleton of a custom Vector class, including the copy constructor and copy assignment:

```
template<class T>
class Vector
{
    T*       myPtr;
    size_t   mySize;

public:

    Vector(size_t size = 0)
        : mySize(size), myPtr(size > 0? new T[size]: nullptr)
    {}

    ~Vector() { delete[] myPtr; }

    Vector(const Vector& rhs)
        :    mySize(rhs.mySize),
             myPtr(rhs.mySize > 0 ? new T[rhs.mySize] : nullptr)
    {
             copy(rhs.myPtr, rhs.myPtr + rhs.mySize, myPtr);
    }

    void swap(Vector& rhs)
    {
        std::swap(myPtr, rhs.myPtr);
        std::swap(mySize, rhs.mySize);
    }

    Vector& operator=(const Vector& rhs)
    {
        if (this != &rhs)
        {
            Vector<T> temp(rhs);
            swap(temp);
        }
        return *this;
    }

    void resize(size_t newSize)
    {
        if (mySize < newSize)
        {
            Vector<T> temp(newSize);
            copy(myPtr, myPtr + mySize, temp.myPtr);
```

```
            swap(temp);
        }
        mySize = newSize;
    }

    T& operator[] (const size_t i) { return myPtr[i]; }
    const T& operator[] (const size_t i) const { return myPtr[i]; }
    const size_t size() const { return mySize; }
    T* begin() { return myPtr; }
    const T* begin() const { return myPtr; }
    T* end() { return myPtr + mySize; }
    const T* end() const { return myPtr + mySize; }
};
```

The copy constructor clones the *rhs* Vector into *this*, implementing a memory allocation followed by a copy of the data.[6] The copy assignment operator does the same thing, but it must also release the data previously managed by *this* (for the copy constructor, *this* is not yet constructed so it doesn't manage any data).

To avoid duplicating the copy constructor code into the copy assignment operator, we applied the well known "copy and swap" idiom. The method *swap*() swaps the pointers (and sizes) of **this* and *rhs* Vectors, effectively swapping the *ownership* of data, without modifying the data itself in any way: after the swap, *this* owns *rhs*'s previous data, and *rhs* owns the data previously managed by *this*. In the assignment operator, we construct a temporary vector *temp* by copy of *rhs*, and swap *this* with *temp*. As a result, *this* holds a copy of the previous contents of *rhs* and *temp* manages the previous data of *this*. When *temp* goes out of scope, its destructor is invoked and the data previously managed by *this* is destroyed and its memory is released. A similar process is implemented in the resizer.

Copy semantics make a copy of *rhs* into *this*, without modification to *rhs*, at the cost of an expensive allocation and an expensive copy of the data. We now implement move semantics. A move is not supposed to copy any data, but transfer the ownership of *rhs*'s resources to *this* and leave *rhs* in an empty state. It follows that *rhs* is *not* a const argument to the move constructor:

```
// ...
Vector(Vector&& rhs)
{
    swap(rhs);
    rhs.myPtr = nullptr;
}
// ...
```

[6] *copy*() is another generic algorithm from <algorithm>; its purpose and interface are self-explanatory from the code.

After the swap, *this*− > *myPtr* points on *rhs*'s former data, *rhs.myPtr* points on *this*'s former, uninitialized memory, since *this* is not yet constructed. No data was copied, no memory allocated, and the ownership of *rhs*'s data was swiftly and efficiently transferred to *this*, in exchange for some uninitialized memory. Therefore, *rhs* is empty after execution and *this* effectively owns its former contents and resources. For avoidance of doubt, we set *rhs.myPtr* to *nullptr* so its destructor would not attempt to deallocate it.

The move assignment is implemented in a similar way, the difference being, *this* may own resources prior to the assignment, in which case they must be released. We move *rhs* into a temporary Vector, and swap *this* with *temp*, so that *this* ends up with the ownership of *rhs*'s previous resources, *rhs* ends up empty after it was moved, and the previous resources of *this*, transferred to *temp* in the swap, are released when *temp* exits scope.

```
// ...
Vector& operator=(Vector&& rhs)
{
    if (this != &rhs)
    {
        Vector<T> temp(move( rhs));
        swap(temp);
    }
    return *this;
}
// ...
```

A move transfer is an order of magnitude faster than a copy. All it does is swap pointers, something called a *shallow copy*, without allocation of memory or copy of data. But it renders the moved object unusable after the transfer.

When the *rhs* object is *unnamed*, for example, returned from a function, then it couldn't possibly be reused in any way. The compiler knows this, so it always invokes move semantics in place of slower copy semantics in these situations. When the *rhs* is a named object, the compiler cannot safely move it, and it would normally make a copy. When *we* know that we don't reuse a named *rhs* object after the transfer, we can explicitly invoke its move semantics with *move*().

In those situations where move semantics are invoked, either automatically or explicitly, on an object that doesn't implement them, the compiler falls back on to more expensive copy semantics. Copy semantics are always implemented: when a copy constructor and assignment operator are not explicitly declared on a class, the compiler generates default ones, by copy of all data members. On the contrary, the compiler doesn't generally produce a default move constructor or move assignment operator. This only happens in restrictive cases. It follows that we must always declare move semantics explicitly in classes that manage memory or other resources, or expensive copies will be executed in situations where a faster move transfer could have been performed safely instead.

Move semantics are implemented in all STL containers and standard library classes out of the box. It is our responsibility to implement them in our own container classes and other classes that manage resources. For example, we update our simple matrix code with move semantics. We will use it in Parts II and III. The code below is in the file matrix.h on our repository.

```
#include <vector>
using namespace std;

template <class T>
class matrix
{
    //  Dimensions
    size_t      myRows;
    size_t      myCols;

    //  Data
    vector<T>   myVector;

public:

    //  Constructors
    matrix() : myRows(0), myCols(0) {}
    matrix(const size_t rows, const size_t cols) :
        myRows(rows), myCols(cols), myVector(rows*cols) {}

    //  Copy, assign
    matrix(const matrix& rhs) :
        myRows(rhs.myRows), myCols(rhs.myCols), myVector(rhs.myVector) {}
    matrix& operator=(const matrix& rhs)
    {
        if (this == &rhs) return *this;
        matrix<T> temp(rhs);
        swap(temp);
        return *this;
    }

    //  Copy, assign from different (convertible) type
    template <class U>
    matrix(const matrix<U>& rhs)
        : myRows(rhs.rows()), myCols(rhs.cols())
    {
        myVector.resize(rhs.rows() * rhs.cols());
        copy(rhs.begin(), rhs.end(), myVector.begin());
    }
    template <class U>
    matrix& operator=(const matrix<U>& rhs)
    {
        if (this == &rhs) return *this;
        matrix<T> temp(rhs);
        swap(temp);
        return *this;
    }
```

```

    //  Move, move assign
    matrix(matrix&& rhs) :
        myRows(rhs.myRows),
        myCols(rhs.myCols),
        myVector(move(rhs.myVector)) {}

    matrix& operator=(matrix&& rhs)
    {
        if (this == &rhs) return *this;
        matrix<T> temp(move(rhs));
        swap(temp);
        return *this;
    }

    //  Swapper
    void swap(matrix& rhs)
    {
        //  Call std::vector::swap()
        myVector.swap(rhs.myVector);
        //  Call free function std:swap()
        ::swap(myRows, rhs.myRows);
        ::swap(myCols, rhs.myCols);
    }

    //  Resizer
    void resize(const size_t rows, const size_t cols)
    {
        myRows = rows;
        myCols = cols;
        if (myVector.size() < rows*cols) myVector = vector<T>(rows*cols);
    }

    //  Access
    size_t rows() const { return myRows; }
    size_t cols() const { return myCols; }
    //  So we can call matrix [i][j]
    T* operator[] (const size_t row) { return &myVector[row*myCols]; }
    const T* operator[] (const size_t row) const
        { return &myVector[row*myCols]; }
    bool empty() const { return myVector.empty(); }

    //  Iterators
    using iterator = typename vector<T>::iterator;
    using const_iterator = typename vector<T>::const_iterator;

    iterator begin() { return myVector.begin(); }
    iterator end() { return myVector.end(); }
    const_iterator begin() const { return myVector.begin(); }
    const_iterator end() const { return myVector.end(); }
};

//  Free function to transpose a matrix
template <class T>
inline matrix<T> transpose(const matrix<T>& mat)
{
    matrix<T> res(mat.cols(), mat.rows());
```

```
        for (size_t i = 0; i < res.rows(); ++i)
        {
            for (size_t j = 0; j < res.cols(); ++j)
            {
                res[i][j] = mat[j][i];
            }
        }

        return res;
}
```

2.4 SMART POINTERS

Smart pointers wrap standard (or *dumb*) pointers and implement the RAII (Resource Acquisition Is Initialization) idiom. RAII is a rather verbal name for an idiom that implements the release of resources in destructors, so that when an object exits scope, for whatever reason (the function returned or an exception was thrown), resources are always automatically released. Smart pointers relieve developers from the concern of explicitly releasing allocated memory, and protect against memory leaks.

Smart pointers are otherwise manipulated just like dumb pointers; in particular they can be dereferenced with operators $*$ and $->$ to read and write the managed object.

C++ developers have been using smart pointers for decades, either hand-crafted or from third-party libraries like Boost. They are part of the standard C++ library since C++11.

A simplistic smart pointer could be coded as follows. This smart pointer cannot be copied, since the memory is owned and released on destruction by a single object. But it can be moved, in which case the *rhs* pointer loses ownership of memory when the *lhs* pointer acquires it.

```
template <class T>
class SmartPointer
{
    //  Wrap a dumb pointer
    T*  myPtr;

public:

    //  Adopt ownership of memory
    SmartPointer(T* ptr = nullptr) : myPtr(ptr) {}

    //  This is the smart bit: release memory on destruction
    ~SmartPointer() { delete myPtr; }

    //  Forbid copy
    SmartPointer(const SmartPointer& rhs) = delete;
    SmartPointer& operator=(const SmartPointer& rhs) = delete;

```

```
    //  Implement move
    SmartPointer(SmartPointer&& rhs) : myPtr(rhs.myPtr)
    {
        rhs.myPtr = nullptr;
    }

    SmartPointer& operator=(SmartPointer&& rhs)
    {
        if (this != &rhs)
        {
            delete myPtr;
            myPtr = rhs.myPtr;
            rhs.myPtr = nullptr;
        }
        return *this;
    }

    //  Swapper
    void swap(SmartPointer& rhs)
    {
        std::swap(myPtr, rhs.myPtr);
    }

    //  Adopt dumb pointer
    void reset(T* ptr)
    {
        delete myPtr;
        myPtr = ptr;
    }

    //  Release ownership
    //      memory is not released and the managed object is not destroyed
    //      ownership is transferred to the returned dumb pointer
    T* release()
    {
        T* temp = myPtr;
        myPtr = nullptr;
        return temp;
    }

    //  Access managed object
    T* get() const
    {
        return myPtr;
    }

    T& operator*() const
    {
        return *myPtr;
    }

    T* operator->() const
    {
        return myPtr;
    }
};
```

The standard library smart pointer *unique_ptr* located in the <memory> header is implemented along these lines.

Importantly, the standard guarantees that *unique_ptr*s are manipulated without overhead compared to dumb pointers. Dynamic memory management with dumb pointers is inconvenient, prone to memory leaks, and without benefit compared to *unique_ptr*s. RAII management is a free benefit, and it is considered best practice to always manage heap memory with smart pointers.

We can create an object on the heap with the traditional operator *new* and assign it to a smart pointer for RAII management:

```
Object* object = new Object;
unique_ptr<Object> ptr(object);
// RAII:
//      object is destroyed and its memory released
//      on destruction of ptr when ptr exists scope
```

or more simply:

```
unique_ptr<Object> ptr(new Object);
```

Since C++14, the free function *make_unique*() offers a terser, potentially more efficient syntax for the creation of objects in dynamic memory managed with a *unique_ptr*:

```
unique_ptr<Object> ptr = make_unique<Object>();
```

or more simply:

```
auto ptr = make_unique<Object>();
```

make_unique() also forwards its parameters to the managed object's contructor:

```
class Object
{
    string myName;

    // ...

public:
    private Object(const string& name) : myName(name) {}

    // ...
};

auto ptr = make_unique<Object>("object");
```

The second breed of standard smart pointers, *shared_ptr*s, also located in the <memory> header, offer further benefits, but not for free. Shared pointers are *reference counted*. They *can* be copied, in which case all copies *share* the ownership of the managed memory, and it is only when the *last* owner exits scope that the managed memory is released and its contents destroyed.

Shared pointers are very powerful because we never worry about memory being released too early or too late. A resource remains alive as long as there are pointers referencing it, and it is released automatically when this is no longer the case. But they are slower than dumb and unique pointers. Reference counting is not free, especially in a concurrent environment. For this reason, we must use *shared_ptr*s parsimoniously, always pass them by reference (to avoid unnecessary reference counting), and do nothing else than dereference them[7] at low level and especially in repeated code. Obviously, we only use *shared_ptr*s when we effectively need reference counting, and *unique_ptr*s otherwise.

Like *unique_ptr*s, *shared_ptr*s can adopt dumb pointers for RAII management, although this is implemented in a more convenient and more efficient manner in the factory function *make_shared*() which syntax is identical to *make_unique*().

The following example demonstrates how the two types of standard library smart pointers work.

```
#include <memory>
using namespace std;

struct MyClass
{

    ~MyClass() { cout << "dtor called" << endl; }

    void read() const
        { cout << "i am being read" << endl; }

    void write()
        { cout << "i am being written" << endl; }
};

void myFunc()
{
    {
        unique_ptr<MyClass> up2;

        {
            //  Create an object of type MyClass in dynamic memory,
            //      get a unique_ptr on the object
            unique_ptr<MyClass> up = make_unique<MyClass>();

            //  Manipulate the managed object through the unique_ptr
            up->read();
            up->write();
```

[7]Dereference does not involve reference count: only the creation, copy, and destruction of *shared_ptr*s are expensive.

```
        // Get a direct reference on the managed object
        MyClass& obj = *up;

        // up2 = up;    ERROR: not copyable
        up2 = move(up); //      OK: move

        // myClass& obj2 = *up; // ERROR:
        //      ownership no longer belongs to up

        // up destroyed here,
        // yet nothing happens: ownership was moved to up2
    }
    cout << "out of inner block" << endl;
    //  up2 destroyed here, object is destroyed and memory is released
  }
  {
    shared_ptr<MyClass> sp2;

    {
        // Create an object of type MyClass in dynamic memory,
        //      get a shared_ptr on the object
        shared_ptr<MyClass> sp = make_shared<MyClass>();
        // sp is the only pointer on the managed object
        //      reference count = 1

        // Manipulate the managed object through the shared_ptr
        sp->read();
        sp->write();
        // Get a direct reference on the managed object
        MyClass& obj = *sp;

        sp2 = sp;              // OK, resource now shared
        // reference count = 2

        MyClass& obj2 = *sp; // OK, sp and sp2 reference the same object

        // sp destroyed here, sp2 still alive
        // reference count = 1
        // managed object remains alive
    }
    cout << "out of inner block" << endl;
    // sp2 destroyed here, reference count goes to 0,
    //      object is destroyed and memory is released
  }
}
```

C++11 comes with many, many more new features, the most useful probably being hash tables and variadic templates. They are covered in many textbooks and online resources. Readers wishing to investigate these matters in further detail are referred to Scott Meyers' [24]. An outstanding innovation is undoubtedly the Standard Threading Library investigated in the next chapter.

CHAPTER 3

Parallel C++

In this chapter, we introduce the C++11 Standard Threading Library, discuss its main components, and develop a basic thread pool. We reuse that thread pool to perform concurrent calculations in the rest of the book.

In order to understand concurrent programming and make sense of this chapter, we must first appreciate what are the different pieces involved, how they interact, and who is responsible for their management.

Cores are the hardware processing units that execute sequences of instructions. Typical laptops and PCs have two to six physical cores today, typical workstations have eight cores and up.

Intel's hyperthreading technology runs multiple (two) *hardware threads* per core. Hardware threads, also called *logical cores*, accelerate execution by switching logical cores on physical cores when beneficial, in accordance with a logic hard coded on the chip. From the application's point of view, hardware threads are seen and treated as physical cores.

Performance benefits are entirely dependent on the number of cores on the executing machine. Multi-threaded applications run no faster (actually, slower) on single-threaded computers.[1]

Threads are the sequences of instructions executed on the machine's available cores. Every application creates at least one thread, called *main thread*, which starts when the application launches. Multi-threaded applications are those who explicitly create additional threads, with a syntax explained in Section 3.2.

[1]Those are not being built anymore. Even phones are multi-core and have been for years.

The threads created by an application are distributed, or *scheduled* for execution, by the OS, and outside of the application's control,[2] over the available cores.[3]

Performance is not the only reason for building multi-threaded applications. Other reasons include multi-tasking (the ability to do several things at the same time is in itself valuable, even when it is the same CPU that performs these tasks: otherwise we couldn't run calculations on Excel and document them on Word simultaneously) or separation of concern (in a video game, the movements of the player and their enemies may be coded on separate threads, which clarifies code, and *may* increase performance if executed on multiple cores).

When we have more threads than cores, the OS emulates concurrency by sharing CPU time, quickly switching execution threads, evidently without a performance benefit. On the contrary, performance typically suffers from context switching. The application may create a lower or higher number of threads than there are available cores, although performance benefits are limited to the number of cores.

The creation of threads involves an overhead, and the management of a large number of threads may slow down the OS. For these reasons, we must carefully choose the number of threads and refrain from starting and ending them repeatedly.

Tasks are the logical slices of a parallel algorithm that may be executed simultaneously, like the transformation of a subset of the source collection, or the processing of a batch of scenarios in a Monte-Carlo simulation. The number of tasks may be very large, and it is the application's responsibility to schedule their execution on the application's threads.

Thread pools are constructs that centralize and encapsulate the scheduling of tasks over threads in a multi-threaded application. Thread pools are a best practice that help produce modular, readable code and improve performance. They are not strictly speaking necessary. The management of threads and tasks in the application could be conducted manually, although this is not recommended.

[2]Unless the application uses a specific API to directly program into the OS, something outside of our scope.

[3]The OS schedules threads "at best," and there is no guarantee that threads will execute in parallel. In practice, we could verify that Windows scheduling is very efficient and uses hardware to its full potential. We have no doubt that this also applies to other major OSs.

The best design for a thread pool depends on the application's specific needs. Therefore, the C++ standard library does not provide thread pools, although it provides all the building blocks for a flexible, custom implementation. Financial applications typically only require basic scheduling, and the simple design implemented in Section 3.18 is sufficient for our needs. Williams develops in [15] more sophisticated thread pools, and introduces advanced scheduling techniques like *work stealing*. That publication, written by a developer of the Standard Threading Library, provides a more comprehensive discussion. It is a recommended reading for anybody wishing to earn a deep understanding of concurrency and parallelism. Our coverage is more condensed, concurrency in itself not being the primary topic of this publication.

3.1 MULTI-THREADED HELLO WORLD

It is customary to begin programming discussions with something that displays "Hello World." To do so concurrently, we implement the following code. It compiles and works out of the box in Visual Studio, without the need for linking to any third-party library or special project settings. All we need do is include the header <thread>.

```
#include <thread>
using namespace std;

void threadFunc()
{
    cout    << "Hello world from worker thread "
            << this_thread::get_id()
            << endl;
}

int main()
{
    const size_t n = thread::hardware_concurrency();
    vector<thread> vt(n);
    for( size_t i=0; i<n; ++i)
    {
        vt[i] = thread(threadFunc);
    }
    cout << "Hello world from main thread " << endl;
    for(size_t i=0; i<n; ++i)
    {
        vt[i].join();
    }
    cout << "Completed " << endl;
}
```

thread : : *hardware_concurrency*() is a static method that provides at run time the number of hardware threads on the machine executing the code. Our code results in interlaced messages on the application window, since multiple threads are writing into the window without any form of synchronization:

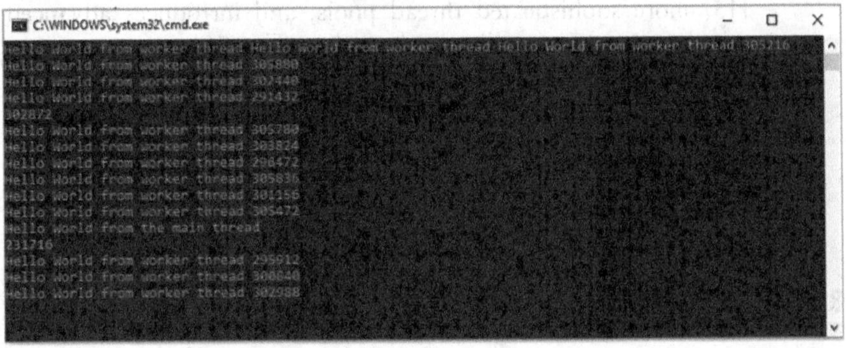

3.2 THREAD MANAGEMENT

The instruction $vt[i] = thread(threadFunc)$ on line 17 creates a new object of a class type *thread* and initializes it with a callable object. This results in the creation of a new thread, which immediately begins the execution of the callable argument, while the main thread continues the execution of *main*().

The *thread* object returned in $vt[i]$ is not an actual thread but a thread *handle*, conceptually similar to a pointer on a thread. Its constructor fires the execution of the callable argument on a newly created parallel thread; the handle provides basic control over that thread. It is this handle and not the thread itself that is represented in the *thread* class.

This is how we start parallel threads. We create new threads and give them a job to do, but the actual *scheduling* (the execution of the threads on CPU cores) is delegated to the OS.[4] We access the number of available hardware threads with *thread* :: *hardware_concurrency*(), which helps write generic code that scales at run time with the numbers of cores. The standard library does not allow to schedule a thread on a specific core.[5] This is called *affinity* and may be implemented with OS specific libraries. We will not use affinity.

[4]With hyperthreading, the OS schedules the execution of threads on hardware threads and the CPU schedules hardware threads on cores.

[5]Or to interfere with scheduling in any way.

The callable argument to the *thread* constructor may be a function, a member function belonging to an object, a function object, or a lambda. It may have parameters, in which case we pass its arguments to the thread constructor after the name of the function. This pattern for callable objects is consistent throughout C++11 and works as follows. It is important to understand this part well, because this pattern for callables is reused everywhere in C++11. In a vast majority of cases of practical relevance, however, the most convenient way to manipulate callables is with lambda expressions.

Free function We give the name of the function followed by its arguments. For instance:

```
int someFunc(const int a, double& b, const string& c);
//  ...
double x;
auto t = thread(someFunc, 2, ref(x), "I am multithreading!");
```

We note that whatever the function returns gets lost in the process, and, indeed, we cannot return results from threads this way. Passing results from worker threads to caller threads is discussed in Section 3.17.

We can pass arguments by reference, but in this case, there is a subtlety we must point out. We don't send arguments directly to our function; we pass them to the constructor of *thread*, which *forwards* them to our function. Our function may take some arguments by reference, but the thread constructor is implemented with (variadic) templates, where type resolution would result in passing the argument by value, unless we explicitly make it a reference with the keyword *ref* (which is actually a function located in the header <functional>). When in doubt, it is best to use pointers to avoid confusion. Even more convenient is to use a lambda that can capture variables by reference instead of taking them as parameters.

Function Object A function object is an instance of a class that defines the operator () and which instances are therefore callable like functions. For example:

```
//  Some class for the representation of implied vol surfaces
class ImpliedVolSurface
{
    //  ...
public:
    //  Picks an implied vol for a given strike and expiry
    void operator() (const double K, const double T, double* ivol);
};
```

In this case, to pick a volatility *asynchronously* on the surface we proceed as follows:

```
//  An instance of ImpliedVolSurface named iVolSurf
ImpliedVolSurface iVolSurf;
//  ...
double strike, mat;
//  ...
double vol;
//  Pick vol from a different thread
//  We pass the object (INSTANCE) as callable
//       followed by arguments
thread t(iVolSurf, strike, mat, &vol);
//  Do something in main thread in parallel
//  ...
//  Wait for parallel thread to finish
t.join();
//  The required volatility is is now computed in vol
//  Do something with the picked implied vol
//  ...
```

The first argument to the thread constructor is the name of the function object (the instance, not the class), followed by a list of the arguments. Note that the argument for the result is passed by pointer to avoid difficulties with references.

Member Function We may also ask the thread to execute a method on a given object. For instance, the previous example could have been written differently:

```
//       Just an object, not a function object
class ImpliedVolSurface
{
//  ...
public:
    //  Just another method, not the operator ()
        void getVol(const double K, const double T, double* ivol);
};

// The instance
ImpliedVolSurface iVolSurf;
//  ...
double strike, mat;
//  ...
double vol;
//  Pick vol on the surface from a different thread
thread t(&ImpliedVolSurface::getVol, &iVolSurf, strike, mat, &vol);
//  Do something in main thread in parallel
//  ...
//  Wait for parallel thread to finish
t.join();
//  Do something with the picked implied vol
//  ...
```

In this case, the first argument to the thread constructor is the name of the *method* to be called, qualified with the name of the class: "&class::method." The second argument is the instance that executes the method. And the rest is the list of the arguments to the method as before.

Now that may be confusing and hard to remember. It is most convenient to use lambdas in all cases. Of course, we can easily replicate all the examples above with simple lambdas, and there may not be any doubt about references or values in this case, as this is clearly specified in the lambda's capture clause. For example, the last example could be rewritten as follows:

```
thread t( [&] () { iVolSurf.getVol(strike, mat, &vol); } );
```

That syntax "thread t(callable, . . .)" creates a new thread and fires the execution of the *callable* on that thread. When the execution of the *callable* completes, the thread is destroyed. The thread handle *t* provides the caller with basic means to inspect and control the thread. More specifically, the caller may use the handle to:

Identify the thread with *t.get_id*().

Wait for the thread to complete with *t.join*(). If the thread already completed, *join*() returns immediately. Otherwise, it causes the calling thread to wait *in idle state*, without consuming any CPU resources, for the thread referenced by *t* to complete. *t.join*() cannot return before "t" completes, hence it is guaranteed that the job that was given to *t* on construction completed after *t.join*() returned. The code that comes after this can rely that the asynchronous job is complete; in particular, it may safely use the results.

Break the handle with *t.detach*(). This breaks the *link* between the handle *t* and the thread it represents. It does not affect the thread itself in any way. This means that we lose control of the thread; we have no longer any means to know what is going on there. In particular, we can no longer wait for the thread to complete its job. The thread will finish its work and terminate behind the scenes, unmanaged and invisible.

Thread handles are *moveable* objects. They cannot be copied (an attempt to do so results in a compile error) but they can be *moved*. When something like "thread lhs = move(rhs)" is executed, *rhs* loses the link to the thread, which ownership is transferred to *lhs*. As a curiosity, the instruction we programmed in our Hello World, *vt*[*i*] = *thread*(*threadFunc*), implements an implicit move from a temporary object. *thread* does not allow copy assignment.

When the thread linked to a handle completes, the handle is still considered *active* until its method *join*() is invoked. After that call, the handle becomes inactive. When the link between the thread and the handle is broken by a call to the handle's *detach*() method, the handle also becomes inactive. The active status of a handle may be checked with a call to its *joinable*() method (returns *true* if the handle is active). We cannot call *join*() on an inactive handle. An active handle, on the other hand, cannot exit scope. If that ever happens, the application *terminates*. This means that either *join*() or *detach*() *must* be called on the handle before it vanishes. We must either wait for the thread to complete or release it to finish its life outside of our control, and we must do so explicitly.

For this reason, it is often advised to manage thread handles with the RAII idiom, similarly to smart pointers. We should wrap a *thread* within a custom class that calls *join*() in its destructor. This would guarantee that an active handle exiting scope would cause to wait for the linked thread to complete rather than crash the application. However, all this doesn't matter that much to us because *we never manage threads directly*.

The creation of a thread comes with a significant overhead, and the OS can only handle so many threads. For these reasons, we limit the number of threads we create (typically to the number of cores *minus one* so that one core remains available for the main thread) and we want to keep the thread logic outside of the algorithmic code, both for performance (no thread creation overhead during the execution of the algorithm) and encapsulation (algorithms shouldn't worry about hardware logic).

For example, in the case of parallel Monte-Carlo simulations, we execute different paths on different threads. If we simulate 65,536 paths, we cannot create 65,536 threads. That would bloat the OS. Instead, we could create *thread* : : *hardware_concurrency*() – 1 threads, divide the 65,536 paths into *thread* : : *hardware_concurrency*() batches, send one batch to each created thread, and keep one batch for execution on the main thread. This whole logic should not be implemented in the Monte-Carlo code. This hardware-dependent division does not belong to simulation logic. Besides, thread creation is expensive so we would rather create the worker threads once and for all (for instance, when the application starts) and have them wait (without consuming any CPU or other resources) to be given jobs to do in parallel, until they are destroyed when the application exits.

When some algorithm such as the Monte-Carlo simulator sends a job for a parallel execution, some kind of scheduler should assign a worker thread to execute it and then send the thread back to sleep, but without destroying it (otherwise we would incur its creation cost again when we need it next). This whole logic is best encapsulated in a dedicated object. An object that works in this manner is called an *active object* because it sleeps in the background and awakens to conduct parallel work when needed. Modern GUI design is based on this active object pattern. The particular active object that dispatches tasks over worker threads is called a *thread pool* and we discuss these in detail in Section 3.18.

In addition, basic thread management does not provide mechanisms to return results to the caller or communicate exceptions in case something goes wrong. We will provide such facilities in our thread pool.

First, we investigate the *synchronization* of threads.

3.3 DATA SHARING

Each thread owns its private execution stack. Concretely, this means that each thread works with its own copy of all the variables declared within the functions and methods it executes, including nested ones. When the type of the variable is a reference or a pointer, however, the thread still exclusively owns its copy of the pointer/reference, but this is not necessarily the case for the referred object. The object could live on the heap, or on the stack owned by another thread. It could be *shared*, referenced, and accessed *concurrently* (meaning at the same time) from multiple threads. Multiple threads reading a shared object causes no trouble, but concurrent writes produce deep, dangerous, hard-to-debug problems called *race conditions*. The same applies to global and static variables shared by all threads.

The demonstration code below shows a few examples.

```
#include <thread>
using namespace std;

// Global
vector<int> v1 = { 1, 2, 3, 4, 5 };

void threadFunc(vector<int>* v)
{
    // Local
    vector<int> v3 = { 5, 4 };

    // OK: reading shared data, writing into own data
    v3[0] += (*v)[0];

    // RACE: changing shared data
    v->push_back(1);

    // RACE: writing into global data
    v1[0] += 1;
}

int main()
{
    // Belongs to the main thread's stack
    vector<int> v2 = { 3, 2, 1 };

```

```
    const size_t n = thread::hardware_concurrency();
    vector<thread> vt(n);
    for (size_t i = 0; i<n; ++i)
    {
        //  v2 passed by reference to other threads
        vt[i] = thread(threadFunc, &v2);
    }
    for (size_t i = 0; i<n; ++i)
    {
        vt[i].join();
    }

    for_each(v1.begin(), v1.end(),
        [](const int& i) {cout << i << " "; }); cout << endl;
    for_each(v2.begin(), v2.end(),
        [](const int& i) {cout << i << " "; }); cout << endl;
}
```

Worst thing is, this code will probably run just fine and display the expected results. We may have to run it many times to see a problem. Races occur when threads read and write into memory *at the same time*. That is a low-probability event. But this does not make it better. It makes it worse. Races cause random, nonreproducible bugs that may crash the application, produce absurd results, or (which is much worse) produce wrong results by small amounts so they may not be noticed.

3.4 THREAD LOCAL STORAGE

The variables that are global (declared outside classes and functions) are shared among threads. That is also true of *static* variables, which, for all intents and purposes, are global variables declared within classes or functions. Concurrent writes into global or static variables cause data races, just as with other shared objects. For this particular case, however, C++11 provides a specific solution and even a special keyword *thread_local*.

This keyword only applies to global and static variables, and when any such variable is declared with the *thread_local* specifier, every thread works with its own copy of it. The code accesses this variable like any other global or static variable, but in doing so, each thread reads and writes to a different memory location.

We will see a simple, yet very useful example in our thread pool implementation of Section 3.18, where a thread local variable represents the index of the threads: 0 for the main thread or 1 to

$$thread :: hardware\_concurrency() - 1$$

for the worker threads. Another example will be given in Part III, where the *tape*, the data structure at the core of AAD that records mathematical operations, is declared thread local, allowing AAD to work with parallel algorithms.

3.5 FALSE SHARING

We briefly introduced (true) sharing and the dangers of writing into shared data, and discussed thread local storage as a solution for the (rare but important) case of global/static data. Before we investigate more general mitigation to (true) sharing, we introduce *false sharing*.

As a first example of a real multi-threaded algorithm, we compute the non-parametric distribution of some data in a vector. The single-threaded code is self-explanatory.

```
#include <vector>
#include <algorithm>
using namespace std;

vector<int> dist(
    const vector<double>& data,
    const vector<double>& knots)
{
    // Vector of results: count data between knots
        const size_t n = knots.size() + 1;
        vector<int> res(n, 0);

    // Loop on knots
        for (size_t i = 0; i < n; ++i)
        {
            //  Lower bound
                const double lb = i == 0 ?
                    -numeric_limits<double>::max()
                    : knots[i - 1];

                //  Upper bound
                const double ub = i == n - 1 ?
                    numeric_limits<double>::max()
                    : knots[i];

        //  Count data
        for (size_t j = 0; j < data.size(); ++j)
            res[i] += (lb < data[j] && data[j] <= ub);
        }
        return res;
}
```

We test the algorithm as follows:

```
#include <iostream>
#include <ctime>
int main()
{
    const size_t n = 50000000;
```

```
    vector<double> data(n);
    srand(12345);

    // Populate data with n uniform numbers in (0, 1)
    generate(data.begin(), data.end(),
        [] () {return double(rand()) / RAND_MAX; });

    // Knots
    vector<double> knots = { 0.1, 0.15, 0.2, 0.21, 0.25, 0.35, 0.4,
        0.5, 0.55, 0.6, 0.7, 0.75, 0.8, 0.85, 0.9, 0.95 };

    // Measure time
    time_t t1 = clock();
    vector<int> result = dist(data, knots);
    time_t t2 = clock();

    // Display results
    cout << "result " << endl;
    for_each(result.begin(), result.end(),
        [=](const int& i) { cout << double(i) / n << " "; });
    cout << endl << "time (ms) " << t2 - t1 << endl;
}
```

We generate a large vector of random uniform data between 0 and 1 (using the STL algorithm *generate*()), send it to our algorithm and display results (with the very useful algorithm *for_each*()) as well as the time spent.

We get the expected results in around 3 seconds on our iMac Pro. To make this algorithm parallel (across knots) is apparently trivial. Note that we create threads inside the algorithm. We are not supposed to do that, but we don't know any other way yet:

```
#include <thread>
vector<int> parDist(
    const vector<double>& data,
    const vector<double>& knots)
{
    const size_t n = knots.size() + 1;
    vector<int> res(n, 0);
    vector<thread> threads(n);

    for (size_t i = 0; i < n; ++i)
    {
        // Exactly the same code, except we send each knot
        //     on a separate thread

        // i is captured by value
        threads[i] = thread([&, i] ()
        {
            // i read here, main thread would interfere
            //     if captured by reference
            const double lb = i == 0 ?
              -numeric_limits<double>::max()
              : knots[i - 1];

            const double ub = i == n - 1 ?
```

```
25              numeric_limits<double>::max()
26              : knots[i];
27
28          for (size_t j = 0; j < data.size(); ++j)
29          res[i] += (lb < data[j] && data[j] <= ub);
30        });
31    }
32    for (size_t i = 0; i < n; ++i) threads[i].join();
33    return res;
34  }
```

Note that the lambda captures the counter *i* by value. The main thread changes *i* in the dispatch loop, so if we captured it by reference, we could read a changing value of *i* from the lambda executed on worker threads. The (const) vectors are captured by reference.

We get the exact same results (thankfully) in around a second. Now this is disappointing. Why only 3 times the speed with 8 cores?

Note that this code does not write to any shared data. On the line 28, each thread writes into its dedicated entry *i* of the result vector, hence, threads don't interfere with one another. Or do they? What if we change this line of code so that within the loop we write into a local variable instead, replacing the code of lines 28–29 by:

```
int count = 0;
for (size_t j = 0; j < data.size(); ++j)
    count += (lb < data[j] && data[j] <= ub);
res[i] = count;
```

Now we get the same results in 2.5 seconds single-threaded, 300 milliseconds multi-threaded. Not only is the serial code faster; because it is faster to access the stack variable *count* than access *res*[*i*] on the the heap, we also get a parallel acceleration of more than 8 times, as expected from multi-threading over 8 hyper-threaded cores.

Note that if we had used the STL algorithm *count_if*() in place of the hand-crafted loop, as is recommended best practice, we would have got this result in the first place:

```
res[i] = count_if(data.begin(), data.end(),
    [=](const double x) { return lb < x && x <= ub; });
```

This illustrates, once again, that we should always prefer STL algorithms to hand-crafted loops. So *count_if*() would have shielded us from the problem by avoiding repeated concurrent writes into the vector *res*. But why was that a problem in the first place? Each thread was writing into its own space within that vector, so there shouldn't have been any interference. And yet they did interfere. While the threads wrote into separate spaces of memory, so there was no "true" sharing; they were writing to locations close enough to one another so they belonged to the same *cache line*. We recall that a cache line is what is transferred back and forth between the RAM and the CPU cache, and its size is generally 64 bytes or 8 integers (we compiled in 64 bits). So the entries *res*[0], *res*[1], …, *res*[7] lived on the same cache line. Threads were writing into separate memory cells but into the same cache line. So in some sense there was sharing involved. This form of sharing through cache lines is called *false sharing* or cache ping-pong for a reason that will be made apparent imminently.

We have seen in Chapter 1 that when a core reads a memory cell, it first looks for that location in the cache. If it is not there, the cache is *missed*, and the cell is fetched from RAM and cached along with its line (the 64 bytes surrounding it). Each core has its own cache. If another core reads another RAM cell on the same line, it also caches the line:

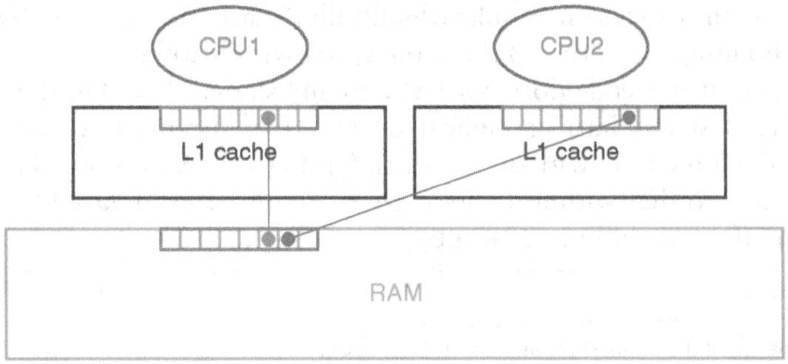

Read

At this point, both cores cached the line. When one core subsequently *writes* into a memory cell on that line, it actually writes into its own cache. This causes an inconsistency between the RAM and the cache. If another core tried to read the memory cell, it would get a wrong read from the RAM because the first core only updated its cache. So the first core must "warn" the system about the discrepancy. It does so by marking the RAM cell invalid, for everyone else to know. But it cannot invalidate a single cell; it invalidates the whole line. When the second core reads another cell on the same line, it cannot read it from its own cache, because it sees the whole line as invalid. It cannot read it from RAM, either, for the same reason. The only place it is guaranteed to find an up-to-date read is in *the first core's cache*. So, the whole line must be transferred from the first core's cache into RAM, and from RAM into the second core's cache, where the second core finally reads it. Then, of course, the second core writes into the same memory line and the whole process is repeated. We know that this is all unnecessary, that each core actually invalidates one cell on the line that is not used by the other cores. But the CPU does not know that. It can only see that lines are valid or not; it doesn't know the status of individual cells. And, when in doubt, it transfers the whole line.

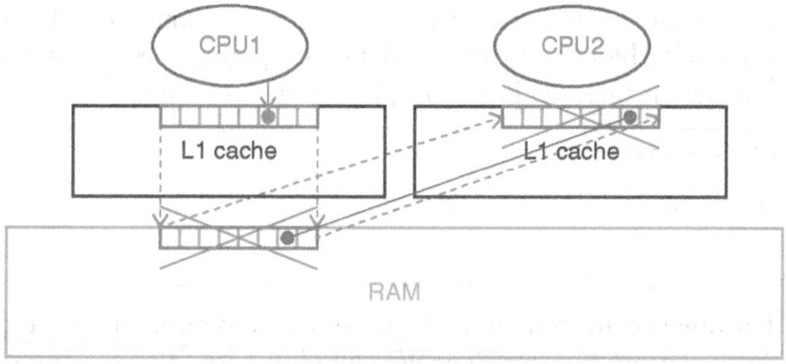

Write

This is false sharing, or cache ping-pong, and this is why our initial concurrent code did not perform as expected. Now we understand the matter, we may even wonder by what miracle it still achieved a decent speed improvement? It could easily have resulted in a parallel code *slower* than its serial counterpart.

The answer is that we simplified the design of CPU caches to avoid confusion in our initial discussion. In reality, modern CPUs typically incorporate a hierarchy of three levels of cache. L1 cache is the smallest and fastest and sits on CPU cores. L2 is bigger and slower and there is also one for every core. L3 is slower, bigger[6] and more importantly shared among all cores on a CPU. RAM is the biggest and slowest of all. These cache levels form a hierarchy, which means that every memory access is first checked in L1, if missed, checked in L2, if missed in L3 and finally in RAM, caching the line all the way up to the core.

That the L3 cache is shared among cores is what mitigates false sharing, because cache ping-pong occurs between L2 and L3 caches, not between cache and RAM. This is why we still saw a decent speed improvement from our initial multi-threaded code.

We noted in Chapter 1 that code *within a thread* should operate on coalescent memory for cache efficiency. Now we see that in addition, codes that execute on separate threads should work with memory distant from one another. We are now aware of false sharing and we learned a few things about hardware in the process, so we can compile a few guidelines to mitigate false sharing, and for the production of effective code in general:

1. Prefer STL algorithms to hand-crafted loops.
2. Cache repeatedly accessed heap data into local stack variables.

[6] L3 is what manufacturers advertise. When they claim that some high-end CPU has 25 MB cache, for instance, what this really means is 25 MB L3 cache. The sizes of the (much smaller) core specific L2 and L1 caches are generally much harder to find.

3. When concurrent writing into close memory is unavoidable, *pad* that memory with blocks of dummy data to separate memory written into from different threads. For example, say we use a data structure:

```
struct SomeObject
{
    double x;
    double y;
};
```

so that one thread frequently writes into x and another into y, producing false sharing interference. We can eliminate false sharing (at some memory cost) by padding the structure with data the size of a cache line:

```
struct SomeObject
{
    double x;
    char pad[64];
    double y;
};
```

3.6 RACE CONDITIONS AND DATA RACES

Race conditions and data races are the most common flaws in multi-threaded programs, caused by incorrectly synchronized data sharing among threads.

A *race condition* is a flaw that occurs when the timing or ordering of events affects a program's correctness. A *data race* is one nasty race condition where a thread reads some memory location *while* it is being written by another thread, which may result in reading invalid or corrupted data.

The textbook example of a race condition is the concurrent access to a bank account. Consider the class:

```
class SharedBankAccount
{
    double myBalance;

public:

    double withdraw( const double amount)
    {
        if( amount <= myBalance)
        {
            myBalance -= amount;
            return amount;
        }
        else return 0;
    }
};
```

When the method *withdraw*() is called concurrently from two threads, its instructions are executed in parallel from both threads. How fast each thread executes the sequence of instructions depends on scheduling and may be different every time the program runs. For instance, two concurrent withdrawal requests for $100 from an account with a balance of $120 could be scheduled as follows:

thread 1	thread 2	balance
withdraw(100)	withdraw(100)	120
amount ≤ balance : true		120
	amount ≤ balance : true	120
balance -= amount		20
	balance -= amount	−80
get 100		−80
	get 100	−80

This is one *possible* order of execution where thread 2 checks that the balance is sufficient after thread 1 conducted the same check, but *before* thread 1 actually updated the balance. Hence, at the exact moment where thread 2 checks the balance, it is still sufficient and thread 2 is authorized to perform the withdrawal. Both threads perform the withdrawal, resulting in a negative balance, something that the program is supposed to prevent.

The order of execution could have been different. The execution of the actual withdrawal on line 11 could have been scheduled on thread 1 *before* thread 2 checked line 9. In that case, thread 2 would have correctly been denied because the balance would have been insufficient at check time. Another possibility is that thread 2 could have been scheduled to conduct the check on line 9 *while* thread 1 was updating *myBalance* on line 11. In that case, thread 2 could have read some invalid, corrupted, partly written data, resulting in a random number or a crash.

Therefore, depending on the exact scheduling of the execution on the two threads, the result could be correct, incorrect, or corrupted. The result may be different every time the program runs *with the same parameters*, depending on *who gets there first*, hence the name *race* condition. When the result is a corruption where a thread reads a location in memory at the exact same time another thread is writing into it, we have a *data race*.

Races are particularly hard to identify, reproduce, debug, and correct, precisely because they may happen or not depending on scheduling. The fact that we inspect or debug a program may in itself affect scheduling and prevent races. This is why some developers call them "quantum bugs."

A lot of concurrent programming is about avoidance of races. One means of avoiding races is by locking parts of the code so they don't execute concurrently.

3.7 LOCKS

C++11 provides a *thread synchronization primitive* called *mutex* in the <mutex> header. A mutex is an object that may be locked (by calling its *lock*() method) by one thread at a time. Once a thread *acquired a lock* on a mutex, the mutex will deny locks to other threads until its owner thread unlocks it (by a call to *unlock*()). In the meantime, other threads will wait for the lock *in idle state*. When the owner thread releases the lock while other threads are waiting on it, one of the waiting threads will awaken and acquire the lock. Therefore, the blocks of code that cannot be safely executed in parallel (we call those *critical sections*) may be surrounded by calls to *lock*() and *unlock*() on some mutex, which guarantees that only one thread can execute this code at a time.

The withdrawal code is fixed as follows:

```
#include <mutex>
using namespace std;

class SharedBankAccount
{
        mutex myMutex;
        double myBalance;

public:
        double withdraw( const double amount)
        {
                myMutex.lock();

                if( amount <= myBalance)
                {
                        myBalance -= amount;
                        myMutex.unlock();

                        return amount;
                }
                else
                {
                        myMutex.unlock();

                        return 0;
                }
        }
};
```

A thread locks the mutex before checking the balance and only releases it after it finished updating the balance. This guarantees that another thread cannot read or write the balance in the meantime. In order to do so, it would need to acquire the lock, and for that, it would have to wait until the lock is released. The code surrounded by *myMutex.lock*() and *myMutex.unlock*() is *atomic*: it is executed entirely on one thread without exterior interference. Other threads may read or write the balance before or after, but not during, the execution of the critical section. The following possible scheduling illustrates this notion:

thread 1	thread 2	balance
withdraw(100)	withdraw(100)	120
lock granted		120
	lock denied	120
amount ≤ balance : true	wait	120
balance -= amount	wait	20
unlock	lock granted	20
get 100	amount ≤ balance : false	20
	unlock	20
	get 0	20

It is clear that we may no longer see data races or negative balances. However, we have *not* eliminated race conditions. Whether it is thread 1 that is served and thread 2 that is denied, or the contrary, still depends on which locks the mutex first, which depends on scheduling and may be different on every run. However, such remaining race is *benign* in the sense that it cannot defeat the balance check or corrupt data. One of the users always withdraws $100, the other user is always denied, and the balance always ends at $20. No corruption may occur.

In order to effectively prevent races, locks must be acquired before the first manipulation of shared data and released after the last manipulation, so as to make the combined manipulation atomic. On the other hand, locks serialize code and defeat the benefits of parallelism, since only one thread can execute it at a time. For this reason, locks should never be acquired earlier than necessary and never released later than necessary. In our example, we acquire a lock before we check the balance and we release it as soon as we are done writing its update.

3.8 SPINLOCKS

How difficult is it to code a mutex? The following code apparently works:

```
class MyOwnMutex
{
    bool myLocked;

public:
    MyOwnMutex() : myLocked( false) {}
    void lock() { while( myLocked); myLocked = true; }
    void unlock() { myLocked = false; }
};
```

but it is effectively flawed: the implementation of *lock*() consists of two *distinct* instructions: check ("while(myLocked)") and set ("myLocked = true"). Because these two instructions are distinct (or *non-atomic*), another thread may step in between the check and the set and execute the critical code along with the thread that locked the mutex, defeating protection.

This is fixed by making the check and set *atomic*, meaning that they are executed in such a way that another thread could see the state of *myLocked* before or after, but not during the whole operation. The class atomic<bool> from header <atomic> defines a method *exchange*() that sets the state to the argument and returns the former state, and does all of this *atomically*, as a single instruction, so that another thread cannot step in or peek into the data during its execution. Therefore, the following code effectively fixes our custom mutex, which we now call a "spinlock":

```
#include <atomic>
using namespace std;

class SpinLock
{
    atomic<bool> myLocked;

public:
    SpinLock() : locked( false) {}
    void lock() { while( myLocked.exchange( true)); }
    void unlock() { myLocked = false; }
};
```

That is a perfectly valid lock and it effectively protects critical code against concurrent execution.

But it is not a mutex.

In fact, we just developed a well-known synchronization object called a *spinlock*. A spinlock works just like a mutex and it even implements the same interface, but it behaves differently when waiting on a lock. A spinlock waits "actively," constantly checking availability, consuming an entire CPU core while waiting, which makes that CPU unavailable to other threads. A mutex puts a waiting thread in idle state, not consuming any resource while waiting, and awakens the thread to resume execution when the lock becomes available. In the meantime, the core is available to conduct work for other threads.

The truth is *we cannot program a mutex in* C++. A mutex really is a primitive, which implementation is delegated to the OS, and ultimately to hardware.

What we can easily program is a spinlock, and we should use one or the other depending on the context. A thread that waits on a mutex releases its CPU for other threads, but a thread that waits on a spinlock resumes faster when the lock becomes available. When the speed of resuming execution on acquiring the lock is critical, we use a spinlock. When we know that the thread may wait a while and need its CPU for other tasks, we use a mutex. Mutexes are generally preferable and spinlocks must be used with care in particular situations only. Multiple spinlocks in waiting state cause a heavy CPU load that may result in a severe performance drag. This is particularly the case with hyper-threading, where threads share CPUs.

We call mutexes and spinlocks and anything that exposes the methods *lock*() and *unlock*() for the purpose of protection of critical code *lockable* objects.

3.9 DEADLOCKS

What happens when a thread acquires a lock and never releases it? Another thread attempting to acquire the lock will wait *forever*. Such situation is known as a *deadlock*. This is obviously something we must avoid.

How could that happen? One textbook situation (not really applicable in contexts of practical relevance in finance) is when thread 1 acquired lock A and needs to acquire lock B to complete a critical operation before it may release lock A. Thread 2 holds lock B and must acquire lock A to complete the execution of its critical code and release lock B. Both threads are waiting for the other to release its lock so they may release their own. Obviously, this never happens and both threads are effectively deadlocked.

This is why we must avoid nested locks. When nested locks cannot be avoided[7] C++11 provides in <mutex> a free function *lock*() that acquires multiple locks atomically with a deadlock avoidance algorithm. It is used as follows:

```
mutex m1, m2, m3;
//  ...
lock( m1, m2, m3);
//  ...
m1.unlock(); m2.unlock(); m3.unlock();
```

[7]We never faced that situation in our years of developing multi-threaded financial code.

The free function *lock*() works not only with mutexes, but all lockable objects, including spinlocks.

3.10 RAII LOCKS

The other cause of deadlocks is much more common. It happens when a thread acquires a lock and "forgets" to release it. Then any other thread that attempts to acquire the lock is deadlocked. This may happen in functions with multiple branches. Our withdrawal code has two branches, depending on whether the balance is sufficient or not. The lock must be released on all branches. This includes "implicit" branches, when the function exits on an *exception*. An exception may be thrown in the function, or in a nested call to another function. Exceptions cause functions to exit immediately, and in that case an acquired lock may not be released.

This situation is similar to memory leaks that occur when a function allocates memory, then exits on an exception without freeing memory. Memory leaks are undesirable, and resolved with smart pointers. Smart pointers release memory in their destructor, so memory is released whenever they exit scope for whatever reason. Forgetting to free memory is a serious nuisance, but forgetting to release a lock is much worse.

For this reason, we *never* manipulate mutexes[8] directly but always through RAII objects that release the locks in their destructors. This ensures that acquired locks are always released when execution exits the block where the RAII object was declared, whatever the reason. This is so important that C++11 provides two such RAII objects out of the box, in the header <mutex>: the *lock_guard* and the *unique_lock*.

A *lock_guard* acquires a lock on a mutex on construction, and releases it on destruction. It is a templated type that works with mutexes, spinlocks, and anything that implements the *lockable* interface. It is used as follows:

```
#include <mutex>
using namespace std;

class SharedBankAccount
{
    mutex myMutex;
    double myBalance;

public:
    double withdraw( const double amount)
    {
        lock_guard<mutex> lk( myMutex);
        //  Acquire lock
```

[8] Or spinlocks or any type of lockable objects.

```
        if( amount <= myBalance)
        {
            myBalance -= amount;
            return amount;
        }
        else
        {
            return 0;
        }
    } // lk goes out of scope, the lock is released
};
```

We acquire the lock through the *lock_guard* and we never release it explicitly. When execution exits the block that declares the *lock_guard*, for whatever reason, its destructor is invoked and the lock is released automatically.

A *lock_guard* acquires a lock on construction, releases it on destruction, and doesn't do anything else. It is typically declared on the first line of a block of code (code surrounded by brackets), which makes this block atomic.

C++11 also provides a more flexible construct *unique_lock*, also in the header <mutex>. That is another RAII lock that offers the functionality of a *lock_guard*, plus:

- A *unique_lock* can *defer* the acquisition of a lock until after construction, when constructed with the *defer_lock* flag, whereas a *lock_guard* always acquires the lock on construction.
- A *unique_lock* can explicitly release the lock *before* destruction with a call to its *unlock*() method. A *lock_guard* only releases the lock on destruction.
- *unique_lock* provides a non-blocking *try_lock*() method. If the mutex is already locked by another thread, *try_lock*() returns false but does not put the caller thread to sleep or stop execution.
- A *unique_lock* can *adopt* a mutex previously locked on the caller thread when constructed with the *adopt_lock* flag.
- And more, like a timer base *try_lock*() that attempts to acquire a lock for a given time.

A *unique_lock* is used as follows:

```
mutex m;
// ...
unique_lock<mutex> lk( m, defer_lock);   //  not locked
// ...
lk.lock();            //  now locked
// ...
lk.unlock();          //  unlocked here, otherwise on destruction
```

```
mutex m2; m2.lock();
// ...
// adopted: released on destruction
unique_lock<mutex> lk( m2, adopt_lock);
```

The flexibility of a *unique_lock* comes with an overhead, whereas a *lock_guard* provides RAII management for free.

RAII locks can be instantiated with all types of lockable objects. A lockable class is any class that defines the methods *lock*() and *unlock*(). Hence, a *unique_lock* is itself lockable (but a *lock_guard* is not). In particular, *lock*() can acquire multiple *unique_locks* at the same time.

None of the RAII locks is copyable. A *unique_lock* is moveable. A *lock_guard* is not.

3.11 LOCK-FREE CONCURRENT DESIGN

It is clear that locks effectively *serialize* critical code, preventing its parallel execution in all circumstances, hence preventing races but also defeating the benefits of concurrency. In the Introduction, we explained how partial parallelism leads to disappointing results in accordance with Amdahl's law. Locks always result in partial parallelism and cause poor parallel efficiency.

In addition, locks are expensive in themselves. Standard mutexes may put a thread in idle state if the mutex is locked by another thread, and awaken it when the mutex is unlocked. This may result in a context switch on the executing core. All of this causes an overhead. Spinlocks, on the other hand, occupy an entire core while waiting on a lock, making the core unavailable for other threads to perform useful operations. Finally, locks may result in deadlocks, and for this reason they must be used with care and parsimony and always with RAII constructs.

Locks are necessary in certain circumstances, but to use them systematically may lead to parallel code both slower and more dangerous than its serial counterpart. The art of concurrent programming is very much about making code thread safe *by design*, without locks.

How do we concretely achieve lock-free thread safe design? First, reading from shared objects is always safe; it is only concurrent *writes* that cause races. For this reason, *const correctness* goes a long way toward thread safety. The compiler guarantees that methods marked *const* cannot modify objects. We may therefore assume that they are safe to execute concurrently. The same applies to *const* arguments. For this reason, we must clearly mark *const* all the methods that don't modify the state of the object, and the arguments that are not modified by functions or methods. This is called *const correctness* and is routinely ignored by many developers. It doesn't matter that much in a single-threaded context, but it is critical with concurrent programs.

In addition, we must absolutely refrain from *false constness*, whereby methods marked *const* somehow modify state nonetheless. Such undesirable behavior is achieved, for example, with the *mutable* keyword. Data members that are *conceptually* not part of the state (such as a cache or some preallocated working memory) may be marked *mutable* so the compiler does not complain when *const* methods modify them. But, of course, that would cause race conditions all the same when such methods are executed concurrently. The *mutable* keyword must be banned from concurrent code. Similar behavior is achieved by casting away constness, something also best avoided for the same reasons.

Another example of false constness is when *const* methods on an given object O effectively don't modify that object, but change the state of another object, that may be referred to by pointer or reference in O. In this case, just like with mutables, *const* methods may no longer be assumed safe, making concurrent programming more difficult and prone to error. Thread safe design starts with const correctness and the avoidance of all types of false constness.

Following this logic to its extreme consequences, we may be tempted to program software where all methods are (truly) *const* and where free functions never modify their arguments. Such a program would ban all kinds of mutable state, forcing functions and methods to compute results out of arguments without side effects. The absence of a mutable state and side effects is a major principle of *functional programming* and is enforced in pure functional languages like Haskell. One reason why functional programming earned so much popularity recently is that stateless programs are thread safe by construction. When the language guarantees that functions and methods cannot modify state or parameters, they are all (truly) *const* and therefore safe for concurrent execution. Races cannot happen and locks are unnecessary.

So functional programming is particularly well suited to concurrency. C++, on the other hand, is a multi-paradigm language, and state is extremely useful in C++ programs. We believe it would be a mistake to forbid state in C++ programs for the benefit of concurrency. Many of the concurrent programs in this book use state. Our methods are not all *const* and they are not all safe for parallel execution. Our free functions occasionally modify their (non *const*) parameters. Yet, all our calculation code is absolutely lock-free. The only place we have locks is in the thread pool of Section 3.18, and even there they are all encapsulated in concurrent data structures.

When our concurrent code calls non-*const* methods or modifies non-*const* arguments, we have *mutable objects*. How we deal with mutable objects is this way: *we make copies of all the mutable objects for every thread before we execute concurrent code*. Our thread pool of Section 3.18 facilitates the management of thread-wise copies. The simple instructions on page 109 allow to determine the number of threads involved so we can make copies, and identify the thread executing concurrent code so it can work with its own copy.

Finally, we reiterate that memory allocation involves hidden locks, unless we use a purposely designed concurrent allocator from a third-party library like Intel's TBB. Therefore, we must also ban memory allocation from concurrent code, preallocating necessary memory before multi-threading, which may be difficult. First, we don't always know how much memory we need ex-ante. Second, whenever we create, copy, or pass to a function or method by value, objects that manage resources, like vectors or other containers, all these operations involve hidden allocations. It takes creativity and careful inspection of the code to move both explicit and implicit allocations out of concurrent code. Our simulation and differentiation code in Parts II and III show some examples. It is not always possible to remove all allocations from concurrent code, but at the very least we must strive to minimize them.

3.12 INTRODUCTION TO CONCURRENT DATA STRUCTURES

We know enough at this stage to start implementing concurrent data structures, classes that store data in a way that is safe to manipulate concurrently. We illustrate our purpose with the design of a basic concurrent queue. A queue is a container that implements a FIFO (first in, first out) logic whereby elements are pulled from the queue in the same order they have been pushed into the queue. The standard STL queue[9] exposes the methods *push*() (push an element into the queue), *front*() (access the element in front of the queue), and *pop*() (removes the front element in the queue), as well as the accessors *empty*() and *size*() and it *cannot* be safely manipulated from different threads at once. None of the STL containers can. The following code, therefore, wraps a STL queue with locks to produce a *concurrent queue*, one that is safe to manipulate concurrently:

```
#include <queue>
#include <mutex>
using namespace std;

template <class T>
class ConcurrentQueue
```

[9]Which is not really a data structure, but an adapter over another data structure, typically a *deque*.

```
{
    queue<T> myQueue;
    mutable mutex myMutex;

public:

    bool empty() const
    {
        //  Lock
        lock_guard<mutex> lk( myMutex);
        //  Access underlying queue
        return myQueue.empty();
    }   //  Unlock

    //  Pass t byVal or move with push( move( t))
    void push( T t)
    {
        //  Lock
        lock_guard<mutex> lk( myMutex);
        //  Move into queue
        myQueue.push( move( t));
    }   //  Unlock

    //  Pop into argument
    bool tryPop( T& t)
    {
        //  Lock
        lock_guard<mutex> lk( myMutex);
        if( myQueue.empty()) return false;
        //  Move from queue
        t = move( myQueue.front());
        //  Combine front/pop
        myQueue.pop();

        return true;
    }   //  Unlock

    void clear()
    {
        queue<T> empty;
        swap(myQueue, empty);
    }
};
```

This implementation is rather basic. All it does is serialize accesses to the underlying queue with locks. It exposes three methods, *empty*(), *push*(), and *tryPop*(), and all three are integrally protected with a *lock_guard*. This is not a particularly efficient concurrent queue. A high-performance implementation would require a re-implementation of the internal mechanics of the queue with minimum granularity locks; [15] provides such an implementation in its Section 6.2.3. Perhaps surprisingly, it is even possible, although very difficult, to implement a concurrent queue, safe for concurrent manipulation, without locks at all! Such high-performance implementations make a significant difference when concurrent data structures are heavily manipulated from within concurrent threads. For the purpose of parallel Monte Carlo simulations, our basic implementation is sufficient, and a more advanced implementation makes no difference in performance.

3.13 CONDITION VARIABLES

Condition variables (CVs), declared in the header <condition_variable>, provide a mechanism for different threads to communicate and synchronize. When a thread calls *cv.wait*() on an object *cv* of type *condition_variable*, the thread is put to sleep (passively, in idle state, without consuming any resources), until *another* thread calls either *cv.notify_one*() or *cv.notify_all*() on the same object. *cv.notify_one*() awakens one thread waiting on *cv* (if any). *cv.notify_all*() awakens all the waiting threads. CVs allow threads to control one another and to react on events occurring on different threads.

CVs and mutexes are the *only* thread synchronization primitives in the Standard Threading Library, in the sense that:

1. CVs and mutexes cannot be developed in standard C++; their implementation is delegated to the OS, and ultimately to hardware. We already discussed this point with mutexes; the same applies to CVs for the same reasons.
2. CVs cannot implemented with mutexes (and mutexes cannot be implemented with CVs).[10]
3. Other classical synchronization constructs, briefly discussed later, can all be implemented with a combination of mutexes and CVs. Hence, mutexes and CVs are the two building blocks of thread synchronization. The Standard Threading Library, following the C/C++ tradition of flexible, close-to-the-metal programming, only provides these two building blocks, and lets developers build their own, higher level synchronization constructs in accordance with the specific needs and design of their application.

Condition variables are extremely useful in concurrent programming. The reactive, event-driven behavior of many popular applications use CVs behind the scenes. As a useful example, we make our concurrent queue *reactive*. Pushing an element onto the queue *automatically* causes

[10]What prevents CVs to be implemented with mutexes is that only the thread that acquired a mutex can release it. Hence, mutexes, contrary to CVs, cannot be used for inter-thread communication.

the element to be processed on a different thread. The threads that push elements are called *producers*; those that process the elements are called *consumers*. Consumers hitting an empty queue are put to sleep. Pushing an element onto the queue causes one of them to awaken and process the element. This design pattern is called the *active object pattern*, because it makes threads react to events occurring on these objects. An active concurrent queue is called a *producer-consumer queue*. Producer-consumer queues are implemented with mutexes and condition variables:

```
#include <queue>
#include <mutex>
#include <condition_variable>
using namespace std;

template <class T>
class ConcurrentQueue
{
    queue<T> myQueue;
    mutable mutex myMutex;
    condition_variable myCV;

public:

    bool empty() const
    {
        //  Lock
        lock_guard<mutex> lk( myMutex);
        //  Access underlying queue
        return myQueue.empty();
    }   //  Unlock

    //  Pop into argument
    bool tryPop( T& t)
    {
        //  Lock
        lock_guard<mutex> lk( myMutex);
        if( myQueue.empty()) return false;
        //  Move from queue
        t = move( myQueue.front());
        //  Combine front/pop
        myQueue.pop();

        return true;
    }   //  Unlock

    //  Pass t byVal or move with push( move( t))
    void push( T t)
    {
        {
            //  Lock
            lock_guard<mutex> lk( myMutex);
            //  Move into queue
            myQueue.push( move( t));
        }   //  Unlock before notification

```

```
        myCV.notify_one();
    }

    //  Wait if empty
    bool pop( T& t)
    {
        //  (Unique) lock
        unique_lock<mutex> lk( myMutex);

        //  Wait if empty, release lock until notified
        while( myQueue.empty()) myCV.wait( lk);

        //  Re-acquire lock, resume, combine front/pop
        t = move( myQueue.front());
        myQueue.pop();

        return true;

    }   //  Unlock

    void clear()
    {
        queue<T> empty;
        swap(myQueue, empty);
    }
};
```

The accessor *empty*() and the non-blocking *tryPop*() are unchanged. The method *push*() gets one additional line of code "myCV.notify_one()" that awakens one thread waiting on the condition variable so it consumes the new element. Note that we unlocked the mutex *before* the notification. When a CV awakens a thread, the thread must acquire the lock before it can do anything else. If the lock is unavailable, the thread goes back to sleep until the lock is released. This would cause an unnecessary overhead unless the lock is released prior to notification.

We have a new method *pop*() that is called by consumer threads to dequeue the front element. The consumer thread starts by acquiring a lock on the concurrent queue's mutex. Multiple threads cannot manipulate the queue at the same time. It locks the mutex with a *unique_lock* rather than a *lock_guard* for a reason that will be explained shortly. If the queue is empty, the consumer thread calls *wait*() on the CV and goes to sleep until notified.

Note that we wrote:

```
while( myQueue.empty()) myCV.wait( lk);
```

and not

```
if( myQueue.empty()) myCV.wait( lk);
```

This is the correct syntax due to *spurious wakes*. Spurious wakes occur when a thread awakens without being notified, something that may happen due to low-level details in the OS and hardware-specific implementation of CVs. Because of spurious wakes, when a thread awakes on a CV, it should always start with a check that the condition for the notification (the queue is no longer empty) is satisfied, and go back to sleep otherwise. In practice, this is achieved by expressing the condition for waiting on the CV with a *while* rather than an *if*.

Calling *wait*() on the CV puts its caller to sleep and *releases its lock*. Evidently, a thread should never hold onto its locks when it sleeps. Nobody could push an element and awaken the waiting thread if it had kept its lock. Therefore, the CV's method *wait*() takes a lockable object as an argument and releases the lock when it puts the calling thread to sleep. A *lock_guard* is not a lockable object. It does not define a method *unlock*(). This is why a *lock_guard* cannot be used with a CV and this is why we used a *unique_lock* instead. The implementation of the CV also ensures that the thread reacquires the lock when it awakens on a notification. This is why we released the lock in *push*() prior to notification, so that the consumer may reacquire the lock with minimum overhead. Once it reacquires the lock, the consumer thread *moves* the front element (we know there is one at this point) into the argument to *pop*() and removes it from the queue, all under the protection of the lock it reacquired on awakening. When *pop*() exits, the destructor of the *unique_lock*() releases the mutex.

A consumer-producer queue is the centerpiece of a thread pool. In this case, the elements in the queue are the *tasks* to be executed in parallel. Worker threads are the consumers and wait for tasks to appear in the queue. Every task pushed into the queue awakens one worker thread to execute the task. When multiple tasks are pushed into the queue, worker threads dequeue and execute the tasks in order and go back to sleep when there are no more tasks to execute. A client pushing a task into the queue is guaranteed that a worker thread will awaken to execute the task, unless all workers are busy executing prior tasks, in which case the tasks are executed by worker threads, in parallel, in the same order they were pushed into the queue.

Our producer-consumer queue is *almost* complete. We need to notify all waiting threads in its destructor, so they may awaken before the CV is destroyed. A CV, like a mutex, should *never* be allowed to die while threads are waiting on it. We define an *interrupt*() method that pulls all consumer threads out of waiting and invokes it on destruction:

```
template <class T>
class ConcurrentQueue
{
    queue<T> myQueue;
    mutable mutex myMutex;
    condition_variable myCV;
    bool myInterrupt;
```

```
public:

    ConcurrentQueue() : myInterrupt( false) {}
    ~ConcurrentQueue() { interrupt(); }

    void interrupt()
    {
        {
            lock_guard<mutex> lk( myMutex);
            myInterrupt = true;
        }

        myCV.notify_all();
    }

    void resetInterrupt()
    {
        myInterrupt = false;
    }

    //  ...
};
```

We also modify *pop*() so the threads awakened by a call to *interrupt*() don't go back to sleep immediately and exit *pop*() instead (hence also releasing the mutex):

```
    bool pop( T& t)
    {
        //  (Unique) lock
        unique_lock<mutex> lk( myMutex);

        //  Wait if empty, release lock until notified
        while( !myInterrupt && myQueue.empty()) myCV.wait( lk);

        //  Re-acquire lock, resume

        //  Check for interruption
        if( myInterrupt) return false;

        //  Combine front/pop
        t = move( myQueue.front());
        myQueue.pop();

        return true;

    }   //  Unlock
```

Below is the final listing for the concurrent producer-consumer queue in ConcurrentQueue.h:

```
#include <queue>
#include <mutex>
```

```
#include <condition_variable>
using namespace std;

template <class T>
class ConcurrentQueue
{

    queue<T> myQueue;
    mutable mutex myMutex;
    condition_variable myCV;
    bool myInterrupt;

public:

    ConcurrentQueue() : myInterrupt( false) {}
    ~ConcurrentQueue() { interrupt(); }

    bool empty() const
    {
        //  Lock
        lock_guard<mutex> lk( myMutex);
        //  Access underlying queue
        return myQueue.empty();
    }   //  Unlock

    //  Pop into argument
    bool tryPop( T& t)
    {
        //  Lock
        lock_guard<mutex> lk( myMutex);
        if( myQueue.empty()) return false;
        //  Move from queue
        t = move( myQueue.front());
        //  Combine front/pop
        myQueue.pop();

        return true;
    }   //   Unlock

    //  Pass t byVal or move with push( move( t))
    void push( T t)
    {
        {
            //  Lock
            lock_guard<mutex> lk( myMutex);
            //  Move into queue
            myQueue.push( move( t));
        }   //  Unlock before notification

            //  Unlock before notification
        myCV.notify_one();
    }

    //  Wait if empty
    bool pop( T& t)
    {
        //  (Unique) lock
```

```
        unique_lock<mutex> lk( myMutex);

        //  Wait if empty, release lock until notified
        while( !myInterrupt && myQueue.empty()) myCV.wait( lk);

        //  Re-acquire lock, resume

        //  Check for interruption
        if( myInterrupt) return false;

        //  Combine front/pop
        t = move( myQueue.front());
        myQueue.pop();

        return true;

    }   //  Unlock

    void interrupt()
    {
        {
            lock_guard<mutex> lk( myMutex);
            myInterrupt = true;
        }

        myCV.notify_all();
    }

    void resetInterrupt()
    {
        myInterrupt = false;
    }

    void clear()
    {
        queue<T> empty;
        swap(myQueue, empty);
    }
};
```

3.14 ADVANCED SYNCHRONIZATION

C++11 provides only two synchronization primitives: the mutex and the condition variable. Parallel programming classically also involves more advanced synchronization constructs such as semaphores, barriers, and shared mutexes. The standard threading library does not provide these constructs.[11] They can all be developed by combining mutexes and condition variables.

[11]Although shared mutexes, and their associated RAII shared locks, are planned for future C++ updates.

Although we don't need these constructs in the remainder of the publication, or in typical financial applications, readers are encouraged to attempt implementing them on their own, as this is an excellent opportunity to manipulate mutexes and condition variables and earn a deeper understanding of concurrent programming.

Semaphores

A semaphore is essentially a mutex with added flexibility.

First, a semaphore may be constructed locked. A mutex is always constructed unlocked. Hence, a semaphore may forbid access to a critical section until the gate is explicitly open, something a mutex cannot do.

Second, a semaphore can grant access to critical code to more than one thread at a time. It has a parameter n so that no more than n threads may acquire a lock and execute the critical section concurrently. When n threads are executing the critical section, another $(n + 1)$th thread requesting the lock waits until one of the threads leaves the critical section and unlocks the semaphore. For example, the management of multiple downloads typically uses semaphores to limit the number of simultaneous downloads and avoid network congestion.

The Semaphore class signature is as follows:

```
template <class LOCKABLE>
class Semaphore
{

public:

    //  Will let n threads past wait()
    //  To construct locked, use n=0, then unlock with post
    Semaphore( const size_t n);

    //  Means unlock p spaces
    //  with p = 1: same semantics as mutex::unlock()
    void post( const size_t p = 1);

    //  Same semantics as mutex::lock()
    void wait();

    //  For compatibility with lock_guard<semaphore>
    void lock() { wait(); }
    void unlock() { post(); }
};
```

For historical reasons, *lock*() is called *wait*() on a semaphore and *unlock*() is called *post*(). We define *lock*() and *unlock*() anyway so the semaphore implements the lockable interface and works with RAII locks. We also template it in the mutex type so it may be instantiated with a mutex or a spinlock.

Barriers

Barriers are another classical synchronization construct. Threads that call *wait*() on a barrier wait there (actively or not depending on whether the barrier is implemented with a mutex or spinlock) until *n* threads checked out, at which point the *n* threads resume execution.

The signature is as follows:

```
template <class LOCKABLE>
class Barrier
{

public:
    Barrier( const size_t n);
    void wait();
};
```

Calling *wait*() puts the thread to sleep or active wait until *n* threads made the call; then all the *n* threads resume. The reduction algorithm described in the Introduction typically applies barriers to synchronize threads so that all threads complete a step in the algorithm before any thread starts working on the next step.

Shared locks

Much more difficult is the correct implementation of a shared mutex, a construct particularly useful in the context of concurrent data structures. Imagine we implement a concurrent container. A simplistic implementation would protect all kinds of data access with locks. Even read access must be protected because a thread cannot be allowed to read data while another thread is writing into the container. So every type of access must be protected, as we did with our concurrent queue, which results in a completely serialized data structure that only one thread may manipulate at a time.

This is obviously inefficient and may be improved with the realization that multiple *reader* threads can access data concurrently, as long as no *writer* thread is manipulating data at the same time. Therefore, reader threads could acquire a *shared* lock, whereas writer threads would need an *exclusive* lock. A *shared mutex* exposes a *lock_shared*() method for the acquisition of a shared lock, in addition to the usual *lock*() for the obtention of an exclusive lock. We would also need a *shared_lock* class that provides RAII management for shared mutexes in the same way a *unique_lock* manages a standard mutex. That part is easy, but the correct development of a shared mutex is more involved.

Shared and exclusive locks are granted according to some policy. One possibility is to prioritize reader threads, since multiple ones may access the data simultaneously. The problem with that is, as long as at least one reader is accessing data, writers are denied access. In a situation where readers and writers come in continuously, writers will *never* be granted access and will wait forever. This is sometimes called *thread starvation*. Similarly, priority might be given to writer threads; for instance when a writer requests access, no further readers are allowed and the writer may come in as soon as the present readers cleared the section. This policy would starve reader threads. Writers need access exclusive from readers and other writers, so they access the structure one at a time. As long as there are writers in the queue, no readers will be allowed.

Hence, some accountancy is necessary here. Priority must depend on the number of readers and writers in the queue, and perhaps how long the threads have been waiting. Like a security employee at the gate of a Disneyland attraction, the shared mutex must balance the shared and exclusive passes as best it can.

3.15 LAZY INITIALIZATION

The lazy initialization idiom is a frequently used programming pattern, including in finance. When some object, like a database connection, or an interpolated curve, is expensive to initialize, and may or may not be used during the execution of the program, it is common sense to postpone the costly initialization to the first use.

Here is an illustration of lazy initialization in the case of a curve. Lazily initialized curves are frequent in finance in the context of splines. The initialization of splines is costly and a natural candidate to laziness.

```
class Curve
{
    //  Initialized?
    bool myInit;

    //  Expensive initialization
    void init();

    //  Pick value on the curve
    //  Curve must be initialized
    double getValPrivate( const double) const;

public:

    //  Constructed uninitialized
    Curve() : myInit( false) {}

    //  High level picking function
    //  Implements lazy initialization
    //  Then calls private getValPrivate
    double getVal( const double x)
    {
```

```
        if( !myInit)
        {   //  Lazy initialization
            init();
            myInit = true;
        }

        return getValPrivate( x);
    }
};
```

The code should be self-explanatory, especially with the comments.

This is a perfectly valid curve object in a serial context. But it cannot be used concurrently. Picking a value on the curve is thread safe, of course, and we note that the related method *getValPrivate*() is *const*. But initialization is not, and we note that the public *getVal*(), where lazy initialization is implemented, is *not const*.

How can we make that curve safe for concurrent use without sacrificing this very useful lazy initialization feature? A natural solution would be to protect initialization with locks:

```
#include <mutex>
using namespace std;

class Curve
{
    mutex myMutex;

    //  ...

public:

    //  ...

    double getVal( const double x)
    {
        unique_lock<mutex> lk( myMutex);
        if( !myInit)
        {   //  Lazy initialization
            init();
            myInit = true;
        }
        lk.unlock();

        return getValPrivate( x);
    }
};
```

(note that we used a *unique_lock* so we can unlock the mutex *before* the call to the thread safe *getValPrivate*(). Locks should always be released ASAP.)

There is, however, a problem with this solution: it is *extremely* inefficient. A lock must be acquired *every time getVal*() is called, although it is necessary only once, for initialization. To acquire a lock is expensive, typically more expensive than reading a curve; hence that solution is not satisfactory at all. In fact, it is worse than the sacrifice of lazy initialization.

Another solution called *double checked locking* has been proposed, and although that solution is *broken*, it may still be found in literature:

```
#include <mutex>
using namespace std;

//  ...

double getVal( const double x)
{
    //  First check unlocked
    if( !myInit)
    {
        //  We never get there after initialization
        //      so we lock only once
        lock_guard<mutex> lk( myMutex);

        //  Check again, under protection of lock
        if( !myInit)
        {
            init();
            myInit = true;
        }
    }

    return getValPrivate( x);
}
```

This code does not work: a thread may be checking *myInit* under the first, unlocked check while another thread is writing into *myInit*, executing "myInit=true" under the protection of a lock. This of course would result in a data race. It could be fixed by making *myInit* an atomic boolean type, and call an atomic *exchange* to conduct checks, like we did for spinlocks, but this is not necessary. Lazy initialization is so common that the standard threading library provides a dedicated solution with the type *once_flag* and the function *call_once*(), both defined in the header <mutex>. This construct resolves the problem and guarantees that the *callable* argument to *call_once*() is called once and once only. One possible implementation is with atomic double checked locking, as discussed above. This is how it is applied to fix our curve class:

```
#include <mutex>
using namespace std;

class Curve
{
    once_flag   myOF;

    void init();
    double getValPrivate( const double) const;
```

```

public:

    double getVal( const double x)
    {
        call_once( myOF, &Curve::init, this);

        return getValPrivate( x);
    }
};
```

The arguments to *call_once*() following the *once_flag* follow the callable pattern described in Section 3.2.

3.16 ATOMIC TYPES

We have briefly introduced atomic types when we coded the spinlock mutex in Section 3.8. More specifically, we introduced the type *atomic* < *bool* > that behaves like a *bool* for all intents and purposes, but provides a method *exchange*() that sets the state of the atomic bool and returns its former state, *atomically*, that is in such a way that another thread cannot read or write the atomic object during the whole *set* + *check* operation, providing a guarantee against data races.

Another typical example is counting the number of times an object is accessed. For instance, a counter could be put on a curve and increase every time the curve is read. We define a basic counter class below:

```
class Counter
{
    int myCount;

public:

    Counter( const int count = 0) : myCount( count) {}

    int get() const { return myCount; }
    void set( const int count) { myCount = count; }
    void increment() { myCount++; }
    void decrement() { myCount--; }
};
```

It should be clear that this counter is not thread safe and cannot be manipulated concurrently. One problem is that operations *a la* ++, −−, + =, − =, etc. are not atomic. For instance, a statement like "x++" really translates into multiple instructions:

1. Read the state of x.
2. Add 1.
3. Set the state of x to the result.
4. Return the *former* state of x.

The problem, of course, is that another thread may step in during that sequence of operations and cause a race condition. Our counter may be made thread safe, at the cost of performance, with locks:

```
#include <mutex>
using namespace std;

class Counter
{
    int myCount;
    mutable mutex myMutex;

public:

    Counter( const int count = 0) : myCount( count) {}

    int get() const
    {
        lock_guard<mutex> lk( myMutex);
        return myCount;
    }

    void set( const int count)
    {
        lock_guard<mutex> lk( myMutex);
        myCount = count;
    }

    void increment()
    {
        lock_guard<mutex> lk( myMutex);
        myCount++;
    }

    void decrement()
    {
        lock_guard<mutex> lk( myMutex);
        myCount--;
    }
};
```

Atomics provide an alternative to locks that may be lighter, faster, and closer to the metal. We can implement our thread safe counter with atomics in place of locks, with a considerable performance gain. In fact, the atomics-based counter is almost identical in performance to its unsafe counterpart:

```
#include <atomics>
using namespace std;

class Counter
{
    //  atomic<int> in place of int
    atomic<int> myCount;

public:

    Counter( const int count = 0) : myCount( count) {}
```

```
    // no locks anywhere

    int get() const
    {
        return myCount.load();
    }

    void set( const int count)
    {
        myCount.store(count);
    }

    void increment()
    {
        myCount++;
    }

    void decrement()
    {
        myCount--;
    }
};
```

The atomic template can be instantiated with any type T: bool, int, double, any struct or class type, or, importantly, a pointer or a reference to any type: *atomic* $<$ *MyClass*$*$ $>$ is an atomic *pointer* to an object of class type *MyClass*, contrary to *atomic* $<$ *MyClass* $>$, which is the atomic equivalent of an object of type MyClass. Atomic types provide a number of methods that are guaranteed to execute atomically. The most common methods are:

store() sets the state of the object to the value of the argument.

load() returns the state of the object.

operators *a la* ++, −−, + =, − =, etc. are guaranteed to execute atomically.

exchange() sets the state of the object and returns its former state, atomically.

What is an atomic object? The generic template *atomic* $< T >$ is a *wrapper* to any type T that uses locks to guarantee the atomicity of certain operations. In this respect, our counter code with atomics should be identical to the version with locks. However, and this is the whole point with atomics:

The template *atomic* $< T >$ is *specialized* for types that support atomic operations on OS or hardware level, including:

All integral types like bool, char, int short, long, unsigned and friends.

Pointers and references to any type pointers are *always* integral types.

Other POD types like float, double, etc. depending on implementation, OS, and hardware.

For those supported types, atomic operations are delegated to the OS/hardware and implemented without locks, at a fraction of the cost of locks. Whether a type T supports lock-free atomic operations may be checked with a call to *is_lock_free*() on an abject of type $atomic < T >$. Again, all pointer and referenced types are lock-free.

Applications include spinlocks, counters, and reference counts for smart pointers. More generally, atomics provide a faster alternative to locks in many situations. It is even possible to program a concurrent queue without locks (!) with atomic pointers, and a detailed example is given in [15]. Such developments, called *lock-free algorithms and data structures*, are *very* advanced and make a difference in specific cases, to our knowledge, not relevant to finance. We only use atomics in simple cases like reference counters or spinlocks.

3.17 TASK MANAGEMENT

The standard threading library also provides, under the header <future>, a number of convenient constructs for sending *tasks* for execution on another thread, and monitor the execution from the dispatcher thread: the *async*() function, the *future* and *promise* classes, and the *packaged_task* construct. The free function *async*() is a one-stop solution that sends a callable to execute on another thread. The other, lower level constructs, provide facilities for inter-thread communication but not the management of the threads themselves. Combined with thread management explained in Section 3.2, these constructs provide a convenient and practical way to manage the execution of tasks on a set of worker threads in a thread pool, of which they constitute an important building block.

Async

The first and simplest construct C++11 provides for asynchronous task execution is the free function *async*(), defined in the header <future>. This function takes a callable argument, sends it to be executed asynchronously, and returns a *future* so the caller thread may monitor its execution.

The free function *async*() offers the simplest form of parallelism in C++11. To some extent, it is comparable to OpenMP's loop parallelism in terms of simplicity: instead of calling a function, wrap the call within *async*() and it executes on a parallel thread. The asynchronous execution is controlled by the implementation, delegated to the OS, and does not offer much control or flexibility over the process. It is perfectly suited in simple cases where we execute a simple function on a parallel thread while performing other tasks on the main thread. A typical application is perform an expensive task in the background without blocking the application.

For example, a database query could be executed asynchronously with *async*() without freezing the GUI. An example is given below:

```
#include <future>
using namespace std;

double someLongComputation( const double);

// ...

double x = 0;

// Sends someLongComputation() for asynchronous execution
//      and returns immediately

future<double> f = async( someLongComputation, x);

// Do something on this thread
//      while someLongComputation() calculates asynchronously

// ...

// Now the result is needed, we wait for completion and get the result
// Note: if someLongComputation() throws an exception
//      future::get() will rethrow it on the caller thread

double result = f.get();

// ...
```

The arguments to *async*() follow the callable pattern described in Section 3.2. We may pass a function, function object, member function, or more conveniently a lambda. The function *async*() also accepts an optional parameter *before* the callable to specify the *launch policy*:

async(launch::async, callable, . . .) forces execution on a parallel thread.

async(launch::deferred, callable, . . .) defers execution until *get*() or *wait*() is called on the returned *future*.

async(callable, . . .) when no launch policy is specified, the implementation "decides" on the best asynchronous execution based on hardware concurrency, how busy the CPUs are, and so forth.

In all cases, the call to *async*() returns immediately and provides the caller with a *future* templated on the type of the result of the task: in our example, the task *someLongComputation*() returns a double; hence the call to *async*() returns a *future* < *double* >.

Futures and promises

The free function *async*() encapsulates thread creation and management; hence its simplicity. But it does not offer sufficient flexibility to achieve maximum performance or multi-thread sophisticated algorithms. In these cases, we must manage the process ourselves, but the standard library does provide the necessary building blocks for task management. Futures and promises, also declared in the header <future>, provide an inter-thread communication mechanism, whereby threads communicate to one another the result of their work. These constructs are not *synchronization primitives*. They allow threads to communicate, but not *control* one another in any way, and their implementation is in OS-independent, standard C++. Developers can implement their own futures and promises with mutexes and condition variables and this is indeed a recommended exercise. With futures and promises, threads can communicate:

1. Whether a computation is complete.
2. Whether the computation completed normally or threw an exception.
3. The result of the computation or the thrown exception.

Futures Futures are on the *receiving* end of the communication and live on the dispatcher thread. We just met them with *async*(). The caller to *async*() (dispatcher thread) starts a computation, possibly on a different thread (computer thread), and receives a future for monitoring this computation. The dispatcher thread can call the method *wait*() on this future and block until the computation completes (or return immediately if the computation already completed). This guarantees that the instructions following a call to *future.wait*() will not execute before the asynchronous task completed.

The same result is obtained by a call to the future's method *get*(), which, in addition to blocking the caller until the asynchronous computation completes, returns the result of the computation[12] if it completed normally, or *rethrows* any exception thrown during the computation.

Futures may be moved but not copied.

Promises The function *async*() communicates its status and the results of the asynchronous computation (including exceptions) to its associated future. The promise, also declared in <future>, is a lower level mechanism to communicate this information to a future. The promise is on the *broadcasting* end of the communication and lives on the computer thread.

A promise object has an associated future, accessed by a call to the promise's *get_future*() method, and that subsequently receives communications from the promise. The promise has two methods, *set_value*() and

[12]Contrary to *wait*(), which return type is *void*, the return type of *get* is the future's template parameter.

set_exception(), to communicate completion, along with either a result or an exception. Prior to a call to one of these methods, the status of the task is incomplete, and a call to *wait*() or *get*() on the associated future from a different thread blocks that thread. A call to either *set_value*() or *set_exception*() on the promise wakes the thread (if any) waiting on the associated future.[13] After such call on the promise, a thread calling *wait*() on the associated future will not block, and a call to *get*() returns the result or rethrows the exception immediately.

The implementation of *async*() generally uses promises. Promises are flexible, low-level constructs that may accommodate many kinds of inter-thread communication architectures. As an example, we develop our own version of *async*() with promises. Our code is simplified in the sense that launch::async is compulsory (it is by far the most frequent use) and the callable takes no arguments (which is not a problem if we use lambdas: lambdas capture their environment):

```
#include <future>
#include <memory>
#include <thread>
#include <functional>
using namespace std;

//  Task executed on a separate thread by our simpleAsync:
//      execute the callable and set result on the promise

//  RTYPE is the return type of the callable argument

template<class RTYPE>
void threadFunc(
    //  Arguments: the callable and the promise
    function<RTYPE(void)> callable,
    //  Promise managed by shared pointer so it is kept alive
    shared_ptr<promise<RTYPE>> prom)
{
    try
    {
        RTYPE res = callable(); //  Execute
        prom->set_value( res);  //  Set value, mark as completed
    }
    catch( ...)
    {
        // Exception thrown: set exception, mark as completed
        prom->set_exception( current_exception());
    }
}

//  simpleAsync implementation

```

[13]If the thread was waiting on *get*(), the result is returned or the exception rethrown.

```
template<class RTYPE>
future<RTYPE> simpleAsync( function<RTYPE(void)> callable)
{
    //  Create a promise<RTYPE> and manage it by shared pointer
    //      so it is kept alive during asynchronous execution
    //      even after simpleAsync() exits
    //  It will be released eventually when threadFunc() completes
    //      and the last shared pointer exits scope

    auto prom = make_shared<promise<RTYPE>>();

    //  Access the future
    future<RTYPE> fut = prom->get_future();

    //  Execute threadFunc() asynchronously on a new thread
    //      threadFunc() executes the callable
    //      and sets the result on the promise
    thread ( threadFunc<RTYPE>, callable, prom).detach();
    //  Detach the thread so it continues the execution of threadFunc()
    //      in the background after simpleAsync() returns

    //  Return is automatically moved in C++11
    return fut;
}
```

The code illustrates how to use promises. This implementation takes a callable argument and wraps it into a *function*; see Section 2.2. The first thing it does is construct a *promise*, templated on the return type of the task, and held by shared pointer; see Section 2.4. Shared pointers automatically destroy objects and release memory when the *last* pointer referencing the object exits scope. Shared pointers is what guarantees that the promise remains alive during the asynchronous execution of the task, even though *simpleAsync*() returns immediately.

The code then invokes the method *get_future*() on the promise. As the name indicates, this method returns the *future* associated to the *promise*, the object that may read communications from the promise from a different thread.

Next, we create a new thread to execute *threadFunc*() in parallel, with the callable and the *shared pointer* on the promise as arguments, and return the *future* to the caller of *simpleAsync*(). Note that we detach the thread immediately, so the execution of *threadFunc*() continues in the background when *simpleAsync*() returns. This breaks the thread handle, so we lose the connection to the computer thread (see Section 3.2), which is fine; we have a future to directly monitor the asynchronous task.

The function *simpleAsync*() completed and returned the future to the dispatcher thread. The execution of *threadFunc*() on the computer thread started and may still be running. The promise is used from within *threadFunc*() so it must remain alive throughout the computation. It was passed as an argument to the thread constructor by shared pointer, and the thread constructor forwarded the shared pointer to *threadFunc*(). Therefore, as long as *threadFunc*() is running, there is one shared pointer that references the promise and the promise remains alive. When *threadFunc*() returns, the last shared pointer on the promise exits scope, the promise is destroyed, and its memory on the heap is released.

The function *threadFunc*() executes the callable in a try-catch block in case the callable throws, and invokes the method *set_value*() on the promise. In case the callable throws, we catch the exception and set it with the promise's method *set_exception*(). The call to either *set_value*() or *set_exception*() on the promise causes *a prior or future call* to either *wait*() or *get*() on the associated future (the one we got earlier with a call to *get_future*() on the promise), maybe from a different thread, to unblock, and in the case of *get*(), return the result of *set_value*() or rethrow the exception of *set_exception*().

Promises provide flexible means to control the state accessed by their associated futures from different threads, and may be applied to communicate between threads in many kinds of parallel designs. We could develop a thread pool with promises, but the perhaps higher level *packaged task* is better suited for this purpose.

Packaged tasks

A packaged task, also defined in <future>, is a convenient construct that encapsulates the promise – future mechanics for us, like *async*(), but, contrary to *async*(), it does not manage threads for us. The type *packaged_task* is a *callable* type. It wraps a callable and defines an operator () to invoke it. It is templated by its return and parameter types, like *std* : : *function*.

A packaged task is constructed with a callable of any type. It provides an associated future (with its method *get_future*()), and when the packaged task is invoked on a thread, its wrapped callable is executed on that thread. When execution completes, however, the packaged task does *not* return a result or throw an exception; instead, it sets the result or the exception on a shared state, like a promise. Packaged tasks are typically implemented with promises. The shared state is accessible through the associated future, possibly from a different thread.

A packaged task, like *std* : : *function*, implements type erasure so it may wrap all types of callables. Its creation involves an allocation. The implications for us are that:

1. We should not use this type at low level, and especially not within repeated code.
2. We should avoid the creation of a large number of small tasks, and instead attempt to combine small tasks into larger ones and send those for parallel execution. We will discuss this point again in Section 3.19.

Packaged tasks are non-copyable, moveable types. We illustrate how they work with an example. Say we have a function that reduces a vector into a number; for instance it could compute its average or standard deviation and throw an exception if something goes wrong:

```
double processVector( const vector<double>&);
//  ...
```

Say that we have a number n of (large) vectors to process:

```
//  ...
vector<vector<double>> vectors2process( n);
//  Initialize and fill the n vectors
//  ...
```

We want to process the n vectors on parallel threads. This is exactly where we would use a thread pool, although we don't know how to do that yet, so, instead, we apply basic thread management for now with the convenience of packaged tasks:

```
//  ...

vector<future<double>> futs( n);

for( size_t i=0; i<n; i++)
{
    //  Create the packaged task
    packaged_task<double(const vector<double>&)> task( processVector);

    //  Get future
    futs[i] = task.get_future();

    //  Send for execution on a different thread
    //      We explicitly move the task, tasks are not copyable
    //      The processed vector is explicitly passed by reference
    //      We detach the thread and monitor the task with its future
    thread( move( task), ref( vectors2process[i])).detach();
}
//  ...
```

The processing of each vector is a separate parallel task. For every vector, we create the packaged task from the callable *processVector*(), which, in this case, is a free function taking a vector as a parameter by const reference, and returning a double as the result of its processing. Hence, the packaged task template is instantiated with $<$ *double*(*const vector* $<$ *double* $>$ &) $>$.

Once the packaged task is created, we keep track of its associated future. Like promises, packaged tasks expose their future with *get_future*(). Finally, we create a parallel thread for the execution of the task. We pass the task itself to the thread as the thread constructor's callable argument, along with the argument of the task, the *i*th vector to process. We note that the task must be explicitly moved, since it cannot be copied, and that the argument must be explicitly passed as a reference, otherwise it would be passed by copy, as explained in Section 3.2.

We detach the new thread straightaway. We don't need a handle on the thread, since we have a future on the task. We don't store the thread handle into a variable; hence, the temporary handle returned by the thread constructor is immediately destroyed, which would terminate the application if we had not detached the thread.

The task is the callable that is passed to the thread constructor for execution, so the task's operator () will be invoked on the parallel thread with its (by const reference) argument, the *i*th vector to be processed. This will execute the wrapped callable *processVector*() on the parallel thread and set its result or exception on a shared state, accessible with the future from the main thread:

```
//    ...
vector<double> res( n);

try
{
    for( size_t i=0; i<n; i++) res[i] = futs[i].get();

}
catch( ...)
{
   // handleException( current_exception());
}
```

Calling *get*() on the future blocks the caller until the execution of the corresponding task completes on the parallel thread: more precisely, it blocks until the packaged task sets the result or the exception on the shared state. If this is already done, *get*() returns the result immediately. Otherwise, the caller waits in idle state. Any exception thrown during the execution of the task is rethrown on the thread calling *get*(). Our code would catch and handle this exception with the hypothetical function *handleException*().

What makes packaged tasks particularly well suited to thread pools is that, as illustrated in the previous examples, packaged tasks provide futures so callers may monitor the asynchronous tasks.

3.18 THREAD POOLS

The singleton pattern

We are finally ready to build the thread pool we will use in the remainder of this publication. A thread pool is designed to encapsulate thread logic in an application. It controls a collection of worker threads. It is convenient to implement the well-known *singleton* pattern (see [25]) so that the application holds one and only one instance of the thread pool. The singleton instance is created when the application starts, destroyed when the application exits, and remains globally accessible in the meantime. One benefit is that the thread pool is globally accessible so functions don't have to pass it to one another as an argument.

```
#include <thread>
using namespace std;

class ThreadPool
{
    //  The one and only instance
    static ThreadPool myInstance;

    //  The constructor stays private, ensuring single instance
    ThreadPool() {}

    //  The worker threads
    vector<thread> myThreads;

    //  ...

public:

    //  Access the instance
    static ThreadPool* getInstance() { return &myInstance; }
    //  Alternatively, we could return a reference

    //  Forbid copies and moves
    ThreadPool(const ThreadPool& rhs) = delete;
    ThreadPool& operator=(const ThreadPool& rhs) = delete;
    ThreadPool(ThreadPool&& rhs) = delete;
    ThreadPool& operator=(ThreadPool&& rhs) = delete;

    //  We start the thread pool once when application starts
    void start();

    //  The destructor is invoked once when application ends
    ~ThreadPool();

    //  ...
};
```

This pattern guarantees that there may be only one instance of a thread pool in the application. It is constructed when the application starts, and destroyed when it exits. This unique instance is accessed with the static method:

$$ThreadPool :: getInstance()$$

which provides a pointer for the manipulation of the thread pool.

Task spawning

The purpose of the thread pool is to encapsulate thread logic, so that algorithms divide into logical slices, or *tasks*, *spawned* for parallel execution into the thread pool, without concern of how the tasks are scheduled over threads. Algorithms have no knowledge or control of what thread executes what task: this is the private responsibility of the thread pool. Algorithms just know that the pool executes the spawned tasks in parallel and in the same order they are spawned into the pool.

Hence, the thread pool exposes a method for spawning tasks. What kind of tasks will clients be spawning? We could design it in a very general manner so that clients may spawn all kinds of callables, but this is unnecessary. It is more convenient to always spawn lambdas, which capture the information they need, hence, don't need arguments, and set results directly in memory, hence, don't need to return anything. However, *void* is not a valid return type for C++11's task management constructs. Therefore, all our tasks return a *bool*, by convention *true* if they execute correctly, throwing an exception otherwise:

```
#include <future>
using namespace std;

using Task = packaged_task<bool(void)>;
using TaskHandle = future<bool>;
```

The thread pool internally manages a (concurrent) queue of tasks, so to spawn a task really means push it into the queue. The code for the concurrent queue designed in Section 3.13 is in the file ConcurrentQueue.h:

```
#include <future>
#include "ConcurrentQueue.h"
using namespace std;

class ThreadPool
{
    ConcurrentQueue<Task> myQueue;

    //  ...

public:

    //  ...

    template<typename Callable>
    TaskHandle spawnTask(Callable c)
    {
        Task t(move(c));
        TaskHandle f = t.get_future();
        myQueue.push(move(t));
        return f;
    }
};
```

Client code calls *spawnTask*() to send a task to the thread pool for a parallel execution, which pushes the task in the pool's queue, and returns a future to monitor progress. Note we *move* tasks around to avoid the potential overhead of copying tasks that capture large amounts of data.

Worker threads

The worker threads pop tasks from the queue and execute them in parallel. We recall that the method *pop*() on the concurrent queue causes the caller to wait *in idle state* until a task is pushed into the queue, which automatically awakens one of the waiting threads to consume the task. This is the keystone of the thread pool's design: its worker threads wait in idle state until client code spawns tasks. The worker threads are created when the application starts, and wait in the background for clients to spawn tasks. Spawning a task does not involve any thread creation overhead, because the threads are already there, waiting for jobs to execute. Further, they don't consume any resources waiting so their presence in the background and readiness to execute the spawned tasks is free.

We can now write the function *threadFunc*(), which each one of the worker threads executes. All they do is repeatedly pull tasks from the queue and execute them. When the queue is empty, *pop*() causes them to sleep until a task is pushed in the queue, causing one waiting thread to awaken, as seen in Section 3.13, and consume the task. It follows that *threadFunc*() never exits so the worker threads remain alive until the application exits.

```
    //  The main function executed by all threads
    void threadFunc()
    {
        Task t;

        //  This is the infinite loop
        while (true)
        {
            //  Pop and execute tasks
            myQueue->pop(t);
            t();
        }
    }
```

The starter of the thread pool (called once when the application starts) creates the worker threads and sets them to execute *threadFunc*(). How many worker threads shall we create? We find it best to create, by default, as many threads as we have hardware threads on the machine running the application, *minus one* so one hardware thread always remains available for the main thread. The number of threads can be changed by the client code, for example, for debugging and profiling, as explained in Section 3.20.

```
class ThreadPool
{
    //  The threads
    vector<thread> myThreads;

    //  Active indicator
```

```
    bool myActive;

    //  ...

    void threadFunc()
    {
        Task t;

        //  This is the infinite loop
        while (true)
        {
            //  Pop and executes tasks
            myQueue->pop(t);
            t();
        }
    }

    //  The constructor stays private, ensuring single instance
    ThreadPool() : myActive(false) {}

public:

    //  Starter
    void start(const size_t nThread = thread::hardware_concurrency() - 1)
    {
        if (!myActive)  //  Only start once
        {
            myThreads.reserve(nThread);

            //  Launch threads on threadFunc and keep handles in a vector
            for (size_t i = 0; i < nThread; i++)
                myThreads.push_back(thread(&ThreadPool::threadFunc, this));

            myActive = true;
        }
    }

    //  ...
};
```

This is all we need. The mechanisms in the concurrent queue ensure that the machinery works as planned. We just need to add a small number of convenient features.

Interruption

First of all, we cleanly dispose of the threads when the pool is destroyed, otherwise the destruction of the pool would cause the application to crash. That would not be a major problem, since the singleton pattern guarantees that the pool is not destroyed before application exit. But to let the application crash on exit is poor development standard, so we program a clean disposal instead. This means that we must interrupt all waiting threads, which somewhat changes our code:

```
class ThreadPool
{
    //  ...

    //  Active indicator
    bool myActive;

    //  Interruption indicator
    bool myInterrupt;

    //  Break infinite loop on interruption so threads can exit

    void threadFunc()
    {
        Task t;

        //  "Infinite" loop, only broken on destruction
        while (!myInterrupt)
        {
            //  Pop and executes tasks
            myQueue.pop(t);
            if (!myInterrupt) t();
        }
    }

    //  The constructor stays private, ensuring single instance
    ThreadPool() : myActive(false), myInterrupt(false) {}

public:

    //  Interrupt starts false so threads don't exit

    //  Starter
    void start(const size_t nThread = thread::hardware_concurrency() - 1)
    {
        if (!myActive)  //  Only start once
        {
            myThreads.reserve(nThread);

            //  Launch threads on threadFunc and keep handles in a vector
            for (size_t i = 0; i < nThread; i++)
                myThreads.push_back(thread(&ThreadPool::threadFunc, this));

            myActive = true;
        }
    }

    //  Destructor
    ~ThreadPool()
    {
        stop();
    }
```

```
    void stop()
    {
        if (myActive)
        {
            //  Interrupt mode
            myInterrupt = true;

            //  Interrupt all waiting threads
            myQueue.interrupt();

            //  Wait for them all to join
            for_each(
                myThreads.begin(),
                myThreads.end(),
                mem_fn(&thread::join));

            //  Clear all threads
            myThreads.clear();

            //  Clear the queue and reset interrupt
            myQueue.clear();
            myQueue.resetInterrupt();

            //  Mark as inactive
            myActive = false;

            //  Reset interrupt
            myInterrupt = false;
        }
    }

    //  ...
};
```

We added a bool *myInterrupt* that breaks the infinite loop in *threadFunc*() when true so threads may exit. The interrupter is false on construction, and only a call to *stop*() makes it true, so the worker threads remain trapped in *threadFunc*() until the thread pool is stopped, either explicitly, or in its destructor, activated when the application exits. The method *stop*() also interrupts the queue, awakening the waiting threads so they may all exit, and waits for them to join.

Thread local indices

It is also useful that each thread knows its own index among the worker threads. We achieve this with thread local storage, as explained in Section 3.4. We number worker threads 1 to n, leaving number 0 for the main thread. We also expose the number of threads in the pool:

```
class ThreadPool
{
    //  Thread number
    static thread_local size_t myTLSNum;
```

```
    //  The function that is executed on every thread
    void threadFunc(const size_t num)
    {
        myTLSNum = num;

        Task t;

        //  "Infinite" loop, only broken on destruction
        while (!myInterrupt)
        {
            //  Pop and executes tasks
            myQueue.pop(t);
            if (!myInterrupt) t();
        }
    }

    //  ...

public:

    //  Starter
    void start(const size_t nThread = thread::hardware_concurrency() - 1)
    {
        if (!myActive)  //  Only start once
        {
            myThreads.reserve(nThread);

            //  Launch threads on threadFunc and keep handles in a vector
            for (size_t i = 0; i < nThread; i++)
                myThreads.push_back(
                    thread(&ThreadPool::threadFunc, this,  i + 1));

            myActive = true;
        }
    }


    //  Number of threads
    size_t numThreads() const { return myThreads.size(); }

    //  The number of the caller thread
    static size_t threadNum() { return myTLSNum; }

    //      ...
};
```

We added a data member *myTLSNum* as a static integer marked as thread local, which means that every thread holds a different copy of it. The function *threadFunc*() sets it to the index (passed from the starter) of the executing worker thread. Since *threadFunc*() runs on the worker threads, each worker thread sets *its own copy* of *myTLSNum* to its own index. As a result, when the static function:

$$ThreadPool :: threadNum()$$

is called on a worker thread, it returns the index of the caller. We also want the main thread to return 0, so we set the initial value (on the main thread) to 0 in ThreadPoool.cpp:

```
#include "ThreadPool.h"
using namespace std;

ThreadPool ThreadPool::myInstance;

thread_local size_t ThreadPool::myTLSNum = 0;
```

Active wait

Finally, we add a very useful feature that allows the main thread to help executing tasks while waiting for others. The typical utilization of a thread pool is one where the main thread spawns a large number of tasks and waits on the corresponding futures for the worker threads to complete the tasks. The main thread occupies an entire hardware thread, and its hardware thread will not be doing anything useful while waiting. That is a waste of CPU. Instead, we implement a method *activeWait*() causing the caller thread to help executing tasks in the queue while waiting on futures:

```
    //  Run queued tasks synchronously
    //      while waiting on a future,
    //      return true if at least one task was run
    bool activeWait(const TaskHandle& f)
    {
        Task t;
        bool b = false;

        //  Check if the future is ready without blocking
        //  The only syntax C++11 provides for this is:
        //      wait 0 seconds and return status
        while (f.wait_for(0s) != future_status::ready)
        {
            //  Non blocking
            if (myQueue.tryPop(t))
            {
                t();
                b = true;
            }
            else // Nothing in the queue: go to sleep
            {
                f.wait();
            }
        }

        return b;
    }
```

This concludes the development of our thread pool. This is a very basic thread pool. It could be improved in a lot of ways, including a high-performance concurrent queue, and *work stealing*, a design whereby each thread works with its own queue and helps other threads with their jobs when their own queue is empty; [15] develops these themes and more and discusses the production of high-performance thread pools, capable of dealing with a large number of tasks, including nested spawning of tasks from within the jobs executed on the worker threads.

Financial algorithms, however, don't typically require such advanced functionality. Our basic thread pool, in particular, provides everything we need to run parallel simulations. We tested more sophisticated, high-performance designs, and they did not make a difference in situations of practical relevance.

The complete code is reproduced below from our repository files ThreadPool.h and ThreadPool.cpp.

```
#include <future>
#include <thread>
#include "ConcurrentQueue.h"

using namespace std;

using Task = packaged_task<bool(void)>;
using TaskHandle = future<bool>;

class ThreadPool
{
    //  The one and only instance
    static ThreadPool myInstance;

    //  The task queue
    ConcurrentQueue<Task> myQueue;

    //  The threads
    vector<thread> myThreads;

    //  Active indicator
    bool myActive;

    //  Interruption indicator
    bool myInterrupt;

    //  Thread number
    static thread_local size_t myTLSNum;

    //  The function that is executed on every thread
    void threadFunc(const size_t num)
    {
        myTLSNum = num;

        Task t;

        //  "Infinite" loop, only broken on destruction
        while (!myInterrupt)
        {
```

```
            //  Pop and executes tasks
            myQueue.pop(t);
            if (!myInterrupt) t();
        }
    }

    //  The constructor stays private, ensuring single instance
    ThreadPool() : myActive(false), myInterrupt(false) {}

public:

    //  Access the instance
    static ThreadPool* getInstance() { return &myInstance; }

    //  Number of threads
    size_t numThreads() const { return myThreads.size(); }

    //  The number of the caller thread
    static size_t threadNum() { return myTLSNum; }

    //  Starter
    void start(const size_t nThread = thread::hardware_concurrency() - 1)
    {
        if (!myActive)  //  Only start once
        {
            myThreads.reserve(nThread);

            //  Launch threads on threadFunc and keep handles in a vector
            for (size_t i = 0; i < nThread; i++)
                myThreads.push_back(
                    thread(&ThreadPool::threadFunc, this,  i + 1));

            myActive = true;
        }
    }

    //  Destructor
    ~ThreadPool()
    {
        stop();
    }

    void stop()
    {
        if (myActive)
        {
            //  Interrupt mode
            myInterrupt = true;

            //  Interrupt all waiting threads
            myQueue.interrupt();

            //  Wait for them all to join
            for_each(
                myThreads.begin(),
                myThreads.end(),
                mem_fn(&thread::join));
```

```
            // Clear all threads
            myThreads.clear();

            // Clear the queue and reset interrupt
            myQueue.clear();
            myQueue.resetInterrupt();

            // Mark as inactive
            myActive = false;

            // Reset interrupt
            myInterrupt = false;
        }
    }

    // Forbid copies etc
    ThreadPool(const ThreadPool& rhs) = delete;
    ThreadPool& operator=(const ThreadPool& rhs) = delete;
    ThreadPool(ThreadPool&& rhs) = delete;
    ThreadPool& operator=(ThreadPool&& rhs) = delete;

    // Spawn task
    template<typename Callable>
    TaskHandle spawnTask(Callable c)
    {
        Task t(move(c));
        TaskHandle f = t.get_future();
        myQueue.push(move(t));
        return f;
    }

    // Run queued tasks synchronously
    // while waiting on a future,
    // return true if at least one task was run
    bool activeWait(const TaskHandle& f)
    {
        Task t;
        bool b = false;

        // Check if the future is ready without blocking
        // The only syntax C++11 provides for that is:
        //     wait 0 seconds and return status
        while (f.wait_for(0s) != future_status::ready)
        {
            // Non blocking
            if (myQueue.tryPop(t))
            {
                t();
                b = true;
            }
            else // Nothing in the queue: go to sleep
            {
                f.wait();
            }
        }
```

```
        return b;
    }
};

//  ThreadPool.cpp

#include "ThreadPool.h"

ThreadPool ThreadPool::myInstance;

thread_local size_t ThreadPool::myTLSNum = 0;
```

3.19 USING THE THREAD POOL

We will use the thread pool in rest of the publication. In fact, the thread pool is the only parallel construct we are going to use in our multi-threaded programs. For future reference, we are providing here a summary of its functionality, and a user's guide for the client code. We also highlight some important properties of the thread pool.

Instructions

The thread pool is designed for the convenience of client code. Client code starts the thread pool when the application opens. The pool sits in the background for the lifetime of the application. It is accessible from any function or method by static pointer, without the need to pass it around as an argument. It is destroyed when the application closes. It doesn't consume any resources unless used.

The instructions for using the thread pool are articulated below:

1. Include "ThreadPool.h" on the files that use it.

```
#include "ThreadPool.h"
```

2. Access the thread pool by static pointer in any function or method.

```
ThreadPool* pool = ThreadPool::getInstance();
```

Start it as follows (when the application opens):[14]

```
ThreadPool* pool = ThreadPool::getInstance();
pool->start();
```

[14]The pool's *start*() method takes the number of worker threads as argument. 0 is legal and indeed helpful; see our comments on debugging and profiling, Section 3.20. Default is the number of logical cores minus one for the main thread. The number of worker threads can be changed at run time by calling *stop*() to close the pool; then call *start*(*nThreads*) with the desired number of threads.

3. Give the pool jobs to conduct in parallel, and remember that tasks must return a bool:

```
ThreadPool* pool = ThreadPool::getInstance();
vector<TaskHandle> futures(nTasks);
for (size_t task = 0; task < nTasks; ++task)
{
    auto spawnedTask = [&, task] ()
    {
        //  Parallel job here
        //  ...

        //  Tasks must return a bool
        return true;
    };

    //  Spawn the task into the pool, get future
    futures[task] = pool->spawnTask(spawnedTask);
}
```

The pool guarantees that task execution *starts* in the spawning order.

4. Wait for asynchronous tasks to complete, generally helping with the tasks while waiting:

```
for (size_t task = 0; task < nTasks; ++task)
{
    pool->activeWait(futures[task]);
}
```

5. After that loop, all the tasks spawned to the pool completed and the worker threads are back to sleep. The main thread can safely use the results.

Mutable objects

The thread pool facilitates the creation and the management of copies, one per thread, of the *mutable objects*, the objects that are not safe for concurrent access. Mutable objects were introduced in Section 3.11, and their management is illustrated in the parallel code of Chapter 7.

The number of worker threads can be accessed with a call to *pool– > numThreads*(), so we know how many copies we need. Every thread can identify its own index in the pool by calling *pool– > threadNum*(), which returns an index that depends on the thread making the call: 0 for the main thread, 1 to *numThreads*() for the worker threads. This allows threads to work with their own copies of the mutable objects.

When concurrent code modifies objects with calls to non-*const* methods, we can avoid expensive and efficiency damaging locks by making copies of these mutable objects before multi-threading:

```
//  The mutable object instance
MutableObject mutableObject;
```

```
//  ...

//  Initialization code executed before multi-threading
ThreadPool* pool = ThreadPool::getInstance();
//  Copy the mutable object
//  One copy per thread, including the main thread (hence +1)
vector<MutableObject> copies(pool->numThreads() + 1, mutableObject);
```

In the concurrent lambda, we work with the copy belonging to the executing thread, which results in a thread safe code without locks:

```
ThreadPool* pool = ThreadPool::getInstance();
vector<TaskHandle> futures(nTasks);
for (size_t task = 0; task < nTasks; ++task)
{
    auto spawnedTask = [&, task] ()
    {
        //  Parallel job here
        //  ...

        //  Index of the currently executing thread
        const size_t thisThread = pool->threadNum();
        //  Mutable object for this thread
        MutableObject& thisThreadObject = copies[thisThread];

        //  ...

        //  Safely modify own copy of the object
        thisThreadObject.modify();

        //  ...

        //  Tasks must return a bool
        return true;
    };

    //  Spawn the task into the pool, get future
    futures[task] = pool->spawnTask(spawnedTask);
}
```

We apply this pattern systematically in concurrent code in the rest of the publication. In addition to mutable objects, the exact same pattern allows to preallocate working memory for different threads. For instance, say the concurrent code needs working memory of size N in a vector *temp*:

```
ThreadPool* pool = ThreadPool::getInstance();
vector<TaskHandle> futures(nTasks);
for (size_t task = 0; task < nTasks; ++task)
{
    auto spawnedTask = [&, task] ()
    {
        //  Parallel job here
        //  ...

        vector<double> temp(N);
```

```
11
12          //  Do something with temp
13          workWithTemp(temp);
14
15          //  ...
16
17          //  Tasks must return a bool
18          return true;
19      };
20
21      //  Spawn the task into the pool, get future
22      futures[task] = pool->spawnTask(spawnedTask);
23  }
```

We have an allocation on line 10 in the concurrent code. As explained in Section 3.11, allocations generally involve hidden locks and damage parallel performance. The solution is to preallocate working memory in the initialization phase, before multi-threading. But the different threads cannot write into the same vector *temp*; that would cause race conditions. On the other hand, to lock the piece of code that consumes *temp* would damage performance:

```
//  Not lock-free
mutex m;

//  Pre-allocated working memory
vector<double> temp(N);

ThreadPool* pool = ThreadPool::getInstance();
vector<TaskHandle> futures(nTasks);
for (size_t task = 0; task < nTasks; ++task)
{
    auto spawnedTask = [&, task]()
    {
        //  Parallel job here
        //  ...

        //  Don't do that
        {
            lock_guard<mutex>(m);
            //  Do something with temp
            workWithTemp(temp);
        }

        //  ...

        //  Tasks must return a bool
        return true;
    };

    //  Spawn the task into the pool, get future
    futures[task] = pool->spawnTask(spawnedTask);
}
```

Instead, we preallocate workspace *for every thread*, like we made copies of mutable objects:

```
//  Lock-free code: no mutex
//  mutex m;

ThreadPool* pool = ThreadPool::getInstance();

//  Allocate one temp vector per thread
vector<vector<double>> temps(pool->numThreads() + 1, vector<double>(N));

vector<TaskHandle> futures(nTasks);
for (size_t task = 0; task < nTasks; ++task)
{
    auto spawnedTask = [&, task] ()
    {
        //  Parallel job here
        //  ...

        //  Do this
        //  Index of the currently executing thread
        const size_t thisThread = pool->threadNum();
        //  Workspace for this thread
        vector<double>& thisTemp = temps[thisThread];
        //  Do something with temp
        workWithTemp(thisTemp);

        //  ...

        //  Tasks must return a bool
        return true;
    };

    //  Spawn the task into the pool, get future
    futures[task] = pool->spawnTask(spawnedTask);
}
```

This is how we concretely achieve lock-free concurrent code, with a thread safe design that does not damage parallel efficiency. The thread pool's dedicated facilities makes the management of thread-wise copies mostly seamless.

Load balancing

Another important benefit of our thread pool is that it implements automatic load balancing. Client codes sends tasks in a queue, and the worker threads pop the tasks and execute then in the same sequence. When a worker thread takes a long time to complete a task for whatever reason[15] the other threads

[15]This thread may land on a slower hyper-threaded core, or it may share its core with a background task on the OS, or some other program the user runs while waiting for the results, or this particular task may be longer than others.

pop and execute tasks in the meantime. As long as there are tasks in the queue, all the threads work at 100% capacity and never wait for one another. In parallel computing, this is called load balancing, and our thread pool implements it by design.

We generally have as many worker threads as we have (logical) cores, minus one for the main thread. Load balancing only works when we have many tasks. Evidently, if we send 16 tasks into a thread pool of 15 worker threads (plus the main thread, who helps), everybody waits for the slowest to complete. If we send a large number of tasks, load balancing comes into play and a linear improvement may be achieved.

To enable load balancing, the number of tasks must significantly exceed the number of threads. So, we should split our parallel algorithms into many tasks. How far should we split? For instance, in a transformation that applies a math function to many numbers, should we spawn each single number as a separate task? The answer is no. There is an overhead (of the order of a few tens of microseconds) to spawn a task. The overhead remains negligible as long as the duration of the task is orders of magnitude longer. This is why, as a rule of thumb, we spawn tasks of around a millisecond. This is not a firm rule. It could be 0.1 ms or 10 ms. But we must refrain from spawning gazillions of microscopic tasks, or our cores spend more time spawning than computing.

Therefore, the rule (of thumb) is to split parallel algorithms into as many tasks as possible, keeping the duration of each task around 1 ms. The exact rule depends on the context. In our simulation code of Part II, we spawn Monte-Carlo paths in batches of 64.

3.20 DEBUGGING AND OPTIMIZING PARALLEL PROGRAMS

Parallel programs are notoriously hard to debug and optimize. In this final section, we briefly introduce some techniques to help identify and eliminate race conditions, and improve parallel efficiency.

Debugging and profiling serial code

Before we discuss the debugging and optimization of parallel code, we briefly introduce the debugger and profiler bundled with Visual Studio in the context of serial code. We expect readers to have some experience with these from lectures and programming. When this is not the case, what follows should help get started.

The Visual Studio Debugger When the program crashes or produces incorrect results, we have a "bug." Bugs can be found by inspection of the code, but we can identify them faster, and more conveniently, with a debugger.

To start Visual Studio's debugger, compile the program in debug mode.[16] Set a breakpoint on the first line of the suspected block of code with the "Toggle Breakpoint" item of the Debug menu, or press F9. Start the program through the Debug menu, Start Debugging or press F5. The program executes and pauses on the breakpoint. Visual Studio displays its state. The code window shows a yellow arrow on the left of the current line of code. The call stack window displays the current sequence of function calls. The watch window shows the current values of variables and expressions that users type in the name column, including classes and data structures.

Navigate the program with the Debug menu: continue to the end or the next breakpoint (F5), advance to the cursor (ctrl + F10), to the next line (F10), step into the function called on this line (F11), execute and exit the current function (shift + F11), and so on. Double click on a function on the call stack to see the current state in that function's environment, with execution still paused on the same line of code. Toggle breakpoints on and off and add new breakpoints while debugging (F9) or create conditional breakpoints by right clicking on the breakpoint, and select "condition...".

The debugger provides a vast, and sometimes overwhelming, amount of information. We briefly described the most common ones. With these basic tools, we can execute the program step by step and examine its state as the execution progresses. The environment is simple, clear, and user friendly, and generally helps find bugs very quickly.

The Visual Studio Profiler After the program produces correct results, we may attempt to increase its execution speed. The first thing to do is compile in release mode with all optimizations turned on. Visual Studio switches most optimizations on by default, but the following may be set manually on the project's properties page,[17] on the release configuration for the 32 bit or 64 bit platform, whichever is relevant:[18]

Optimization in C/C++

Optimization = Maximum Optimization (Favor Speed)
Inline Function Expansion = Any Suitable
Favor Size of Speed = Favor fast code

---

[16]Visual Studio's toolbar has a scroll-down menu to switch configuration between release (compile code for practical use, with all optimizations switched on) and debug (compile code for debugging, without the optimizations).

[17]Right-click the project in the solution explorer (the project, not the solution), and select "properties."

[18]We generally compile programs in 64 bit, except when we export code to 32-bit Excel. The project in our repository builds a 32-bit xll and therefore compiles on the 32-bit platform. The file xlCpp in our repository includes a tutorial ExportingCpp2xl.pdf that explains how to export C++ code to Excel, and the project in the repository is built this way.

Code Generation in C/C++
  Enable Enhanced Instruction Set = Advanced Vector Extension 2[19]
  Floating Point Model = Fast

Language in C/C++
  Open MP Support = Yes
  C++ Language Standard = ISO C++ 17 Standard

To accelerate the program, we must first identify *where* most execution time is spent. This is the purpose of a profiler, to measure CPU time spent in various parts of the code during an execution.

To start profiling, compile the program in *release* mode with debug information switched on, both on the compiler and the linker. It should be the case by default. For avoidance of doubt, check the project properties page for the release configuration. "C/C++ / General / Debug Information Format" should be set to "Program Database," and "Linker / Debugging / Generate Debug Info" should be set to "Generate Debug Information."

Start the program from the Debug menu, profiler, performance profiler. Select performance wizard, press start. Skip through the four pages of the wizard window, leaving all settings to defaults, press finish. The program starts. When it completes and exits, Visual Studio displays various statistics. Press "Show Just My Code" in the upper-right corner, and start navigating the functions that consume most CPU time. The profiler shows CPU time spent on every line with a color code from white (negligible CPU time) to red (major CPU time).

The profiler holds much more information than what we introduced. Knowing exactly where the program spends time is enough to get started, and it is priceless information. Armed with these statistics, we can investigate and attempt to accelerate those functions and blocks of code that consume most CPU time, ignoring others:

Is the algorithm implemented in this function optimal?

Do we make unnecessary copies? Are we passing arguments by value? Did we forget to implement move semantics?

Do we allocate memory at low level, explicitly or by construction or assignment of containers and data structures?

Do we correctly call STL algorithms in place of hand-crafted loops?

Do innermost loops work on coalescent memory? Are they vectorized?[20]

And of course: is this time-critical code a candidate for multi-threading?

---

[19]Provided the program runs on a recent chip, otherwise it crashes when started.

[20]Write "/Qvec-report:2" in the "Configuration Properties/ C/C++ / Command Line/ Additional Options" box of the project's properties to find out.

It is important to make these investigations with the measures provided by the profiler and not developer guesses. Even the most experienced programmers are often unable to correctly predict the bottlenecks in a C++ program. With modern hardware and compiler optimization, bottlenecks often land in unexpected places, due to cache, memory, and so forth while the compiler optimizes the trivial bottlenecks away on its own.

Debugging and optimizing parallel code

To debug and optimize parallel code is *orders of magnitude* harder. We don't cover this vast subject in detail, instead offering general advice based on our experience, and tips that helped us develop stable, efficient parallel programs, including those in Parts II and III.

Debugging parallel code If the serial code returns the correct results and the parallel code doesn't, the cause has to be one of the following two: either the parallel logic is flawed, or we have race conditions. To find out, *run the parallel code serially*. Assuming the program uses the thread pool, start the pool with zero worker threads. All the tasks will be processed sequentially on the main thread, resulting in a serial execution of the parallel logic. With concurrency and associated problems out of the way, we can debug the parallel logic on its own.

If the results are still incorrect, the parallel code is bugged. Debug a serial run of the parallel code, same as serial code, following the steps of the previous paragraph.

If the results are correct on a serial run, but incorrect with multiple threads, we have race conditions. Race conditions can also be diagnosed directly in what they produce different results on successive executions. This is, however, not entirely reliable, because other flaws, like uninitialized memory, cause similar symptoms.

Debugging race conditions When tests confirm race conditions, we must locate the responsible pieces of code. To automatically detect race conditions is almost impossible. Visual Studio has no such tool. The few free and commercial tools available are not reliable in our experience. The only way we found to debug races is with *manual code bisection*.

Serialize your entire concurrent code[21] by locking a static mutex on the first line and unlocking it on the last line. The code is now executed on one thread at a time, even in a multiple thread environment, so races cannot occur, and the results must be correct. Now, move the lock somewhere in the middle of the code. Are the results still correct? If so, we have no race

[21]The part of the code that is executed concurrently, written in the lambda spawned into the thread pool.

in the first half of the code. Otherwise, we do have races there, and perhaps also in the second half.

Implement a classic bisection, narrowing down the protected part of the code, possibly with multiple mutexes protecting multiple regions, until all the races are identified one by one and corrected. This is a slow and frustrating process. Code must be recompiled between steps. It is well worth striving to write const-correct, thread safe code in the first place, as explained in Section 3.11, rather than spending hours and days hunting race conditions after the fact.

Profiling parallel code Once the parallel program returns the correct results, we can measure its parallel efficiency, which we recall is the ratio $E = T_s/pT_p$ of the parallel acceleration T_s/T_p to the number of physical cores p, and attempt to improve it if sub-par, that is, materially under 100%. For instance, our initial parallel histogram code from Section 3.5 completed in $T_p = 1s$ over $p = 8$ physical cores, compared to $T_s = 3s$ serial, a parallel efficiency of only 37.5%.

First, separate the parallel logic overhead from the concurrent inefficiency. Run the parallel program in serial mode, as explained earlier, and measure its execution time T_{ps}. How does it compare to T_s? The *logic efficiency* $L = T_s/T_{ps}$ measures (the inverse of) the overhead caused by the additional logic necessary to implement the parallel code, irrespective of the number of threads or cores executing it. The *concurrent efficiency* $C = T_{ps}/pT_p$ measures the acceleration per core of a parallel run *over a serial run of the same parallel code.*

Evidently, $E = LC$, but we now have two measures of parallel efficiency so we know where to focus our efforts: logic, concurrency, or both. Our parallel histogram code (upgraded with the thread pool) runs in $T_{ps} = 3.5s$ over a single thread, hence $L = 86\%$ and $C = 44\%$.

The parallel histogram is an embarrassingly parallel algorithm, multi-threaded at a high level. Its only logic overhead comes from task spawning and the general administration of the thread pool, the concurrent queue, the tasks, and the futures. In such cases, the expected overhead is around 10%. The expected logic efficiency is around 90%. A materially lower efficiency would indicate an incorrect application of the thread pool, probably too many small tasks with duration below 1 ms so the administration overhead makes a measurable difference.

Other parallel algorithms may require further additional logic. For instance, the parallel reduction algorithm, discussed in the introduction, implements recursion logic in addition to the reduction logic of the serial version. That may cause substantial overhead, and result in a lower logic efficiency.

Parallel logic may be profiled, and perhaps optimized, in the exact same way as serial code, profiling a serial run of the parallel code and repeating the steps of page 114. To what extent parallel logic may be optimized depends on the algorithm. In many instances, logic overhead is part of the algorithm's nature and cannot be improved.

The expected concurrent efficiency also depends on the parallel algorithm. For instance, the synchronization logic necessary for the parallel reduction, where threads must wait on one another to complete a step before the next step begins, is bound to reduce efficiency. For an embarrassingly parallel algorithm multi-threaded at a high level, we expect efficiency around 100%. For the parallel histogram, 44% clearly indicates a flaw in the parallel code. We now know the problem of that code is false sharing, of course, but in general, the presence of a problem is easily diagnosed with simple efficiency measures.

It is worth considering trivial causes first, and inspect CPU utilization. On Windows, CPU utilization is displayed on the Task Manager. *Before* the parallel program is executed, CPU utilization must be close to 0. Otherwise, other processes are running and the parallel algorithm will share cores with them, obviously reducing efficiency. All the running processes must be stopped in order to measure efficiency accurately.

When the parallel part of the algorithm starts, all CPUs should hit 100% utilization and stay there continuously until it completes. At that point, the utilization of all CPUs should simultaneously drop to 0, safe for one, where the main thread aggregates and displays the results. It is worth running the algorithm for a long time (large data set, many Monte-Carlo paths, . . .), to obtain a clear CPU utilization chart. Those charts must *always* reproduce the pattern below:

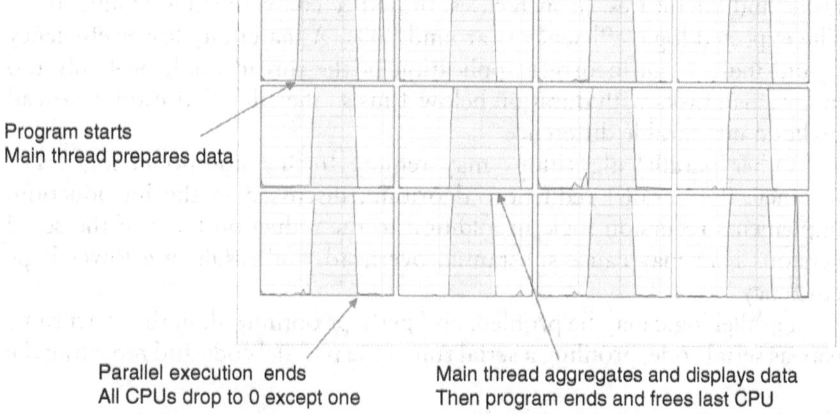
CPU
Intel(R) Xeon(R) W-2140B CPU @ 3.20GHz
% Utilization over 60 seconds
100%
Parallel execution starts
All CPUs rise to 100%
Program starts
Main thread prepares data
Parallel execution ends
All CPUs drop to 0 except one
Main thread aggregates and displays data
Then program ends and frees last CPU

When the chart looks any different, we have a problem so severe that it should be easy to find. If only one CPU is utilized, the program does not run in parallel. The thread pool did not start correctly or the tasks were not properly spawned. When multiple, but not all, CPUs are working, (assuming the thread pool was set to fully utilize hardware concurrency), we probably didn't fire enough parallel tasks. As explained on page 112, an insufficient number of parallel task, also damages load balancing, causing utilization to drop *one CPU at a time*, and not simultaneously, when the parallel part completes.

Try increasing the number of parallel tasks, splitting the algorithm into a larger number of smaller tasks.

In rare occasions, we have seen all CPUs at work but capped around 50–80%. Every time, this curious phenomenon was caused by severe lock contention, or repeated allocations, usually in the innermost loops. On the other hand, false sharing or a more limited contention leave CPUs working at 100% capacity, but damage efficiency all the same.

More generally, an incorrect CPU utilization always reveals a problem, but a correct utilization pattern absolutely doesn't indicate the absence thereof. If concurrent efficiency remains sub-par after the initial checks, we may have a deeper concurrent flaw. It helps to make more precise measures: run the parallel code on 1, 2, 3, ... threads up to the number of logical cores. Measure execution times T_n with n threads and draw T_1/T_n as a function of n. The chart below displays this information for the parallel histogram, before and after fixing the false sharing problem.

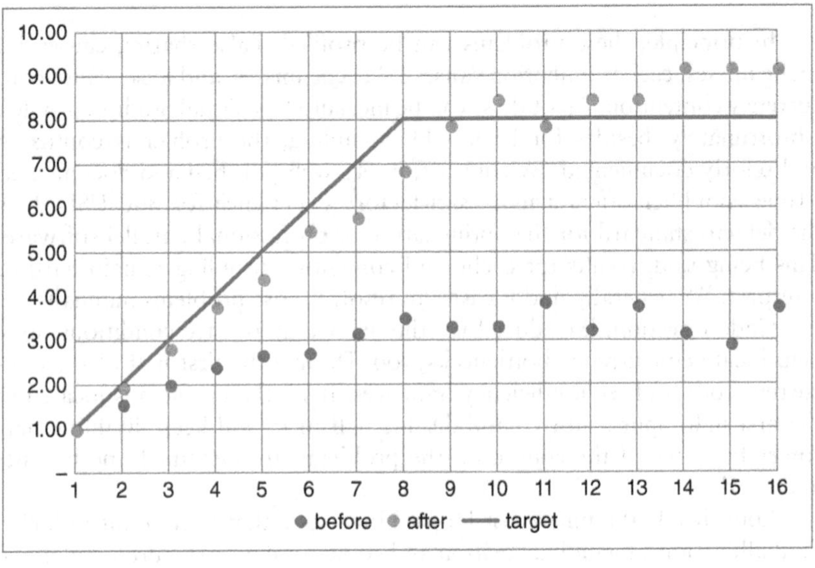

With the correct code, we get a straight 45deg line up to the number 8 of physical cores, followed by a flattening when we hit the hyper-threaded cores. With false sharing, we obtain the same pattern, but with a more than twice lower slope around 20deg. With locks or allocations, we could obtain a *decreasing* curve.

When tests confirm concurrency flaws, we must identify and locate them. The most common concurrent inefficiencies include:

Locks as explained in Section 3.11, including hidden ones, like memory allocation. The remedy is to remove all locks from the concurrent code, making copies of mutable objects instead, as explained on page 109, and move allocations out of the concurrent code, pre-allocating all necessary memory in the initialization phase before multi-threading.

False Sharing as explained in Section 3.5, along with remedies.

L3 cache contention for memory intensive parallel algorithms, like AAD, see Part III. Multi-threaded code processing vast amounts of memory may cause multiple threads to compete for the shared L3 cache. L3 cache contention is remedied by splitting processed data over a larger number of smaller tasks, so that each task works in parallel with a smaller amount of data. This may be hard to implement in practice, and may even require a redesign of the parallel algorithm. The parallel histogram would have to be completely rewritten in a way more complicated manner for the data to be processed over each interval by multiple tasks. In the context of AAD, this is achieved with check-pointing; see Chapter 13.

In principle, these problems can be profiled. False sharing causes L2 cache misses, cache contention causes L3 cache misses, and locks cause concurrency contention. All of these can be measured by Visual Studio's profiler. Unfortunately, besides for basic CPU sampling, the profiler is confusing and poorly documented. By contrast, Intel's well-polished and documented vTune Amplifier offers a more satisfactory experience for 800 USD. It is the defacto standard for the optimization of professional parallel software. This being said, results for cache and contention profiling remain hard to interpret. We generally find it easier to resolve those problems manually.

Code bisection, introduced for the debugging of race conditions, may help locate concurrency bottlenecks, too. Disable the first half of the concurrent code and see if efficiency improves. If so, the problem is located in the first half. Split it in two and disable half of it, and keep disabling narrower fractions of the code until the problems are identified one by one, and fixed.

Look inside the innermost loops. Hunt code that may result in locks. Find allocations, including creation and copies of data structures, and move them outside of the concurrent code. Look for code that writes into the heap and may cause false sharing.

To rule out allocations, install a concurrent lock-free allocator, such as the one bundled with Intel's free TBB library, and set it as the application's global allocator. See Intel's documentation on how to do that exactly.

To confirm L3 cache contention, try running the same algorithm on a smaller data set. If the problem persists, cache contention is not the problem. Cache contention problems grow with the size of the working memory per task. As mentioned above, it is resolved by splitting the algorithm into a larger number of smaller tasks, each manipulating smaller memory. This is generally hard to implement, and may take patience and creativity. Parallel AAD, in particular, manipulates vast amounts of memory and is vulnerable to cache contention. We apply a number of techniques in Part III to reduce the size of the tapes (working memory for AAD) and mitigate cache contention.

This completes our general introduction to parallel programming in C++, and the general part about programming. The next part applies these concepts and techniques, and our thread pool, in the context of Monte Carlo simulations.

PART II

Parallel Simulation

Introduction

In this part, we cover the valuation of derivatives transactions (including regulatory calculations like xVA) with Monte Carlo simulations. We develop a generic simulation library in C++ and accelerate it with parallelism, building on the notions and constructs introduced in Part I.

Derivatives mathematics revolve around *pricing*, the determination of the value of transactions as a function of the current market, and *risk*, the differentiation of that function. Numerical finance consists in the production of pricing and risk algorithms converging with minimum asymptotic complexity. Computational finance brings the pieces together to develop valuation and risk software that is both practical and convenient, and produces accurate results quickly, not in a theoretical, asymptotic context, but in situations of practical relevance.

Pricing is a prerequisite to risk. Evidently, a function must be defined before it is differentiated. This part deals with pricing. Differentiation in general, and risk in particular, are covered in Part III.

Chapter 4 reviews the asset pricing theory, introduces the theoretical context for Monte-Carlo simulations, and establishes the necessary abstractions, definitions, and notations. Chapter 5 covers financial Monte Carlo simulations. Chapter 6 brings these ideas to practice and builds a generic (sequential) Monte Carlo library in C++. In Chapter 7, the library is extended with parallel simulations using the notions and constructs of Part I. The serial and parallel programs are differentiated in Chapter 12, after we cover differentiation in general terms.

The source code is available in our repository.

CHAPTER 4

Asset Pricing

The next chapters cover the development of Monte-Carlo simulations. This chapter introduces the theoretical context. We cover asset pricing in general terms, introduce derivatives markets and the asset pricing theory, and establish the necessary formalism, abstractions, definitions, and notations for the rest of the publication. Derivatives markets and asset pricing are vast subjects, covered in a multitude of publications. Our short introduction is necessarily partial and dense. Derivatives markets are covered in detail in Hull's classic [26], updated in 2017. Asset pricing theory is covered in many textbooks, like [27].

4.1 FINANCIAL PRODUCTS

In order to design *generic* pricing libraries, we must abstract the notions of financial products and models and define them in a general manner. Starting with financial products, we need a formalism encompassing underlying assets (shares, currencies, zero-coupon bonds, . . .), linear transactions (forward contracts, forward rate agreements, interest rate swaps, . . .), European options (calls and puts, caps and floors, swaptions, . . .), path-dependent and early exerciseable options (collectively known as *exotics*), multi-underlying options, including derivatives written on assets of different classes (known as *hybrids*), as well as portfolios of transactions and regulatory amounts like xVA.

Products and cash-flows

We therefore define a financial product as a collection of cash-flows paid over a given schedule. The payment schedule can be either discrete, in which case the product is a set $\{CF_p, 1 \leq p \leq P\}$ of cash-flows paid on dates $\{T_p\}$, or continuous, in which case the product is a continuous set of cash-flows

($dCF_u = cf(u)du$) paid over the time interval $[0, T]$, T being the payment date of the last cash-flow, called the product's maturity.

Valuation is linear, so the value V_t of a product at a time $t \leq T$ is the sum of the values of its subsequent cash-flows: $V_t = \sum_{T_p \geq t} V_t^p$ or $\int_t^T v_t^u du$.

Cash-flows may be deterministic, linear in underlying asset prices, optional, or exotic. In its most general form, a cash-flow CF_T paid at time T is a $\mathcal{F}_T$-measurable random variable. This means that it may depend on the state of the market on and before its payment date T, and is fully determined at T. Formally, $CF_T = h(S_t, 0 \leq t \leq T)$ is a functional h of the *path* (which we also call *scenario*) and denote:

$$\Omega = (S_t, 0 \leq t \leq T)$$

where S_t is the "state of the market" at time t (which we call *sample*): the values at time t of the market variables that affect the cash-flow.

A European call of strike K and maturity T on some underlying asset s pays one cash-flow:

$$CF_T = (s_T - K)^+$$

at maturity. A call K up and out on a barrier B also pays one cash-flow but it is contingent on not breaching the barrier before maturity:

$$CF_T = 1_{\{\max(s_t, 0 \leq t \leq T) < B\}}(s_T - K)^+$$

The floating leg of an interest rate swap (IRS) pays N Libor coupons over a discrete schedule; for instance, a $10y$ quarterly floating leg pays 40 cash-flows:

$$CF_i = L\left(T_i^{start}, T_i^{end}\right)\left(T_i^{end} - T_i^{start}\right)$$

where $L(T_1, T_2)$ is the Libor rate of matuirty T_2 fixed at T_1.[1]

Scenarios

A cash-flow CF_T paid at time T depends on the scenario, generally defined as the evolution of the market up to the maturity, but its specific meaning depends on the cash-flow: for a European call, the scenario is the underlying asset price at maturity. For a barrier option, it is the continuous path of the underlying price series from now to maturity. For an interest rate swap (IRS),

[1]The dates T_i^{start} and T_i^{end} are the start and end dates in the ith period in the floating schedule, and we ignore conventional day counts and fixing lags.

the scenario is the discrete set of Libor fixings on the floating schedule. Cash-flows don't depend on the complete state of the world market at all times in $[0, T]$. Depending on the particular cash-flow, the scenario is *specified* by two characteristics:

1. The subset of $TL \subset [0, T]$ of dates where the state of the market affects the cash-flow, called *timeline*. For a European call, the timeline is the maturity date: $TL = \{T\}$. For a barrier, it is the entire interval $[0, T]$.
   A product's timeline is the union of its cash-flows timelines. The dates $t \in TL$ in the timeline are called *event dates*; they are the contractual dates where something meaningful happens to determine the cash-flows, in reference to the state of the market on those dates: the exercise of a call, the monitoring of a barrier, the fixing of a Libor, or the payment of a cash-flow. For the floating leg of an IRS, the timeline is the discrete set of fixing and payment dates on the floating schedule.
2. The nature of the sample S_t for every event date $t \in TL$ on the timeline. For a European call or a barrier option, this is the underlying asset price on time t. For a floating leg cash-flow, it is the Libor of a given maturity fixed at t. In general, a sample s_t for the event date t is a collection of market variables (asset prices, rates,...) observed at time t. It is the snapshot of the fraction of the market at time t that is useful for the determination of the cash-flows. In principle, its dimension could be infinite or even continuous. The dimension and contents of different samples on the timeline may be different from one another.
   The time t sample for a product is the union of the time t samples of its cash-flows. The collection of all the samples $\{S_t, t \in TL\}$ across the timeline is called *scenario*, and denoted Ω.

The scenario is therefore a multidimensional, and sometimes continuous, collection of samples over a timeline. The timeline, and the nature and the dimension of its samples, depend on the product. The cash-flows are a function(al) of the scenario, more precisely, of the scenario before the payment date:

$$CF_T = h(\{S_t, t \in TL, t \leq T\})$$

Underlying assets

The notions of products, cash-flows, timelines, samples, and scenarios are the building blocks of the product side of a generic financial library (the other sides being models, discussed next, and algorithms that articulate the two to produce values and risk, like the Monte-Carlo simulations of the next chapter). We illustrate these notions with a few examples, starting with the underlying assets. Underlying assets, like stocks or currencies, are not exactly financial products, but they can be modeled as financial products with the introduction of a holding horizon T. A stock is not a financial product, but the strategy of buying the stock and holding it to some horizon T is one.

A stock without dividends, with value s_t at time t, can be assimilated, given a holding horizon T, to a financial product with a single cash-flow $CF_T = s_T$ on the horizon date T. The timeline is therefore the singleton $\{T\}$ and the scenario is $\Omega = \{S_T = s_T\}$.

With a discrete schedule of dividends $\{d_k\}$ paid on times $\{T_k\}$, a stock is modeled as a schedule of cash-flows $CF_p = \{d_p, T_p \leq T\}$, plus a final cash-flow $CF_T = s_T$. The timeline is the set of ex-dividend dates before horizon $\{T_k < T\}$, plus the horizon date T, and the scenario is $\Omega = \{S_{T_k} = d_k\} \cup \{S_T = s_T\}$.

A stock or stock index with a continuous dividend yield (y_t) is modeled as a continuous schedule of cash-flows $(cf(t) = s_t y_t, t < T)$ and a final cash-flow $CF_T = s_T$. The timeline is the interval $[0, T]$; the scenario is $\{(s_t, y_t), t \in [0, T]\}$.

A foreign currency, which is economically identical to a stock with dividend yield the foreign short rate, is modeled as a financial product in the same way.

In interest rate markets, the notion of "underlying assets" is typically associated with zero-coupon bonds, although they are not directly traded instruments. The zero-coupon bond of maturity T is a financial product that pays a unique cash-flow of one monetary unit at time T. Its timeline is $\{T\}$. Its cash-flow is deterministic and independent of a scenario.

The value of a zero-coupon bond of maturity T at a time $t < T$, denoted $DF(t, T)$, acts as a conversion rate between payments at t and T, the value at t of a cash-flow X_t known at t ($\mathcal{F}_t$-measurable) and paid at T being $DF(t, T)X_t$. For this reason, zero-coupon bond prices are also called *discount factors*. Evidently, $DF(T, T) = 1$.

The collection $DF(t, T)$ of discount factors of all maturities T at time t is called the *discount curve* at t. The discount curve is part of the market state at t. Discount factors are market primitives and may be part of samples in a scenario. What discount maturities T are included in what samples depends on the product and its cash-flows. In general, a cash-flow doesn't depend on the whole collection of discount factors $DF(t, T)$ on an event date t, but on a smaller number of discounts of a specific set of maturities $T \in \Theta_t^{DF}$. The set Θ_t^{DF} of discount maturities in the sample S_t is part of the product-specific definition of the scenario.

Finally, the discount curve can be expressed in price units, discount factors, or in rate units,

$$R(t, T) \equiv -\frac{\log[DF(t, T)]}{T - t} \Longleftrightarrow DF(t, T) = \exp[-R(t, T)(T - t)]$$

is called *discount rate* of maturity T. The quantity:

$$f(t,T) \equiv \frac{\partial}{\partial T}[R(t,T)(T-t)]$$

is called *instantaneous forward rate* (IFR).[2] It follows immediately that:

$$f(t,T) = -\frac{\partial \log[DF(t,T)]}{\partial T} \Longleftrightarrow DF(t,T) = \exp\left[-\int_t^T f(t,u)du\right]$$

and the short rate at time t is defined as $r_t \equiv f(t,t)$.

Linear transactions

A *forward contract* with strike K and maturity T on an underlying asset s is a financial product paying a unique cash-flow $s_T - K$ at date T. Its timeline is the singleton $\{T\}$ and the time T sample is the scalar $\{s_T\}$.

On a date $t \leq T$, the unique strike K such that the value of the corresponding forward contract is 0 is called the *forward price* of s and denoted $F(t,T)$. Forward prices $F(t,T)$ are part of the market state at time t, and it is best, in order to accommodate a wide range of products and models, to consider forwards as market primitives and parts of samples in the scenario. In this case, we don't need spot prices in the samples, since $s_t \equiv F(t,t)$.

What forwards of what maturities are part of the samples depends on the exact nature of the cash-flows, same as discount factors. For instance, a forward contract depends on the sample $S_T = s_T = F(T,T)$ so in this case, $\Theta_T^F = \{T\}$.

In interest rate markets, a forward rate agreement (FRA) pays $[L(T_1,T_2) - K](T_2 - T_1)$ on date T_2,[3] where $L(T_1,T_2)$ is the *Libor* rate fixed at T_1 for maturity T_2. The FRA's timeline is the union of its fixing and payment dates $\{T_1,T_2\}$ and its dependency is on the sample $S_{T_1} = L(T_1,T_2)$ on the fixing date T_1. Why T_2 is on the timeline without a sample will be clarified when we move on to pricing, and realize that we are missing "something" on the payment dates samples.

On a date $t \leq T$, the unique strike K such that the value of the corresponding FRA is 0 is called forward rate, or *forward Libor*, and denoted $F(t,T_1,T_2)$. We consider forward Libors as market primitives and part

[2]We should call it "instantaneous forward *discount* rate" to avoid confusion with Libor or OIS rates, which generally trade with a significant difference called *basis* to discount rates.

[3]Ignoring conventional day counts and fixing lags.

of samples in the scenario. Spot Libors are therefore not necessary, since $L(T_1, T_2) \equiv F(T_1, T_1, T_2)$.[4]

What forward Libors of what start and end dates and what index are included in the samples is something specified from the nature of the cash-flows. For instance, a FRA of maturity T on a $3m$ Libor ("$3m$" referring both to the Libor's duration and to its index) pays $CF_{T+3m} = 0.25[F(T, T, T + 3m, 3m) - K]$ on time $T + 3m$. Its timeline is $\{T, T + 3m\}$, and it is determined by the time T sample $S_T = F(T, T, T + 3m, 3m)$. It follows that $\Theta_T^L = \{(T, T + 3m, 3m)\}$, one Libor being defined, in addition to its fixing date, by two dates and an index.

Since the cash-flow is known at time T and paid $3m$ later, we can equivalently model this FRA as the cash-flow:

$$CF_T = 0.25[F(T, T, T + 3m, 3m) - K]DF(T, T + 3m)$$

paid at time T. We used the discount factor to *convert* an amount paid in $3m$ into an amount paid immediately. This allows to reduce the timeline to the singleton $\{T\}$ but increases the sample to include the discount factor: $S_T = \{F(T, T, T + 3m, 3m), DF(T, T + 3m)\}$. In this case, $\Theta_T^{DF} = \{T + 3m\}$ and $\Theta_T^L = \{(T, T + 3m, 3m)\}$.

An interest rate swap (IRS) is a financial product that exchanges a fixed leg for a floating leg. The payment schedules of both legs share a common start date T_s and end date T_e but they typically have different periods δ_{fix} and δ_{float}, hence a different number of *coupons* N_{fix} and N_{float}. In a receiver swap, the fixed leg pays the coupons $K\delta_{fix}$ on dates $T_0 + i\delta_{fix}$ and the floating leg

---

[4] "Old school" interest rate modeling uses the "identity":

$$F(t, T_1, T_2) = \frac{1}{T_2 - T_1}\left(\frac{DF(t, T_1)}{DF(t, T_2)} - 1\right)$$

to "deduce" Libors from discount factors. It would follow that Libors are not additional market primitives, but aggregates of discount factors. However, this formula ignores discount basis and assumes Libor discounting, something completely wrong in modern markets. Since the explosion of basis adjustments, the advent of collateral discounting, and the artificial stabilization of interest rates around zero in 2008–2011, Libor basis (roughly, the difference of interest rates picked on the Libor curve or on the discount curve) largely exceeds interest rates, both in value and daily moves, and accounts for a major fraction of interest rate risk. This is why Libors and discounts are different market primitives. In addition, we have multiple Libor indices per currency, with large, fast-moving bases between indices, so a complete definition of a forward Libor at time t includes its index: $F(t, T_1, T_2, idx)$, although we often drop the index here to simplify notations.

pays the coupons $-L(T_0 + (j-1)\delta_{float}, T_0 + j\delta_{float})\delta_{float}$ on dates $T_0 + j\delta_{float}$. A payer swap pays the opposite coupons. The timeline of an IRS is the union of the fixed and floating schedules. The scenario that determines an IRS is the sequence of Libor fixings over the floating schedule.

The fixed coupon K such that the value of the swap is 0 on a date $t \leq T_s$ is called "par swap rate," or sometimes "forward swap rate." Par swap rates are not market primitives; we will see shortly that they are expressed in a simple, universal, and model-independent manner from forward Libors and discount factors.

Forward contracts, FRAs, and IRS are all examples of a particular type of financial products called *linear* products. Linear products are defined as schedules of linear cash-flows. Linear cash-flows are often defined as linear functions of the scenario, although this definition is approximate and in some cases incorrect. The definition of linear cash-flows is related to valuation, as we will see shortly.

European options

Traditionally, European options are those that depend on a single observation of an underlying market variable on a single event date. We go with this definition for now, although it is somewhat blurry in our context. We will correctly define European products later, when we move on to valuation.

For now, a cash-flow CF is called European if it depends on a single observation, on a single event date, of a single market variable: $CF = h(s_T)$ where s_T is a scalar market variable. The timeline of a European cash-flow is a singleton $\{T\}$, called "maturity," "expiry," or "exercise date" depending on the product. The cash-flow is a function (not a functional) h of the scalar s_T.

When $h(x) = (x - K)^+$, the cash-flow is a *European call.* When $h(x) = (K - x)^+$, it is a European put. More generally, any smooth function h can be written as:

$$h(x) = h(k_0) + h'(k_0)(x - k_0) + \int_{-\infty}^{k_0} h''(k)(k - x)^+ dk + \int_{k_0}^{\infty} h''(k)(x - k)^+ dk$$

This is known as Carr-Madan's formula [28] and shows that *all* European cash-flows are combinations of calls and puts. Carr-Madan's formula is demonstrated with simple calculus and remains valid when h is discontinuous, as long as derivatives are read in the distributional sense of Laurent Schwartz.

A European call (respectively put) on a Libor is called a caplet (floorlet). A succession of caplets (floorlets) over a floating schedule is called a cap (floor).

A cash-settled swaption pays a single cash-flow on its exercise date T:

$$h(F_T) = \frac{1}{F_T}\left(1 - \frac{1}{(1+\delta_{fix}F_T)^{N_{fix}}}\right)(F_T - K)^+$$

for a payer swaption, or replace the right-hand side with $(K - F_T)^+$ for a receiver swaption, where F_T is the par swap rate on the exercise date. A swaption's cash-flow is a function of the par swap rate, which is not a market primitive, but an aggregate of a number of discount factors and forward Libors as we will see shortly. A swaption may therefore appear not *stricto sensu* European.[5] We will revisit swaptions shortly.

A financial product is called European if all its cash-flows are European (European products may also include multiple European cash-flows on the same date, on different underlying variables). All the linear products we introduced previously are European products. Nonlinear European products are called European options.

Path-dependent exotics

Cash-flows that are not linear or European depend on multidimensional samples on multiple event dates, in a nonlinear manner. Products that involve such cash-flows are collectively known as *exotics*. Cash-flows that depend nonlinearly on multidimensional samples are called "basket options." Those depending on multiple samples prior to their payment dates are called "path-dependent." When a party holds the contractual right to cancel, redefine, or alter cash-flows in any way, the product is said to be "callable" or "early exerciseable." Those categories are not mutually exclusive.

The timeline of path-dependent cash-flows includes, by definition, multiple event dates. A textbook example is a barrier option[6] on some underlying asset s. A barrier delivers the cash-flow of a European call or put at maturity, but only when the underlying asset price remained below (up and out) or above (down and out) a predefined level called *barrier* and denoted B, at all times before maturity. The cash-flow of an up-and-out call, for example, is:

$$CF_T = 1_{\{M_T<B\}}(s_T - K)^+$$

[5]A swaption is an option on a combination of forward Libors, an exotic product known as a "basket option." The Libor Market Model of [29] builds on this observation to develop an approximate closed-form valuation formula for swaptions as a function of the volatilities and correlations of forward Libors.

[6]Options on average, called "Asian options," also often illustrate the notion of path-dependent products, although they rarely trade in practice, contrary to barriers.

where $M_T = \max(s_t, 0 \le t \le T)$. Its timeline is therefore the continuous interval $[0, T]$ and it is a functional of the continuous scenario $(s_t, 0 \le t \le T)$. Many other variations exist, like knock-in barriers, which pay only if the barrier *was* breached, *American digitals*, which pay a different amount depending on whether the barrier was breached or not, double barriers (those with a down barrier and an up barrier, sometimes called *corridors*), and all the imaginable combinations of these.

One important variation for Monte-Carlo simulations is the *discretely monitored* barrier. In this case, the payment is subject to not breaching (knock-out) or breaching (knock-in) the barrier over a *discrete* timeline, irrespective of what happens in between the observation dates. The timeline is the discrete union of the monitoring schedule with the fixing and payment dates of the cash-flow. The scenario is the spot price on the event dates in the timeline.

American options and cancellable products

Moving on to callable products, we introduce the two most common types of early exerciseable transactions, American options and cancellable products.

American options are those exerciseable at any time t *before maturity* T into a cash-flow $(s_t - K)^+$ for an American call or $(K - s_t)^+$ for an American put. The party who holds the right to exercise does so on the first occasion when the exercise cash-flow exceeds the *continuation value* of the option (the value immediately after if not exercised). An American option's payment schedule and its timeline are therefore both the continuous interval $[0, T]$ and its cash-flows are defined by:

$$a_0 = 1$$
$$CF_t = a_t h(s_t) 1_{\{h(s_t) > V_t\}}$$
$$da_t = -a_t 1_{\{h(s_t) > V_t\}}$$

where a tracks whether the option is still alive and h defines the exercise cash-flow. We have a product which cash-flow at t depends on the product's value on and before t. We have seen that the product's value at time u is the sum of the values at u of all the cash-flows paid after u. Hence, values depend on cash-flows and cash-flows depend on values. In addition, the continuation value V_t depends not only on the state of the market at t, but also on the option pricing model.

American options therefore involve an undesirable coupling between the product and the model and a recursive relationship between values and cash-flows.

Cancellable products are those where a party has the right to cancel all the subsequent cash-flows on a set of exercise dates. A cancellable product's timeline is the union of the timeline of its non-callable counterpart with the exercise dates. When the non-callable product pays the cash-flows CF_p^{NC} on times $\{T_p\}$, the callable product's cash-flows are defined with:

$$CF_p = a_{T_p} CF_p^{NC}$$

where a is 1 before cancellation and 0 thereafter, hence $a_0 = 1$ and a is updated on exercise dates T_e with:

$$a_{T_e^+} = a_{T_e^-} 1_{\left\{V_{T_e} > 0\right\}}$$

where V_t is the time t value of all subsequent *cancellable* cash-flows. Again, cash-flows depend on future values in a recursive manner, and the definition of the cash-flows depends on the model. This is a defining characteristic of callable products; it is what makes them special and the reason why they require special modeling.

In the somewhat simplified case of a cancelable product with *one* exercise date, sometimes called "one-time callable," the cash-flows are the cash-flows of the non-callable counterpart before the exercise date $T*$, and:

$$CF_p = 1_{\left\{V_{T*}^{NC} > 0\right\}} CF_p^{NC}$$

after the exercise date, so the cash-flows are still dependent on a future value, but this is the future value of the non-callable counterpart, not the value of the callable product itself. The recursive relationship is gone, but the dependency on a future value and a model remains.

For instance, a physically settled swaption is the right to enter the underlying swap, or, equivalently, the right to cancel its cash-flows. It is therefore a one-time callable swap, which pays the cash-flows of the underlying swap, provided its present value on the exercise date T is positive:

$$CF_p = 1_{\left\{V_T^{swap} > 0\right\}} CF_p^{swap}$$

We will see shortly that V_T^{swap} is an explicit function v of the forward Libors $F(T, T_i, T_i + \delta_{float})$ on the exercise date with maturities the start and end dates on the floating schedule, and the discount factors $DF(T, T_p)$ to the payment dates of both legs. It follows that:

$$1_{\left\{V_T^{swap} > 0\right\}} = 1_{\left\{v(F(T,T_i,T_i+\delta_{float}),DF(T,T_p),1\le p\le P,1\le i\le N_{float})>0\right\}}$$

All these discounts and forward Libors are therefore included in the swaption's sample on the exercise date: $\Theta_T^{DF} = \{T_p, 1 \le p \le P\}$, $\Theta_T^L = \{(T_i, T_i + \delta_{float}) 1 \le i \le N_{float}\}$, and:

$$S_T = \left\{F(T, T_i), DF(T, T_p), 1 \le p \le P, 1 \le i \le N_{float}\right\}$$

so:

$$1_{\left\{V_T^{swap} > 0\right\}} = 1_{\{v(S_T) > 0\}}$$

Seen from this angle, the swaption is no longer a callable product, but fits the definition of path-dependent products. We will see shortly that we can also consider it a European option that delivers the present value of the underlying swap on the exercise date.

Turning callables into path-dependents

Callable products are therefore those whose cash-flows depend on the future values of themselves, and/or the future values of financial products other than primary market variables. How algorithms deal with them depends on the algorithm. We will see shortly that finite difference methods (FDM) cope with callable features in a native and natural manner. On the contrary, Monte-Carlo naturally handles path dependency, but doesn't know future transaction values and cannot handle callable features. In order to price a callable transaction with Monte-Carlo, we must first transform the transaction into an equivalent path-dependent, like we just did for a swaption.

In general, unlike for the swaption, we don't have an analytic expression of the future values, so we must somehow estimate them as a function of the market:

$$V_{T_e} \approx \tilde{v}(S_{T_e})$$

where $\tilde{v}$ is called a *proxy* function and the sample S_{T_e} is the set of market variables on T_e it accepts as arguments, which must be included in the scenario in order to use the proxy.[7] Using proxies, the cash-flows of the American option simplify into:

$$a_0 = 1$$

[7]In general, for example when the callable product has path-dependent cash-flows, its value depends not only on the current market, but also on the past:

$$V_{T_e} \approx \tilde{v}(S_t, t \in TL, t \le T_e)$$

In what follows, we only consider the simple case. Path-dependency does not materially modify the reasoning.

$$CF_t = a_t h(s_t) 1_{\{h(s_t) > \tilde{v}(S_t)\}}$$
$$da_t = -a_t 1_{\{h(s_t) > \tilde{v}(S_t)\}}$$

and the cash-flows of an early cancellable transaction simplify into:

$$a_0 = 1$$
$$CF_p = a_{T_p} CF_p^{NC}$$
$$a_{T_e^+} = a_{T_e^-} 1_{\{\tilde{v}(S_{T_e}) > 0\}}$$

In both cases, the explicit dependency on future values vanished and the modified cash-flows fit the definition of classic path-dependents. The injection of proxies effectively turns callables into path-dependents.

It is also noticeable that proxies only appear in exercise indicators. On an exercise date, the set of sample values that cause early exercise is called the exercise region. The border of the exercise region is called the exercise boundary. The problem of finding effective proxies boils down to the estimation of exercise boundaries.

In the context of FDM, exercise boundaries are directly estimated on the FDM grid, as we will see shortly. In the context of Monte-Carlo, proxies are estimated by regression over pre-simulations, with an algorithm known as the *least squares method* or LSM, pioneered by Carriere in [13] and Longstaff and Schwartz in [14]. Proxies obtained with LSM are called *regression proxies*. The LSM algorithm is the established best practice for pricing callables with Monte-Carlo since the early 2000s. We briefly introduce it in the next chapter.

CVA

Finally, it is important to understand that counterparty value adjustment (CVA), along with most other value adjustments (xVA) and other bank-wise regulatory calculations, can be modeled as a financial product. This is crucial, because it follows that the models we develop for financial products are also suitable for these calculations. CVA is a real option a bank gives away whenever it trades with a defaultable counterparty. If the counterparty defaults on a future date t when the net sum of all ongoing transactions with that counterparty (called *netting set*) NS_t is positive, the bank loses the positive value of the netting set. This is a one-way street, hence, really an option: if $NS_t < 0$, the liquidator will insist that the bank refunds its value. CVA is therefore a zero-strike put, contingent to default, on the entire netting set, consisting itself in thousands of transactions, each one of which is a financial product in its own right, perhaps exotic or hybrid. The difficulty of computing CVA in a practical manner, and in reasonable time, is directly related to the high dimension of the "underlying" netting set.

In its simplest form, an uncollateralized CVA, defined in a discrete manner and without recursion, is modeled as a financial product that pays, over an *exposure schedule* $\{T_e, 0 \le e \le N\}$, where $T_0 = 0$ and $T_N = T$, the payment date of the last cash-flow in the netting set, the cash-flows:

$$CF_e = D\left(T_{e-1}, T_e\right)\left(NS_{T_e}\right)^+$$

where $D(T_{e-1}, T_e)$ is the proportion defaulted between T_{e-1} and T_e.[8]

The cash-flows of a CVA therefore depend on the future values of the netting set, so everything we said about callable products applies to CVA, too.[9] In addition, CVA cannot be valued with FDM due to its high dimension, so proxies are necessary there, too.[10]

In addition, CVA depends on credit, so the scenarios must be enlarged to incorporate credit variables and credit events, something we are not doing in this publication. The future values of the netting set typically depend on a vast number of market variables from different asset classes: stocks, rates, currencies, etc. CVA is therefore a *hybrid* option, which valuation necessitates a hybrid model (essentially, an assembly of multiple models driving the different underlying markets, brought together and made consistent and arbitrage-free; see [7]).

Due to the dimension of the model and the size of the market samples, Monte-Carlo is the only practical means of pricing CVA. To generate high-dimensional market samples in reasonable time may be challenging; see [7]. The netting set's timeline, which is the union of the timelines of all the transactions, is typically dense, further slowing down evaluation, unless compression techniques, introduced in [30] and [11], are implemented. A reasonably fast generation of scenarios is not enough. To evaluate the large number of cash-flows in the netting set in every scenario is another bottleneck. Our publication dedicated to scripting [11] covers general, efficient means of representing and manipulating cash-flows, which helps with the aggregation and compression of the cash-flows, as well as a fast evaluation over the generated scenarios.

---

[8]To model credit this way is more general than with a default time τ, in which case $D(T_{e-1}, T_e) = 1_{\{\tau \in [T_{e-1}, T_e]\}}$.

[9]And most other regulations.

[10]Many banks use rough closed-form approximations to estimate the future values of transactions for counterparty credit adjustment (CCR). Analytic approximations are other types of proxies, although regression proxies are vastly superior in accuracy when correctly implemented; see [31].

It follows that even though CVA may be seen as another financial product and evaluated in a hybrid platform, it remains that its size raises specific problems when we wish to price it in a practical manner and in reasonable time. Although this publication covers major ground for the efficient management of CVA, including generic simulation, parallelism, and AAD, it does not cover *all* the pieces, in particular, the representation and manipulation of cash-flows, the application of regression proxies, and some important algorithms specifically designed for the acceleration of xVA and explained in [31].

4.2 THE ARBITRAGE PRICING THEORY

Having defined financial products in an abstract manner suitable for the development of generic libraries, and having explored in detail the key notions of cash-flow, timeline, scenario, and sample,[11] we are ready to discuss their valuation. Valuation is permitted by the keystone theorem of modern finance, underlying financial theory and practice for almost 50 years, established by Harrisson, Kreps, and Pliska in [32] and [33], following the ground-breaking work of Black and Scholes in [22].

The Asset Pricing Theorem

We state the fundamental asset pricing theorem here, referring to the original articles or textbooks for demonstrations. In the absence of arbitrage, and in complete markets (both of which we assume in all that follows):

1. Financial cash-flows can be replicated by *self-financing strategies*.

   A self-financing strategy is a dynamic trading strategy over a time interval, without injection or withdrawal of cash or assets (hence, self financing), that consists in holding, at time t, the underlying assets s that constitute the market, of prices $s_t = \left(s_t^1, s_t^2, ..., s_t^n\right)$, in the amounts $\Delta_t = (\Delta_t^1, \Delta_t^2, ..., \Delta_t^n)$.

   A self-financing strategy is specified by the vector process (Δ_t). The vector Δ_t, the amounts of assets held at time t, must be $\mathcal{F}_t$-measurable, which means that they are determined at time t, from the state of the market on a before t, but not after.[12]

[11]These notions will be further clarified with concrete code in Chapter 6.

[12]A process (X_t) where X_t is $\mathcal{F}_t$-measurable is said to be *adapted*. Self-financing strategies trade adapted amounts. A strategy that trades non-adapted amounts is said to be *forward looking* and does not qualify as a self-financing strategy.

It follows from this definition that the value series of a self-financing strategy is an adapted process satisfying:

$$V_{t_2} = V_{t_1} + \int_{t_1}^{t_2} \Delta_t \cdot ds_t$$

with $t_1 \leq t_2$. One trivial, yet important, self-financing strategy is leave \$1 compounding at the short rate (r_t) on a bank account. This strategy is traditionally called "bank account," and its value, denoted β, is of course:

$$\beta_t = \exp\left(\int_0^t r_u du\right)$$

What the theorem says is that for any financial cash-flow CF_T (defined as a $\mathcal{F}_T$-measurable variable) paid at time T, there exists a self-financing strategy such that the value of the strategy replicates the cash-flow at time T with probability 1:[13]

$$V_T = CF_T$$

The amounts of assets (Δ_t) held in the replication strategy are called replication coefficients, *hedge coefficients*, or just "deltas."

The value of the cash flow at a time $t \leq T$ is the value of the replicating strategy.

2. There exists a unique probability measure, equivalent to the "real" probability measure,[14] denoted Q and called *risk-neutral measure*, such that the discounted values V_t/β_t of all financial cash-flows are martingales under Q:

$$\frac{V_{t_1}}{\beta_{t_1}} = E_{t_1}^{Q}\left[\frac{V_{t_2}}{\beta_{t_2}}\right]$$

for any $t_1 \leq t_2 \leq T$.

One consequence is that the value processes (V_t) of all cash-flows have a drift equal to r_t under Q.[15] It is also the case for financial

[13] In theory, in continuous time, and provided that the model parameters are accurate.

[14] Which essentially means that the same events have zero probability under the two measures.

[15] In proportional terms, meaning that

$$\frac{dV_t}{V_t} = r_t dt + ...dW^Q$$

where W^Q is a standard Brownian motion under Q.

products, in between the payment of cash-flows (the value of a product dropping by the cash-flow on a payment date).

Another consequence is the fundamental pricing formula:

$$V_0 = E^Q\left[\exp\left(-\int_0^T r_t dt\right) CF_T\right]$$

3. The value V_t is a function of the market state on and before t, including the current values s_t of the underlying assets in the current market. The replication strategy, the amounts of underlying assets to trade at time t to replicate a cash-flow $CF_T, T \geq t$, are the derivative sensitivities of V_t to these underlying prices:

$$\Delta_t^k = \frac{\partial V_t}{\partial s_t^k} = \frac{\partial E_t^Q\left[\exp\left(-\int_0^T r_t dt\right) CF_T\right]}{\partial s_t^k}$$

4. More generally, the drift of the value process of any self-financing strategy under Q is r_t, and conversely, any adapted process (X_t) with drift r_t, under Q is the value process of a self-financing strategy, and therefore, the price series of a tradable asset.

This theorem guarantees that the market risk of financial products can be covered with trading strategies, justifies the universal practice of measuring risk by taking derivatives of transaction values to current market prices, and provides a general pricing formula for all financial products.

Later work precised, confirmed, and extended this fundamental theorem. For instance, it was no earlier than 2009 that it was finally demonstrated, indisputably and in the most general case, that derivative sensitivities effectively always correspond to hedge coefficients. The demonstration is due to Dupire, who had to develop a new branch of stochastic mathematics, called "stochastic functional calculus," to obtain this result [34].

Pricing with a numeraire

Another extension, extremely useful for analytic calculations and numerical implementations, including Monte-Carlo, was introduced in 1995 by El-Karoui, Geman, and Rochet [35]. The bank account β is not the only choice of *numeraire*: any (strictly positive) self-financing strategy may be used as numeraire, including positive financial cash-flows and products, before payment.

For every such choice of a numeraire N, there exists a unique probability measure Q^N, equivalent to Q, such that all asset values (in between payments) in terms of N are martingales under Q^N:

$$\frac{V_{t_1}}{N_{t_1}} = E^N_{t_1}\left[\frac{V_{t_2}}{N_{t_2}}\right]$$

where $t_1 \le t_2 \le T$. In particular, the value of a cash-flow CF_T paid at time T satisfies:

$$V_t = E^N_t\left[\frac{N_t}{N_T} CF_T\right]$$

and with $N_0 = 1$ without loss of generality:[16]

$$V_0 = E^N\left[\frac{CF_T}{N_T}\right]$$

Remembering that cash-flows are defined as functionals of scenarios:

$$CF_T = h(\Omega) = h(\{S_t, t \in TL\})$$

and that product values are the sum of the values of their cash-flow, it follows that the value V_0 of any financial product with cash-flows $\{CF_p, 1 \le p \le P\}$ paid on dates $\{T_p\}$ satisfies:[17]

$$V_0 = E^N\left[\sum_{p=1}^{P} \frac{h_p(S_t, t \in TL_p)}{N_{T_p}}\right]$$

We augment the definition of scenarios and samples[18] to include the numeraire on payment dates, so that:

$$\sum_{p=1}^{P} \frac{h_p(S_t, t \in TL_p)}{N_{T_p}} = g(S_t, t \in TL)$$

[16]Martingale measures are unchanged when numeraires are scaled, so numeraires can always be normalized by their initial value and $Q^{\frac{N}{N_0}} = Q^N$.

[17]This is the formula for a discrete set of cash-flows; the extension to a continuous set is trivial and not necessary for what follows.

[18]We will not extend the definition of scenarios any further.

The functional g of the scenario, including the numeraire on payment dates, is called *payoff*, and we get the classic result that the value of a financial product is its expected payoff:

$$V_0 = E^N[g(S_t, t \in TL)] = E^N[g(\Omega)]$$

Forward measures

Besides the bank account β, a number of useful numeraires have been widely used in investment banks and universities to resolve analytic and numerical problems. We will see a few in this chapter, and demonstrate a simple change of numeraire in the code of Chapter 6.

For now, we introduce a particularly useful family of numeraires: the zero-coupon bonds. A zero-coupon bond of maturity T satisfies the criteria of a numeraire. The associated martingale measure is called *forward measure* for a reason that will be made apparent shortly. The T forward measure is often denoted Q^T and expectations taken under this measure are denoted E^T.

One of the multiple benefits of forward measures is that they provide the means to price linear products directly and independently of a model.

Forward prices and forward contracts Let us revisit a forward contract of maturity T and strike K on an asset s, and price it under the Q^T measure. From the fundamental pricing formula, we immediately get:

$$V_t = DF(t,T)E_t^T\left[\frac{s_T - K}{DF(T,T)}\right] = DF(t,T)E_t^T[s_T - K]$$

since $DF(T,T) = 1$. Solving for $V_t = 0$ in the strike K to find the forward price $F(t,T)$, we get:

$$F(t,T) = E_t^T[s_T]$$

Forwards of maturity T are expectations under the martingale measure associated with the zero-coupon bond of maturity T, hence its name "forward measure." It also follows that forwards of maturity T are martingales under Q^T, in particular:

$$F(0,T) = E^T[F(t,T)]$$

These formulas are not particularly helpful for pricing, since today's forwards trade in the market. We know their prices. Where they are helpful is the other way around: the forwards that trade on the market today *imply the means of forward-neutral measures*.

Another trivial consequence is that the value of a forward contract is $V_t = DF(t,T)(F(t,T) - K)$. Setting $K = 0$, we get the time t value of an asset delivered at time T:

$$V_t = DF(t,T)F(t,T)$$

Forward rates and FRAs The same can be said of FRAs in interest rate markets. A FRA over the period $T_1 - T_2$ pays a cash-flow on date T_2, hence, we value it under Q^{T_2}:

$$V_t = DF(t,T_2)E_t^{T_2}[L(T_1,T_2) - K](T_2 - T_1)$$

Solving $V_t = 0$ in K, we find that the forward Libor is the forward-neutral expectation of the Libor:

$$F(t,T_1,T_2) = E_t^{T_2}[L(T_1,T_2)]$$

and a martingale under Q^{T_2}. Today's Libor curve implicitly defines the expected values of future Libors under the forward-neutral measures for their end dates. In addition, the value of a FRA is $V_t = DF(t,T_2)(F(t,T_1,T_2) - K)$. Setting $K = 0$, we see that the time t value of a $Libor(T_1,T_2)$ paid at T_2 is:

$$V_t = DF(t,T_2)F(t,T_1,T_2)$$

Forward discount factors and instantaneous forward rates Consider now a product that delivers a zero-coupon bond of maturity T at a time $t < T$. Its value is obviously $DF(0,T)$. From the valuation formula under Q^t, we get:

$$DF(0,T) = DF(0,t)E^t[DF(t,T)]$$

We know, now, that forward-neutral expectations are forwards. It follows that the forward price $DF(0,t,T)$ of maturity t of the zero-coupon bond of maturity T satisfies:

$$DF(0,t,T) = \frac{DF(0,T)}{DF(0,t)}$$

In addition, the value series $DF(t,T_1,T_2)$ of a forward discount factor of maturities T_1, T_2 is a martingale under Q_{T_1}. It follows that today's discount curve fixes the forward-neutral expectations of future discount factors.

The amount

$$R(t,T_1,T_2) \equiv -\frac{\log[DF(t,T_1,T_2)]}{T_2 - T_1} = -\frac{\log[DF(t,T_2)] - \log[DF(t,T_1)]}{T_2 - T_1}$$

is called *forward discount rate*, and effectively corresponds to the rate of a loan from T_1 to T_2 that may be locked at time t by selling zero-coupon bonds of maturity T_1 and buying zero-coupon bonds of maturity T_2 for the same total price.

Denoting $T \equiv T_1$ and considering the limit case where $T_2 \to T_1$, we get the time t instantaneous forward rate of maturity T:

$$f(t,T) \equiv \lim_{T_2 \to T} R(t,T,T_2) = -\frac{\partial \log[DF(t,T)]}{\partial T}$$

$$\Leftrightarrow DF(t,T) = \exp\left[-\int_t^T f(t,u)du\right]$$

which we had defined before, only now we can see where this definition comes from and why the short (discount) rate is effectively $r_t = f(t,t)$.

Forward bank account Another example is the bank account β. $\frac{\beta_t}{DF(t,T)}$ is a martingale under Q^T, therefore:

$$\beta(t,T) \equiv E_t^T[\beta_T] = \frac{\beta_t}{DF(t,T)} = \exp\left(\int_0^t r_u du + \int_t^T f(t,u)du\right)$$

Since forward-neutral expectations are forwards, it follows that $\beta(t,T)$ is the forward value of the bank account β at time t for the maturity T. In particular, we have today's value of bank account forwards of all maturities:

$$\beta(0,T) = \frac{1}{DF(0,T)} = \exp\left(\int_0^T f(0,u)du\right)$$

Interest rate swaps It immediately follows from what precedes that the value on an IRS (a receiver swap in this instance) is:

$$V_t = K\delta_{fix} \sum_{i=1}^{N_{fix}} DF(t, T_0 + i\delta_{fix}) - \delta_{float} \sum_{j=1}^{N_{float}} DF(t, T_0 + j\delta_{float})$$
$$\times F(t, T_0 + (j-1)\delta_{float}, T_0 + j\delta_{float})$$

and, solving for $V_t = 0$ in K to get the par swap rate F_t, we get:

$$F_t = \frac{\delta_{float} \sum_{j=1}^{N_{float}} DF(t, T_0 + j\delta_{float}) F(t, T_0 + (j-1)\delta_{float}, T_0 + j\delta_{float})}{\delta_{fix} \sum_{i=1}^{N_{fix}} DF(t, T_0 + i\delta_{fix})}$$

This formula is universal and model independent, which implies that the par swap rate is a redundant market primitive, hence, not directly included in scenarios. The denominator is called the *annuity* of the fixed leg. The annuity is a combination of zero-coupon bonds, hence, a tradable asset, with time t value:

$$A_t = \delta_{fix} \sum_{i=1}^{N_{fix}} DF(t, T_0 + i\delta_{fix})$$

On the par swap rate formula, we recognize that F_t is the value at time t of the floating leg *in terms of the annuity*:

$$F_t = \frac{V_t^{float}}{A_t}$$

It follows that F_t is a martingale under the martingale measure Q^A associated with the annuity.

Also note that the formula of the value of a (receiver, in this instance) IRS can be rearranged into:

$$V_t = A_t(K - F_t)$$

Swaption pricing A *physically settled* payer swaption (right to enter the underlying payer swap) of strike K on the exercise date T is the value, if positive, of the underlying swap on the exercise date:

$$CF_T = A_T(K - F_T)^+$$

We can therefore price it under the annuity-neutral measure:

$$V_t = E_t^A\left[\frac{CF_T}{A_T}\right] = A_t E_t^A[(K - F_t)^+]$$

where the payoff is that of a standard put (for a payer swaption, it is a call) on the swap rate, which itself is a martingale. This is the theoretical basis for pricing swaptions under models like Black-Scholes [22] or SABR [36]. These established market practices therefore implicitly, and correctly, manage swaptions under the annuity-neutral measure. Assuming a constant volatility σ as in Black and Scholes's specification, $E_t^A[(F_T - K)^+]$ is given by Black and Scholes' formula [22]. Assuming F_t is a stochastic volatility martingale as in SABR's specification, $E_t^A[(F_T - K)^+]$ is given by Hagan's formula [36].

Natural numeraires We demonstrated that forward prices $F(t, T)$, observed at t with maturity T, are martingales under the forward measure Q_T associated with the zero bond of maturity T, the maturity date of the forward. This zero bond and forward measure are respectively called the *natural numeraire* and the *natural measure* of the forward $F(t, T)$.

In particular, the spot price s_t on time t coincides with its forward $F(t, t)$. Its natural numeraire is therefore the zero bond of maturity t, and its natural measure is Q^t. *Forwards are expectations of asset prices under their natural measure.*

This is true of all asset prices. In particular, the forward price of the bank account β_t on time t is its expectation under its natural measure Q^t.

The forward discount factor:

$$DF(t, T_1, T_2) = \frac{DF(t, T_2)}{DF(t, T_1)}$$

is a martingale under Q^{T_1}. Therefore:

$$E^{T_1}[DF(T_1, T_2)] = DF(0, T_1, T_2)$$

The natural measure of the discount factor $DF(T_1, T_2)$, observed at time T_1 with maturity T_2, is therefore Q^{T_1}. The expectation of a discount factor under its natural measure is the corresponding forward discount factor observed today. Its natural numeraire is the zero bond of maturity T_1.

A forward Libor $F(t, T_1, T_2)$ observed at time t with maturities T_1, T_2 is a martingale under its natural measure Q^{T_2}, associated with its natural numeraire, the zero bond of maturity T_2. In particular, a spot Libor observed at T_1 with maturity T_2 $L(T_1, T_2) = F(T_1, T_1, T_2)$ coincides with the forward on the fixing date. It follows that the expectation of a Libor under its natural measure is the forward Libor. Its natural numeraire is the zero bond of maturity the Libor's payment date.

Finally, we have seen that the par rate F_t of a given swap, paid over the annuity A of the swap's fixed leg, is a martingale under Q^A. Q^A is therefore the swap rate's natural measure, and the fixed leg's annuity A is its natural numeraire.

Every observation of a market variable s_T fixed at time T admits a natural measure, such that conditional expectations $E_t(s_T)$ of s_T under the natural measure are, by definition, the forward values observed at time t of the cash-flow s_T at time T. It follows that those forward values are martingales under s_T's natural measure. The numeraire associated to the natural measure of an observation s_T is called its natural numeraire. The notion of natural numeraire and measure depends on the nature of the market variables, and we have seen that prices, Libors, discounts, bank accounts, and swap rates define their natural measures in a different way. In addition, natural measures relate to observations: two observations on different dates of the same variable have different natural measures and numeraires.

In economic terms, the observation of a market variable always expresses the value of an asset in terms of a unit. This unit is the observation's natural numeraire. The market price of an Apple share on 30 September 2025 expresses the value of a fraction of Apple in terms of 30/09/25 USD. Its natural numeraire is the USD denominated zero bond of maturity 30/09/25.

In formal terms, the natural numeraire N of an observation s_T is the numeraire such that under any complete and arbitrage-free probabilistic specification with correct initial values, $E_t^N(s_T)$ corresponds to the forward price observed at t of the cash-flow s_T. This is rather abstract, but in practical terms, for a given observation of a market variable, it is generally quite clear what its natural numeraire and measure are. In particular, for the market variables that populate market samples:

1. The natural measure for a forward price $F(T, T_2)$ is Q^{T_2}, including spot prices $s_T = F(T, T)$ with natural measure Q^T.
2. It directly follows that the natural measure for a discount factor $DF(T, T_2)$, or the bank account β_T observed at T, is also Q^T.
3. The natural measure for a forward Libor $F(T, T_1, T_2)$ is Q^{T_2}, including spot Libors $L(T_1, T_2) = F(T_1, T_1, T_2)$ with natural measure Q^{T_2}.

Constant Maturity Swaps We computed the values of all the linear transactions introduced earlier: forward contracts, FRAs, IRS, without specifying a model, directly, as a function of the forward prices, forward rates, and discount factors trading on the current market. This is a defining characteristic of linear products: linear products are financial products whose value depends on the current market, independently of a model. It follows that today's value of linear products is the same in any arbitrage-free model that respects today's market. We just saw that forwards are expectations under natural measures, so we have a clean definition of linear products:

Linear products are those whose value depends on *expectations*, but not *distributions*. It follows that the value of any linear product is a linear combination of the present values of a number of forward contracts, hence the name "linear" products.

This (correct) definition of linear products conflicts with the (approximate but widely accepted) one of products where cash-flows are linear functions of scenarios. Consider a cash-flow $CF_{T_p} = \alpha s_T$, linear in the market variable s_T fixed on T. We now know that its value satisfies:

$$V_t = \alpha DF(t, T_p) E_t^{T_p}[s_T]$$

When Q^{T_p} is s_T's natural measure, $E_t^{T_p}[s_T] = F^s(t, T)$ is its forward, therefore:

$$V_t = \alpha DF(t, T_p) F(t, T)$$

and it follows that the cash-flow is indeed a linear one. When this is not the case, and s_T's natural measure $Q \neq Q^{T_p}$:

$$E_t^{T_p}[s_T] \neq E_t^Q[s_T] = F^s(t, T)$$

The difference between the two expectations $E_t^{T_p}[s_T]$ and $E_t^Q[s_T]$ is called *convexity adjustment* and can be calculated with Girsanov's theorem, but only under dynamic assumptions on the future behavior of s, in particular its volatility. It follows that the cash-flow is *not* a linear one, but a European one (because it depends on a single variable on a single date).

To see that clearly on an example, a Libor $L(T_1, T_2)$ is a linear cash-flow when paid at T_2. Its value is then:

$$V_t = DF(t, T_2)E_t^{T_2}[L(T_1, T_2)] = DF(t, T_2)F(t, T_1, T_2)$$

as seen earlier. When paid at T_1, its value is:

$$V_t = DF(t, T_1)E_t^{T_1}[L(T_1, T_2)]$$

and:

$$E_t^{T_1}[L(T_1, T_2)] \neq E_t^{T_2}[L(T_1, T_2)] = F(t, T_1, T_2)$$

Ignoring discount basis and assimilating:

$$DF(T_1, T_2) = \frac{1}{1 + (T_2 - T_1)L(T_1, T_2)}$$

the cash-flow $L(T_1, T_2)$ paid at T_1 is equivalent to a T_2 payment, under the underlying Libor's natural numeraire, of:

$$\frac{L(T_1, T_2)}{DF(T_1, T_2)} = L(T_1, T_2) + (T_2 - T_1)L(T_1, T_2)^2$$

which clearly identifies the cash-flow as a nonlinear quadratic function of the underlying Libor. Its linearity was only apparent. It is actually a European cash-flow. Under a simple model a la Bachelier [37] with Gaussian first increments with annual standard deviation σ, its value is:

$$V_t = DF(t, T_2)E_t^{T_2}\left[L(T_1, T_2) + (T_2 - T_1)L(T_1, T_2)^2\right]$$

$$= DF(t, T_1)\left[\underbrace{F(t, T_1, T_2)}_{linear} + \underbrace{\sigma^2(T_1 - t)(T_2 - T_1)}_{adjustment}\right]$$

(where we skipped trivial calculation steps). In a more general, non-Gaussian context, we can statically replicate this cash-flow, like any European cash-flow written on a Libor, with combinations of caps and floors, by an application of Carr-Madan's formula.

This is the cash-flow of a well-known, actively traded financial product called "Libor in arrears." More generally, an actively traded family of financial products, called constant maturity swaps or CMS, pay cash-flows linked to libor or swap rates, over schedules different than their natural numeraire, resulting in nonlinear products, which value depends on volatility and dynamic assumptions. Like Libor in arrears, swap-based CMS are statically replicated with combinations of swaptions, with techniques similar to Carr-Madan's, as explained in [38].

Nonlinear products, like swaptions or CMS, cannot be valued out of the underlying market alone. Their value depends on volatility assumptions, and more generally, dynamic assumptions of how the market is going to evolve. This is the purpose of derivatives *models*.

4.3 FINANCIAL MODELS

Having defined financial products in a general manner, we repeat the exercise for financial models. The asset pricing theory gave us the pricing formula for all financial products:

$$V_0 = E^N[g(\Omega)]$$

where the payoff function g and the definition of the scenario (what market primitives, including numeraire, on what dates) are fully specified by the product.[19] In order to apply this formula and produce prices, we must specify the distribution of Ω under Q^N. This is the defining purpose of a model.

We call "model" a specification of the probability distribution of the scenario Ω. This specification may either be direct, or implied from dynamic assumptions regarding future market behavior, in which case we say that the model is *dynamic*.

More precisely, the scenario:

$$\Omega = \{S_t, t \in TL\}$$

is a collection of market samples S_t on a collection of event dates, and each sample is itself a collection:

$$S_t = \{s_t^i\}$$

[19]The model decides what the numeraire is. The product determines when the numeraire is needed and how it intervenes in the payoff. This will be clarified in code.

of market variables on the event date t. A sample may be a singleton, or a discrete, infinite, or continuous collection of market variables. Hence, the scenario Ω is a collection of market *observations*:

$$\Omega = \{s_k\}$$

where the s_ks are all the individual market variables on all the event dates. Different market variables on the same date are different ks and the same market variable on different dates are also separate ks. Each s_k belongs to one sample S_{T_k}, and we denote T_k its event date, also called observation date or *fixing date*. We now have a more precise definition of models.

The defining purpose of a model is to specify the joint distribution of the observations s_k in the scenario Ω.

Note the separation of concerns: the payoff function g and the definition of the scenario Ω are specified by the product outside of any kind of model logic. The model specifies the probability distribution of Ω outside of any cash-flow logic.[20]

A model is not free to specify *any* joint distribution for the s_ks. Q^N must respect its defining property that asset prices in terms of the numeraire are martingales; otherwise the model is arbitrageable, produces inconsistent prices, and is generally not acceptable. All common models introduced shortly are arbitrage-free. In addition, the model's specification of Q^N must respect the current market prices. For instance, we have modeled underlying stocks, indices, and currencies as financial products. Their value under Q^N as financial products must correspond to their current market price. More generally, we have seen that the forward prices and rates, traded on today's market, fix expectations under forward-neutral measures, imposing a constraint on the model, called *initial value conditions*.

We shall review linear and European models, appropriate for the valuation of respectively linear and European transactions, before we discuss dynamic models, the most general class of models, those appropriate for the pricing of all products.

Linear models

Linear models refer to what we have been calling so far "today's market." They contain all the underlying asset prices, including dividend and repo curves, all the discount and Libor curves, and other variables that define the current state of the market. They "know," among other things, the forward prices $F(0, T)$ and discount factors $DF(0, T)$ of all maturities T, as well as

[20]This separation between models and products is somewhat muddied in the context of callable products and regulatory calculations, although it is reestablished with proxies.

forward Libors $F(0, T_1, T_2, idx)$ of all maturities T_1 and T_2, for all Libor indices.

Linear markets achieve this capability by interpolation of the available market information: spots, futures, par swap rates, basis and currency swaps, and so forth. This is not as easy as it may sound, especially since the multiplication of basis and discount curves in 2008–2011. To properly construct a linear market is a sophisticated, challenging exercise that occupies some of the greatest minds in the industry. We refer to [39] for a review of linear markets, including a presentation of the difficulties and the sophisticated solutions implemented in modern systems.

Mathematically, linear markets, who know all forwards, define the *expectations* $E^k(s_k)$ of the s_ks under their respective natural measures, but they know nothing else of their *distribution* Q^k, because they have no knowledge of their future dynamics. This transpires in the notations: linear markets know all the $F(0, T)$ for all maturities T, but nothing of the $F(t, T)$ when $t > 0$. Linear markets are initial value conditions. They are the modern counterpart of the "initial spot S_0" found in traditional literature.

Why then do we call them linear *models* and discuss them in the model section here? We have defined a model by its ability to assign a probability distribution to a scenario. Linear models assign probability 1 to the *forward scenario* Ω_F where every market variable s_k lands, on its event date, on its forward value seen from today, which we have seen is its expectation under its natural measure Q^k:

$$\Omega_F = (E^k[s_k])$$

This "probability distribution" of the scenario obviously respects initial conditions, and is trivially arbitrage-free, the discounted price series of all assets being (constant) martingales between payments. This is actually the simplest possible arbitrage-free model consistent with today's market.

It follows that the price of any financial product defined by its payoff g and scenario Ω in the linear model is:

$$V_0 = E^Q[g(\Omega)] = g(\Omega_F)$$

which may explain why those models are called linear, although:

$$\Omega_F = (E^k[s_k]) \neq (E^Q[s_k]) = E^Q(\Omega)$$

so the linearity of linear models is an approximate notion.

We priced a number of linear products in the linear model in Section 4.2, although we did not call it linear model at the time. The value in the linear model is called the *intrinsic value* of the product. For European options, this definition coincides with the traditional definition.

For instance, for a European call, $IV = DF(0,T)(F(0,T) - K)^+$. Our definition extends the notion of intrinsic value to all financial products. The difference between the actual value and the intrinsic value is called *time value*: $V_0 = IV + TV$. Linear products are those whose time value is zero.[21] The intrinsic value prices the linear part of a product and is independent of volatility or dynamic assumptions. The time value prices the nonlinear part of a product and depends on volatility and other dynamic assumptions.

From a practical point of view, pricing in a linear model is extremely fast, usually a fraction of a millisecond, because it takes a unique scenario. It is also very accurate because no numerical algorithm is involved. Linear models are suitable for Monte-Carlo simulations; in particular they fit the simulation model interface we build in Chapter 6, albeit in a trivial manner, since they generate only one scenario: the forward scenario Ω_F. Generic, well-designed valuation systems price linear products (and the intrinsic value of other products, an interesting information in its own right) as a simulation in a single scenario.

More sophisticated models used for the valuation of nonlinear transactions can also price linear products, as a special case of options or exotics. Provided that the models respect today's market, the prices coincide with the linear model, but with lower speed and accuracy. Finally, note that those more sophisticated models coincide with linear models, not only for linear products, but for all products, when their volatility parameters are set to zero.[22] Therefore linear models are also zero-volatility arbitrage-free dynamic models.

European models

Moving one step up to European models, we discuss those models appropriate for the pricing of European products, but not exotics.

We have seen that a European cash-flow of maturity T is a combination of calls and puts on the same market variables, same maturity, and same payment date, as demonstrated by Carr-Madan's formula. It follows that a dynamic model is not necessary for the valuation of European cash-flows; all we need are the call prices $C_T(K)$ of all strikes K for the maturity T.[23]

[21]Technically, this is not a sufficient condition: linear products are those whose time value today *and in the future* is always zero. Exceptions are rare but they exist, and are sometimes asked in interviews.

[22]As long as they respect today's market.

[23]Put prices are given by the *parity*

$$C_T(K) - P_T(K) = DF(0,T)[F(0,T) - K]$$

which is an immediate consequence of $(x-k)^+ - (k-x)^+ = x - k$ where we note that European models must rely on linear models for discounts and forwards to respect today's market.

European models, by definition, "know" these prices in the same way linear models "know" forward prices: by interpolation of available market data, in this case, the prices of calls and puts that trade on today's market.[24]

Linear models generally interpolate discount factors not in price, but in rate. Similarly, options are typically interpolated not in their price $C_T(K)$, but in their Black-Scholes *implied volatility* $\hat{\sigma}$, defined by:

$$DF(0,T)BS(F(0,T),K,T,\hat{\sigma}_T(K)) = C_T(K)$$

where BS is Black and Scholes's formula [22]. We call *implied volatility curve*, or IVC, the continuous collection of Black-Scholes implied volatilities of a given maturity. An IVC of maturity T, *combined with a linear model* for the forward and discount in Black and Scholes's formula, can price all European cash-flows of maturity T. It follows that an interpolated collection of IVCs for all maturities T, called *implied volatility surface* or IVS, can price all European cash-flows, hence, all European products.[25] In case a European product includes European cash-flows on different underlying assets, we need multiple IVS, one per underlying asset, although, in the interest of clarity, we just call IVS the collection of all call prices on all underlying assets.

European models, or IVS are we call them now, therefore build on the information contained in linear models, and inject additional information in the form of a continuous, bi-dimensional surface of Black-Scholes implied volatilities. We have seen that linear models define the *expectations* of the observations s_k in a scenario under their natural measure. European models, or IVS as we call them now, define the *marginal distributions* $\widetilde{\varphi}_k$ of these observations under their natural measure. To see this, we apply the valuation formula to a call of strike K, maturity T:

$$C(K,T) = N_0E[(s_T - K)^+] = N_0 \int (x-K)^+\widetilde{\varphi}_k(x)dx$$

Differentiating twice on both sides against the strike K (in the sense of Laurent Schwartz's distributions,[26] we obtain:

[24]In the very rare cases where no such information is available, dynamic models (see next) may be used instead.

[25]To obtain an arbitrage-free IVS from the interpolation of a limited number of traded option prices is a challenging, specialized exercise outside of our scope. Gatheral's SVI [40] and Andreasen and Huge's LVI [41] are the current best practice solutions. In addition, IVS often interpolate not Black and Scholes's volatility, but the dynamic parameters of more sophisticated models, like SABR [36] or Heston [42], see [43] for details.

[26]Generalized derivatives are not really needed to demonstrate this result; it can also be demonstrated by more traditional means. However, Schwartz's generalized derivatives allow to demonstrate many financial results, including Dupire's formula and extensions with stochastic volatility, in a natural, terse, and sharp manner. See [44] for a tutorial and applications in finance.

$$\varphi_k(K) \equiv N_0 \widetilde{\varphi}_k(K) = \frac{\partial^2 C(K, T)}{\partial K^2}$$

where N is the natural numeraire for the observation s_T, and the normalized distribution φ_k is the price of the Dirac mass of s_k in K, called *Arrow-Debreu security*. The call prices of all strikes K for a maturity T are equivalent to the natural probability distribution of the underlying observation.[27] Hence, the IVS, which knows the call prices of all strikes and maturities on all relevant assets, also knows the *marginal* probability distributions of all the market variables s_k in the scenario, called *marginal* distributions.

Because the IVS delegates forwards to the linear model, the means of the natural distributions $\widetilde{\varphi}_k$s are the forwards of the observations s_k. It follows that the IVS respects today's market and produces arbitrage-free marginal distributions. It does *not* produce arbitrage-free *joint* distributions between underlying asset prices on different dates, but, for European products, all that matters are the marginal distributions of the s_ks.

The price of a European cash-flow CF_p paid on the natural payment date T_p of an underlying observation s_p is $\int g_p(x)\varphi_p(x)dx$, the integral of the cash-flow's payoff against the distribution of its underlying observation specified in the IVS. In practice, this integral may be evaluated by numerical integration. Chapter 4 of *Numerical Recipes* [20] covers some efficient procedures and provides source code. Numerical integration consists in the approximation:

$$\int g_p(x)\varphi_p(x)dx \approx \sum_i w_i g_p(x_i)\varphi_p(x_i)$$

where the weights w_i and the knots x_i are carefully chosen by the numerical integration scheme. To price a European *product*, we may price all its cash-flows and sum them up, or we can price it all together with the help of a change of variable (assuming a discrete schedule of cash-flows and a discrete scenario here to simplify notations, although everything applies in the continuous case too):

$$\int g_p(x)\widetilde{\varphi}_p(x)dx = \int_0^1 g_p\left[\phi_p^{-1}(y)\right] dy$$

[27] In particular, the variance of the distribution is related to the at-the-money (ATM) Black-Scholes implied volatility, its skewness is related to the slope of the implied volatility function of the strike around the money, and its kurtosis is related to the curvature (called "smile") of the implied volatility curve.

where $\phi_p(y) = \int_{-\infty}^{y} \widetilde{\varphi}_p(x)dx$ is the cumulative probability distribution of s_p, known to the IVS, as well as its inverse. The value of the product is the sum of the values of its cash-flows:

$$V_0 = N_0 \int_0^1 \sum_p g_p \left[\phi_p^{-1}(y)\right] dy$$

The payoff function g of the product is the sum of the payoff functions g_p of its cash-flows, hence:

$$\frac{V_0}{N_0} = \int_0^1 g\left[\phi_1^{-1}(y), \phi_2^{-1}(y), ..., \phi_P^{-1}(y)\right]dy = \int_0^1 \psi(y)dy \approx \sum w_i \psi(y_i)$$

where, again, the weights w_i and the knots $y_i \in (0, 1)$ are determined by the numerical integration routine, we used the property of European cash-flows where they each depend on exactly one s_p, and the definition of ψ should be clear for the context.

A numerical integration is typically extremely fast, of millisecond order, and very accurate, in particular compared to Monte-Carlo. This procedure can also price linear products (although slower and less accurately than linear models). But it cannot correctly price exotics.

The IVS correctly values European products because it knows the (natural) *marginal* distributions of all the observations s_k in a scenario, but it knows nothing of their *joint* distribution, including the dependence of different underlying assets on the same date, or the dependence of the same underlying asset on different dates. The mathematical mapping of a number of marginal distributions into a single joint distribution is called a *copula* and there exists infinitely many copulas that join the same set of marginals into different joint distributions.

A defining characteristic of European products, and a better definition than the one stated earlier, is that these products depend on marginal distributions, but not on copulas (in the same way that linear products only depend on expectations but not distributions). *European cash-flows are those whose payoff can be written as a function of one observation, under the natural numeraire of the observation.* It follows that a defining characterization of non-European exotics is that they *do* depend on copulas. To see that on a simple example, consider the twice-monitored digital barrier that pays \$1 if two fixings of the underlying asset price on two given dates T_1 and T_2 are both below a barrier B:

$$g(s_{T_1}, s_{T_2}) = 1_{\left\{s_{T_1} < B\right\}} 1_{\left\{s_{T_2} < B\right\}}$$

Then:

$$\begin{aligned}
E[g] &= E\left[1_{\{s_{T_1}<B\}}1_{\{s_{T_2}<B\}}\right] \\
&= E\left[1_{\{s_{T_1}<B\}}\right]E\left[1_{\{s_{T_2}<B\}}\right] + cov\left(1_{\{s_{T_1}<B\}}, 1_{\{s_{T_2}<B\}}\right) \\
&= E\left[1_{\{s_{T_1}<B\}}\right]E\left[1_{\{s_{T_2}<B\}}\right] \\
&\quad + corr\left(1_{\{s_{T_1}<B\}}, 1_{\{s_{T_2}<B\}}\right)\sqrt{Var\left[1_{\{s_{T_1}<B\}}\right]Var\left[1_{\{s_{T_2}<B\}}\right]} \\
&= P\left[s_{T_1} < B\right]P\left[s_{T_2} < B\right] \\
&\quad + corr\left(1_{\{s_{T_1}<B\}}, 1_{\{s_{T_2}<B\}}\right) \\
&\quad \sqrt{P\left[s_{T_1} < B\right]\left(1 - P\left[s_{T_1} < B\right]\right)P\left[s_{T_2} < B\right]\left(1 - P\left[s_{T_2} < B\right]\right)}
\end{aligned}$$

In this formula, the terms $P[s_{T_1} < B]$ and $P[s_{T_2} < B]$ are European digitals that only depend on marginal distributions, but the term

$$corr\left(1_{\{s_{T_1}<B\}}, 1_{\{s_{T_2}<B\}}\right)$$

clearly refers to their dependence.

To price this exotic with the IVS would result in a random and incorrect value. Random because the implementation of the IVS implicitly and arbitrarily chooses a copula: in our numerical integration the same quantiles were sampled simultaneously for all the marginal distributions. And incorrect because a random copula almost certainly results in an arbitrageable joint distributions, even when the marginal distributions are arbitrage-free.

Linear models define the expectations of underlying variables and correctly price linear products. European models inject additional marginal distributions around these expectations, and correctly price the time value of European products, in addition to linear products. The purpose of dynamic models is to produce an arbitrage-free, credible copula between the marginal distributions of the IVS, so as to correctly price not only linear and European products, but also exotics.

Dynamic models

To directly specify the copula is a perilous and probably impossible task. The copula must be specified in a way to produce an arbitrage-free, credible joint distribution, and it is unclear how this can be done, working directly on distributions.

To achieve this, dynamic models specify an arbitrage-free stochastic dynamic for the underlying market variables. They fit expectations by respecting the initial values given by the linear model. They fit marginal distributions by *calibration* to the IVS.[28] And they inject copulas, implicitly, with a credible, realistic, and arbitrage-free stochastic dynamics.

All dynamic models calibrated to the IVS produce the same marginals as the IVS, hence the same values for all European transactions. But models also implicitly specify a copula between these marginal distributions. It follows that two different models (say, one with local volatility and one with both stochastic and local volatility) both calibrated perfectly to the same IVS (which in this case is feasible) produce the same marginals but different copulas. It follows that they assign the exact same value for Europeans (modulo calibration and numerical error) but a potentially very different one for exotics. The sensitivity to the copula is a measure of *model risk*. It may be estimated by pricing an exotic or exotic book with different models calibrated to the same IVS. The price dispersion measures "how exotic is this particular exotic."

For example, it was discovered by Dupire in [45] that a variance swap, which pays the realized statistical variance of an underlying asset price over a period, can be perfectly replicated by a combination of Europeans and a self-financing strategy, irrespective of any dynamic assumption. This theory is the basis of the VIX index published by the CBOE and widely considered as a primary measure of market nervousness. The details are explained, for instance, in the fourth part of [46]. It follows that the variance swap is completely insensitive to copulas, and priced identically (modulo numerical errors) in different models, as long as they calibrate to the same IVS.

As another example, Peter Carr's well-known static replication of barrier options [47] shows that (under some pretty strong assumptions) simple barrier options are also replicated by combinations of European options and trading strategies. It follows that barriers like down-and-out calls are not that exotic (with the barrier on the strike, the value of a down-and-out call is even linear in the underlying asset price in all continuous models) and their price in different models calibrated to the same IVS is very similar.

On the contrary, tight double no touches (options that pay $1 if the underlying asset price remains in tight corridor at all times for a period) have high *volga* (second derivative to volatility) and are therefore very sensitive to stochastic volatility. A stochastic volatility model may price such products orders of magnitude higher than models without stochastic volatility, calibrated to the same IVS. It follows that double no touches are very exotic products, with a massive dependency on copulas, and material model risk. Financial institutions should always classify exotics in accordance to model risk, since this is one of the most difficult risks to cover.

[28] Calibration is the process of tuning the dynamic parameters of the model so that the model matches the IVS prices of European options, hence, marginal distributions.

The authors of dynamic models generally deliver exact or approximate closed-form solutions for European options, which helps calibration.[29] More sophisticated models like Heston's stochastic volatility model [42] don't offer analytic formulas but can value Europeans very quickly with numerical integrals; see the dedicated chapters of [6]. In the simplest models like Black Scholes and, some exotics like barrier options also admit closed-form solutions. But in a vast majority of contexts, the pricing of an exotic necessitates a numerical algorithm. The ultra-efficient FDM, briefly introduced later in this chapter, can be applied in a limited number of situations, other contexts requiring the slower, but almost universally applicable, Monte-Carlo algorithm covered in more detail in the rest of this publication.

Note that the different types of models introduced in this chapter form a hierarchy: linear models are self-contained and correctly, accurately, and efficiently price all linear products. European models (IVS) depend on linear models, so they also correctly price linear products. They are also implemented with accurate and efficient numerical integration algorithms, although not as fast or as accurate as linear models. But they correctly price European products in addition to linear products. Dynamic models depend on linear models for their initial conditions, and on IVS for their calibration. Mainly implemented with Monte-Carlo, they are many orders of magnitude slower and less accurate than European models. But they may value exotics in addition to pricing European and linear products, and they do so consistently with European and linear models. Hybrid models are assemblies of single underlying dynamic models, typically joined with a correlation matrix and made consistent and arbitrage-free by drift adjustments in their various components: for example, the risk-neutral drift of underlying assets that don't pay dividends is the short rate. When a stochastic price model is coupled with a stochastic rate model, the drift of the price process is the short rate produced by the rate model. Another example is that in a multi-currency context, the risk-neutral drift of assets denominated in a foreign currency must incorporate a convexity adjustment known as the quanto adjustment. See [27] for details. It follows that in a hybrid model, each (sub-) model is the same exotic model used for single underlying exotics, and that hybrid models value those exotics consistently with exotic models, hence, also, consistently with IVS and linear models. Hybrid models are slower than exotic models, but they can value hybrids and conduct bank-wide regulatory calculations.

---

[29] Although one-dimensional FDM prices Europeans, one by one or in bulk with the forward FDM technique explained in [44], with speed and accuracy that rivals closed-form formulas, and calibrations are further greatly accelerated with AAD; see Part III.

This hierarchy is fundamental. The intrinsic value of Europeans, exotics, and hybrids is produced in the exact same model that prices linear products. Calibration guarantees that the European part of exotic products is valued consistently with the IVS used for the management of Europeans. Hybrids and institution-wide regulatory calculations are conducted in assemblies of the same exotic models that are used to manage exotics. Not only does this provide consistent risk management across an investment bank's trading book, it also satisfies the regulator's justified insistence that regulatory calculations, like credit counterparty risk (CCR), revalue transactions over simulated scenarios *in a manner absolutely consistent with front office pricing.*

We don't implement a model hierarchy in this publication (actually, we don't implement interest rate models at all, not even linear ones, and keep the initial state as parameters in exotic models for simplicity) because that would double the size of the book and because it is not the primary topic here. Model hierarchies, and the consistency they offer, are nonetheless an extremely important subject in modern finance, and something we intend to address in a forthcoming publication.

The rest of this chapter briefly describes some of the most popular dynamic models. Our introduction is brief and without much detail, with the exception of Dupire's model [12], which we use as the main example to demonstrate parallel simulations and AAD, and which is therefore covered in depth in the rest of this publication. We recommend the three volumes of Andersen and Piterbarg's [6] for a deep and broad review of many financial models.

The Black and Scholes model

The Black and Scholes (BS) model [22] is a single underlying price model on an underlying asset s, like a stock, currency, or index, whose value today is s_0. BS copes with a deterministic short rate $r(t)$, which, in a model hierarchy, should be set to the IFR $f(0,t)$ to respect today's market.

The model also allows a deterministic dividend yield $y(t)$. To hold a dividend-paying asset is *not* a self-financing strategy. What is a self-financing strategy is to hold the asset and reinvest dividends into more units of the asset. Denoting b_t the number of assets held at time t, $b_0 = 1$ and it follows from the reinvestment of the dividend $b_t y(t) s_t$ paid at t that:

$$db_t = b_t y(t) dt$$

so $b_t = \exp\left(\int_0^t y(u)du\right)$ and the value of the strategy is:

$$V_t = s_t \exp\left(\int_0^t y(u)du\right)$$

It follows that

$$\frac{dV_t}{V_t} = \frac{ds_t}{s_t} + y(t)dt$$

and that the drift of V is the drift of s plus y. Under the risk-neutral measure, associated with the numeraire $\beta_t = \exp\left(\int_0^t r(u)du\right)$, the value of the self-financing strategy has a drift $r(t)$; hence the risk-neutral drift of the price series (s_t) is:

$$\mu(t) = r(t) - y(t)$$

The defining assumption of the Black and Scholes model is that asset prices are diffusions with a deterministic volatility $\sigma(t)$. It follows that the risk-neutral process of s is:

$$\frac{ds_t}{s_t} = \mu(t)dt + \sigma(t)dW$$

where W is a standard Brownian motion under Q.

Change of numeraire

Models can be equivalently rewritten under different measures, associated with different numeraires than the classic bank account. A wise choice of numeraire may simplify analytic calculations and improve the efficiency of numerical algorithms. We demonstrate this technique in the simple case of the Black and Scholes model, although it applies in a vast number of more sophisticated contexts; see [35] for some classic applications.

Forward measure First, we note that with deterministic interest rates, all forward-neutral measures are identical and coincide with the risk-neutral measure. To see this, we price a cash-flow X_T paid at date T under the risk-neutral measure:

$$V_t = \beta_t E_t^Q \left[\frac{X_T}{\beta_T}\right] = \frac{\beta_t}{\beta_T} E_t^Q [X_T]$$

where we moved the deterministic β_T out of the expectation. Repeating the exercise under the forward measure Q^T, we get:

$$V_t = DF(t, T) E_t^T [X_T]$$

Trivially:

$$\frac{\beta_t}{\beta_T} = DF(t, T) = \exp\left(-\int_t^T r(u)du\right)$$

and it follows that $E_t^Q[X_T] = E_t^T[X_T]$, for any $\mathcal{F}_T$-measurable variable X_T. Hence, by definition, the two measures coincide. This is all true for any maturity T. Hence, all Q_T measures coincide with the risk-neutral measure Q and are therefore identical.

It follows that forwards are also risk-neutral expectations in this case:

$$F(t,T) = E_t^Q[s_T] = s_t \exp\left(\int_t^T \mu(u)du\right)$$

and that all forwards of all maturities are martingales under Q. In addition, the volatility of a forward is:

$$\sigma_F(t,T) = \frac{\partial F(t,T)}{\partial s_t} \frac{s_t}{F(t,T)} \sigma(t) = \sigma(t)$$

so the risk-neutral dynamics of all forwards is simply:

$$\frac{dF(t,T)}{F(t,T)} = \sigma(t)dW$$

Applying Ito's lemma to the logarithm, we get:

$$d\log F(t,T) = -\frac{\sigma^2(t)}{2}dt + \sigma(t)dW$$

and it follows that for all times $t_1 < t_2$:

$$s_{t_2} = F(t_2,t_2) = F(t_1,t_2)\exp\left[-\int_{t_1}^{t_2} \frac{\sigma^2(t)}{2}dt + \int_{t_1}^{t_2} \sigma(t)dW\right]$$

Hence, the *distribution* of s_{t_2} conditional to $\mathcal{F}_{t_1}$ is:

$$s_{t_2} = F(t_1,t_2)\exp\left[-\frac{V(t_1,t_2)}{2} + \sqrt{V(t_1,t_2)}N\right]$$

where $V(t_1,t_2) = \int_{t_1}^{t_2} \sigma^2(t)dt$ and N is a standard Gaussian variable.

Spot measure An algorithm implementing the model under the risk-neutral measure must track the spot price s_t and the bank account numeraire β_t.[30] Alternatively, we can equivalently rewrite the model using the underlying

[30]Which is not such a drag in this model, the bank account being deterministic, but the method also applies in more sophisticated contexts where it may be beneficial.

asset itself as a numeraire, something known in literature as "spot measure," so we no longer need to additionally track the numeraire, the underlying asset *being* the numeraire.

With dividends, however, the asset price does not qualify as a numeraire. We fix a horizon T on or after the last date of interest in our working context, a technique known as "terminal measure." We consider the present value of a forward of maturity T:

$$f_t = DF(t,T)F(t,T) = s_t \exp\left(-\int_t^T y(u)du\right)$$

where we dropped the reference to the fixed horizon to simplify notations, and f trivially qualifies as numeraire. We call spot measure the associated measure Q^f, even though f is not exactly the spot price s_t of s.

Pricing a $\mathcal{F}_T$-measurable cash-flow X under the spot measure:

$$V_0 = f_0 E^f\left[\frac{X}{f_T}\right]$$

and comparing with the risk-neutral measure:

$$V_0 = E^Q\left[\frac{X}{\beta_T}\right] = \frac{E^Q[X]}{\beta_T}$$

(where we moved the deterministic bank account out of the expectation) implies that for any $\mathcal{F}_T$-measurable variable X,

$$E^Q[X] = E^f\left[\frac{\beta_T f_0}{f_T}X\right]$$

It follows that the Radon-Nikodym derivative of the spot-neutral to the risk-neutral measure, conditional to $\mathcal{F}_T$, is:

$$\frac{dQ^f}{dQ}|\mathcal{F}_T = \frac{f_T}{f_0\beta_T} = \exp\left[-\int_0^T \sigma^2(u)du + \int_0^T \sigma(u)dW^Q\right] = \zeta_T$$

where

$$\zeta_t = \exp\left[-\int_0^t \sigma^2(u)du + \int_0^t \sigma(u)dW^Q\right]$$

is an exponential martingale under Q with volatility $\sigma(t)$. Therefore, by Girsanov's theorem:

$$dW^f = dW^Q - \sigma(t)dt$$

is a standard Brownian motion under Q^f, and we get the dynamics of the forward (*any* forward) under the spot measure:

$$\frac{dF}{F} = \sigma(t)dW^Q = \sigma(t)dW^f + \sigma^2(t)dt$$

Hence, the *distribution* of s_{t_2} conditional to $\mathcal{F}_{t_1}$ under the spot measure is:

$$s_{t_2} = F(t_1, t_2)\exp\left[+\frac{V(t_1, t_2)}{2} + \sqrt{V(t_1, t_2)}N\right]$$

where we note that the only difference with the risk-neutral measure is the sign of the left term in the exponential.

The Black and Scholes formula

Having investigated some fundamental properties of the Black and Scholes model, we can easily derive the analytic formula for call options (the formula for puts following from the call–put parity):

$$C_t = DF(0, T)E_t^T[(F_T - K)^+]$$

and

$$\begin{aligned}
E_t^T[(F_T - K)^+] &= E_t^T\left[F_T 1_{\{F_T>K\}}\right] - KE_t^T\left[1_{\{F_T>K\}}\right] \\
&= F_t E_t^f\left[1_{\{F_T>K\}}\right] - KE_t^T\left[1_{\{F_T>K\}}\right] \\
&= F_t P\left(F_t \exp\left(\tfrac{V(t,T)}{2} + \sqrt{V(t,T)}N\right) > K\right) \\
&\quad - KP\left(F_t \exp\left(-\tfrac{V(t,T)}{2} + \sqrt{V(t,T)}N\right) > K\right) \\
&= F_t P\left(N > \frac{\log\left(\frac{K}{F_t}\right) - \frac{V(t,T)}{2}}{\sqrt{V(t,T)}}\right) - KP\left(N > \frac{\log\left(\frac{K}{F_t}\right) + \frac{V(t,T)}{2}}{\sqrt{V(t,T)}}\right) \\
&= F_t N\left(\frac{\log\left(\frac{F_t}{K}\right) + \frac{V(t,T)}{2}}{\sqrt{V(t,T)}}\right) - KN\left(\frac{\log\left(\frac{F_t}{K}\right) - \frac{V(t,T)}{2}}{\sqrt{V(t,T)}}\right) \\
&= F_t N(d_1) - KN(d_2)
\end{aligned}$$

where we used the notation N to refer to a standard Gaussian variable or the cumulative normal distribution, which is clear from context, and the traditional d_1 and d_2 notations.

One singular beauty in the Black and Scholes formula is that it exposes the hedge coefficients, too. In principle, d_1 and d_2 are functions of F_t and K so the differentiation of the formula could be a painful, if trivial, exercise. Noting that an option price is homogeneous in the forward and strike (an option on n assets for n times the strike is like n options on one asset for the strike price), it follows by Euler's theorem that the price is the sum of its derivatives, hence:[31]

$$\frac{\partial C_t}{\partial F_t} = DF(t,T)N(d_1) \quad \text{and} \quad \frac{\partial C_t}{\partial K} = -DF(t,T)N(d_2)$$

It follows that the replication strategy at time t is hold $N(d_1)$ forwards and borrow an amount $KDF(t,T)N(d_2)$ of cash, the value of the replication portfolio being trivially given by Black and Scholes's formula. It follows that the formula does not only express a price: it expresses the *replication strategy*, of which the price is a byproduct.

Dupire's model

The Black and Scholes model, despite its simplicity, remains a strong reference in the theory and practice of option trading, and not only because of its historical importance or its remarkable analytic tractability. The Black and Scholes model remains a keystone derivatives risk management instrument, useful in many important ways besides converting prices to volatilities and back. To appreciate the robustness of the model and its relevance for risk management, see the recent [48].

But the model suffers from one major flaw: the same assumption of a deterministic volatility that gave it such remarkable analytic tractability prevents it from correctly calibrating to option markets. It should be clear that with a single time-dependent free parameter $\sigma(t)$, the model may hope to hit, at best, one option price (one strike) per maturity. It follows that the *implied volatility* (obtained by solving in volatility so Black and Scholes's price corresponds to the market price) may be different for different strikes, a phenomenon commonly known as "smile," the curve of implied volatility as a function of the strike generally rising away from the money on both sides,[32] looking like a smile (with a bit of imagination).

[31]This is a very common application of Euler's theorem to Black and Scholes, although it takes a shortcut as noticed by Rolf Poulsen in http://web.math.ku.dk/~rolf/BS_clarification.pdf.

[32]For good reason outside of our scope here; see [46].

In the early 1990s, Bruno Dupire worked out the simplest possible extension of the Black and Scholes model that can be calibrated to option prices of all strikes and maturities. Like most ground-breaking ideas, Dupire's specification looks trivial after the fact: make volatility a function of the underlying asset price. This is called *local volatility* and presented here under Q, with zero rates and dividends for simplicity:

$$\frac{ds_t}{s_t} = \sigma(s_t, t)dW$$

To calibrate Dupire's model means to solve $E[(s_T - K)^+] = C(K, T)$ for all strikes and maturities, in the *local volatility surface* $\sigma(s, t)$. We are calibrating a continuous, two-dimensional local volatility surface, to a continuous, two-dimensional IVS. Intuitively, we could find a unique, perfect fit. Dupire's formula [12] confirms the intuition and even provides an explicit formula for $\sigma(s, t)$ as a function of today's options prices $C(K, T)$.

We discuss, demonstrate, implement in code, and apply Dupire's formula to risk management in Chapter 13. Dupire's formula is widely considered one of the three most important formulas in finance (the other two being Black and Scholes and Heath-Jarrow-Morton [49], which we introduce shortly). Dupire's model, or its stochastic volatility extensions by Dupire himself [50] and [45] and Bergomi [51], power almost all modern exotic risk management platforms.

Stochastic volatility models

The milestone stochastic volatility models, where volatility itself is a random process, were developed in 1987 by Hull and White [52], 1993 by Heston [42], and 2002 by Hagan [36]. They gained traction in the early 2000s, when the industry realized that "options are hedged with options," and that stochastic volatility is necessary to measure the consequences on valuation and risk management; see [46].

We briefly introduce Hagan's SABR, which specification is simpler than Heston, although both models produce a similar behavior and mainly differ in their implementation. Heston's model and its implementation are explained in detail in the first volume of [6].

SABR is written directly under a forward or annuity measure, depending on context[33] where the forward is a martingale. With the addition of stochastic volatility, it implements the following dynamics:

---

[33]SABR is a strong market practice for interest rate options, where caplets are priced under a forward measure and swaptions under the annuity measure as explained earlier.

$$dF_t = \sigma F_t^{\beta} dW$$
$$\frac{d\sigma}{\sigma} = \alpha dZ$$
$$\langle dW, dZ \rangle = \rho dt$$

The first line is a trivial extension of Black and Scholes with a power coefficient. The rest specifies the dynamics of volatility. The first thing to note is that this model is incomplete, in the sense that two sources of risk, represented by the two Brownian motions, cannot be hedged by trading one underlying forward. In addition, the initial value σ_0 is undefined.[34] It follows that we can't apply the asset pricing theorem unless *we complete the model by calibration*. After we calibrate σ_0 to hit the market price of some option $C(K, T)$,[35] the model is complete (because we now have *two* hedge instruments: the forward and the option), with well-defined initial conditions and an arbitrage-free dynamics (the forward is a martingale, and so is the numeraire deflated option price, by construction, because $C_t = N_t E_t[(F_T - K)^+]$).

A strong dependency on calibration is a defining characteristic of stochastic volatility models. Calibration is always a concern, but stochastic volatility models, unless calibrated, are ill-defined and fundamentally incorrect. The purpose of these models is to measure the impact of hedging with options on values and risk, hence the strong reliance on the initial values of options.

The parameters α (volatility of volatility), β (an additional local volatility parameter), and ρ (correlation between the forward and its volatility) are the free parameters of SABR. Note that the acronym SABR comes from the initials of its parameters: Sigma-Alpha-Beta-Rho.

Hagan produced a closed-form *approximation* for European options under the SABR dynamics using small noise expansion techniques. The result is a very precise approximation. It may be found in the original paper, a number of textbooks, or Wikipedia's SABR article. Hagan's solution involves steps specific to the SABR model and is not easily reusable. For a general presentation of expansion techniques and applications outside SABR, see Andreasen and Huge's [53].

Stochastic volatility is covered in a vast amount of literature, including Lipton's textbook [54], Gatheral's lecture notes [43], and Bergomi's recent [51]. It is also covered in detail in the lecture notes [46].

---

[34] Volatility is a free parameter in Black and Scholes, too, but in SABR, volatility is a process, not a parameter, which is the whole point of stochastic volatility models.

[35] Which means we solve $E[(F_T - K)^+] = \frac{C(K,T)}{N_0}$ in σ_0.

Interest rate models

The Heath-Jarrow-Morton framework Despite several early attempts to implement models *a la Black and Scholes* in the interest rate world, it was not before 1992 and the ground-breaking work of Heath, Jarrow, and Morton (HJM, [49]) that a consistent, general, and correct theoretical framework was established for modeling stochastic interest rates.

In interest rate markets, the "underlying asset" is the collection of zero-coupon bonds of all maturities, their value at time t being the *continuous curve* $DF(t, T)$. Ignoring stochastic basis for now, the state of the model at time t is the entire discount curve, or, equivalently, the instantaneous forward rate (IFR) curve, introduced earlier:

$$f(t,T) = -\frac{\partial \log[DF(t,T)]}{\partial T} \Leftrightarrow DF(t,T) = \exp\left[-\int_t^T f(t,u)du\right]$$

Under the risk-neutral measure, the drift of all discount factors must be the short rate $r_t \equiv f(t,t)$:

$$\frac{dDF(t,T)}{DF(t,T)} = r_r dt - \sigma_{DF}(t,T)dW$$

where σ_{DF} is the volatility of discount factors, it is conventionally negative, so that Brownian increments that positively affect rates, negatively affect bonds, and we introduced a one-factor version (one Brownian motion) for simplicity, although the original paper works with the general multifactor case, where equations remain essentially unchanged with matrices in place of scalars.

The celebrated Heath-Jarrow-Morton formula gives the dynamics of the IFR under the risk-neutral measure:

$$df(t,T) = \sigma_f(t,T)\int_t^T \sigma_f(t,u)dudt + \sigma_f(t,T)dW$$

where where σ_f is the volatility of the IFR's increments (not returns). From the definition of the IFR it immediately follows that:

$$\sigma_f(t,T) = -\frac{\partial \sigma_{DF}(t,T)}{\partial T}$$

and that

$$f(t,T) = -\frac{\partial DF(t,T)}{DF(t,T)\partial T}$$

is a martingale under Q^T. Further, we prove below that:

$$dW^T = dW - \sigma_{DF}(t,T)dt$$

is a standard Brownian motion under Q^T, and the HJM formula immediately follows.

Pricing a $\mathcal{F}_T$-measurable cash-flow Y under Q^T:

$$V_0 = DF(0,T)E^T[Y]$$

and comparing with the risk-neutral measure:

$$V_0 = E^Q\left[\frac{Y}{\beta_T}\right]$$

implies that for any $\mathcal{F}_T$-measurable variable Y,

$$E^T[Y] = E^Q\left[\frac{Y}{DF(0,T)\beta_T}\right]$$

It follows that the Radon-Nikodym derivative of Q^T to Q, conditional to $\mathcal{F}_T$, is:

$$\frac{dQ^T}{dQ}|\mathcal{F}_T = \frac{1}{DF(0,T)\beta_T} = \frac{DF(T,T)}{DF(0,T)\beta_T} = \zeta_T$$

where:

$$\zeta_t = \frac{1}{DF(0,T)}\frac{DF(t,T)}{\beta_t}$$

is an exponential martingale under Q[36] with volatility, the volatility $\sigma_{DF}(t,T)$ of the zero-coupon bond of maturity T.[37] Therefore, by Girsanov's theorem:

$$dW^T = dW^Q - \sigma_{DF}(t,T)dt$$

is a standard Brownian motion under Q^T, which completes the demonstration.

---

[36]This is the price at t of the zero-coupon bond of maturity T, discounted by the numeraire, hence a martingale; the normalization by a constant $DF(0,T)$ does not affect this result.

[37]The numeraire β is a finite variation process without a diffusion term; hence, the volatility of the ratio is the volatility of the numerator.

When σ_f is deterministic, the distributions of the IFRs are Gaussian and those of the discounts are log-normal. This model, despite being subject to a single Brownian motion, is generally *non-Markov*. This means that there exists no deterministic relationship between $f(t, T_1)$ and $f(t, T_2)$ or between $DF(t, T_1)$ and $DF(t, T_2)$ when $T_1 \neq T_2$. The Markov dimension of the model is infinite: its state at time t is the whole discount or forward curve and it cannot be reduced without loss of information.

Infinite dimension does not lend itself to an effective practical implementation. Therefore, HJM remains to date the theoretical reference for stochastic interest rate modeling, but what is effectively implemented in modern systems, for the risk management of interest rate exotics, are variations and extensions purposely developed for algorithmic tractability. The most successful variations, developed shortly after HJM and still in activity today with substantial upgrades, are the Libor Market Model (LMM) of Brace, Gatarek, and Musiela (BGM, [29]) and Markov versions of HJM, independently established by many researchers in the mid-1990s, most prominently Cheyette [55].

Libor Market Models Libor Market Models partition the rate curve up to a final horizon $T*$ into a finite number of forward Libors of duration δ: $F_i(t) = F(t, t + \delta i, t + \delta(i + 1))$. The original paper postulates a log-normal dynamic a la Black-Scholes with volatility $\sigma_i^F(t)$ and deduces an HJM-like formula for the risk-neutral dynamics of the forward Libors (which we don't reproduce here, referring readers to the original paper or the dedicated chapter of [6]).

LMM are also generally non-Markov, although their dimension is finite. With horizon 50 years and $3m$ Libors, its dimension is 200, which is certainly better than infinite, but too elevated for performance. LMM remain popular nonetheless, partly because they directly model market traded Libors, as opposed to theoretical, shadow variables likes discount factors or IFR, and partly because the large number of free parameters, the collection of curves $\sigma_i^F(t)$, allows them to correctly calibrate to interest rate option markets, even in their one-factor form.

Importantly, the original paper also delivered a closed-form formula for caps and swaptions in the LMM framework.[38] At the heart of this exercise was the determination of the volatility of swap rates as a function of the volatility and correlation of forward Libors. It later turned out that this new way of reasoning about the dynamics of interest rates was very useful in itself, with applications beyond the implementation of LMM. For instance, closed-form approximations for European options could be found for the general HJM model, something that had not been done before. The mapping

[38]The formula is exact for caps, approximate but very precise for swaptions.

of Libor to swap rate dynamics also helped with the structuring and risk management of interest rate exotics during the 1998–2008 decade when rate exotic businesses were growing exponentially in size and complexity.

Markov specification of HJM LMM resolved the infinite dimension problem by implementing what is essentially a discrete version of HJM. Another, diametrically opposite approach is to specify HJM in a way to lower its Markov dimension. In its one-factor Gaussian form presented here, HJM is Markov in dimension one if (and only if) for any couple of maturities $T_1 < T_2$, there exists a deterministic function α such that:

$$f(t, T_2) = \alpha(f(t, T_1), t, T_1, T_2)$$

When this is the case, all the forward rates at time t are deduced from one another, so when we know one, for example, the short rate $r_t = f(t, t)$, we know them them all, and it follows that the state of the model at time t is contained in one scalar variable, for example r_t, so the Markov dimension of the model is, by definition, one. This would allow to model interest rate curves in the same way we model asset prices, with one-dimensional dynamics, and apply the efficient algorithms designed for these, FDM whenever possible, low-dimension Monte-Carlo simulations otherwise.

Although this is somewhat outside of our scope, we demonstrate here how this is achieved, because this is an interesting exercise, and because it sheds light on *why* HJM was infinite dimensional to start with, even in its one-factor form.

Integrating the HJM formula (and dropping the f subscript on the IFR volatility) we find:

$$f(t, T) = \underbrace{f(0, T) + \int_0^t \sigma(s, T) \int_s^T \sigma(s, u) du ds}_{E^Q[f(t,T)]} + \underbrace{\int_0^t \sigma(s, T) dW_s}_{N\left(0, \int_0^t \sigma^2(s,T)ds\right)}$$

The left-hand side is a deterministic, known risk-neutral expectation and does not matter for our purpose. The right-hand side is the unknown, stochastic part of the IFR, a centred Gaussian variable. The stochastic parts of two IFRs of maturities T_1 and T_2 are:

$$\int_0^t \sigma(s, T_1) dW_s \quad \text{and} \quad \int_0^t \sigma(s, T_2) dW_s$$

and we see why the model is non-Markov: the random increments dW_s affect the IFRs with different weights $\sigma(s, T_1)$ and $\sigma(s, T_2)$, and it is, in general, impossible to retro-engineer their aggregated impact on one $\int_0^t \sigma(s, T_2)dW_s$ out of their aggregated impact on the other $\int_0^t \sigma(s, T_1)dW_s$. The model is Markov if and only if one is somehow a deterministic function of the other. Since both are centered Gaussian variables, if such function exists, it has to be linear:

$$\int_0^t \sigma(s, T_2)dW_s = \alpha(t, T_1, T_2) \int_0^t \sigma(s, T_1)dW_s$$

This must be true for all times t, so we can differentiate the equation in t:

$$\sigma(t, T_2)dW_t = \frac{\partial \alpha(t, T_1, T_2)}{\partial t} \int_0^t \sigma(s, T_1)dW_s + \alpha(t, T_1, T_2)\sigma(t, T_1)dW_t$$

For this equation to be satisfied irrespective of the Brownian path, the derivative in t of the right-hand side's left term must be zero, so α is not a function of t after all, and the equation simplifies into:

$$\sigma(t, T_2) = \alpha(T_1, T_2)\sigma(t, T_1)$$

which implies that for any T, $\alpha(T, T) = 1$ and for any triplet $T_1 < T_2 < T_3$:

$$\alpha(T_1, T_2)\alpha(T_2, T_3) = \alpha(T_1, T_3)$$

It follows that there must exist a function $k(u)$ such that

$$\alpha(T_1, T_2) = \exp\left(-\int_{T_1}^{T_2} k(u)du\right)$$

in particular:

$$\alpha(t, T) = \exp\left(-\int_t^T k(u)du\right)$$

and denoting $\sigma(t) \equiv \sigma(t, t)$, we find that the model is Markov if and only if:

$$\sigma(t, T) = \sigma(t) \exp\left(-\int_t^T k(u)du\right)$$

This requirement restricts the specification of HJM to a so-called *separable* volatility specification, because $\sigma(t, T)$ is the product of a function of only t: $\sigma(t) \exp\left(\int_0^t k(u)du\right)$ and a function of only T: $\exp\left(-\int_0^T k(u)du\right)$. It is also clear that the dimension of the free parameters in HJM is reduced by an order of magnitude from a two-dimensional surface $\sigma(t, T)$ to two curves $\sigma(t)$ and $k(u)$. The resulting model is orders of magnitude more tractable in a practical implementation, but its ability to calibrate to the market prices of interest rate options is limited due to the small number of free parameters.[39]

In addition, going back to the integral form of the IFR and setting $T = t$, we find the integral form of the short rate:

$$r_t = f(t,t) = f(0,t) + \int_0^t \sigma(s,t) \int_s^t \sigma(s,u)duds + \int_0^t \sigma(s,t)dW_s$$

Injecting the Markov volatility form, differentiating, and rearranging, we quickly find the dynamics of the short rate (we make k a constant here to simplify the equations):

$$dr_t = \left[\frac{\partial f(0,t)}{\partial t} + \varphi(t) - k(r_t - f(0,t))\right] dt + \sigma(t)dW$$

where $\varphi(t) = \int_0^t \sigma^2(u)\exp[-2k(t-u)]du$ is deterministic. The short rate follows a *mean-reverting* process a la Vasicek (1977) [56], so one of this model's many names is "extended Vasicek." It is also widely known as Hull and White's interest rate model [57], although, like Vasicek whose work precedes HJM by 15 years, Hull and White postulated a Markov dynamic for the short rate and worked their way up to the dynamics of the curve, not the other way around. We shall call this model by another of its popular names, Linear Gauss Markov (LGM). Its parameters are $\sigma(t)$, the volatility of the first differences (not returns) of the short rate, and its mean-reversion $k(u)$ (even though we kept k constant to simplify the presentation, it doesn't have to be).

It is more convenient to work with the variable $X_t \equiv r_t - f(0,t)$, which dynamics is the same as r_t but drops the reference to today's forwards in the drift:

$$dX_t = [\varphi(t) - kX_t]dt + \sigma(t)dW$$

where $X_0 = 0$. The following results may be found with simple (if somewhat painful) calculus.

---

[39]However this limitation dissipates in a multifactor context. When this model is written with multiple Brownian motions, the number of free parameters increases linearly with the number of factors. A typical modern implementation in production uses three to five factors, allowing for a very decent calibration to market prices. The Markov dimension increases to the number of factors, but remains low. The main benefit of a multifactor model, however, is to produce a realistic correlation between rates of different maturities: in a one-factor model, with the same Brownian motion driving the entire curve, the correlation between rates is necessarily 100%.

1. All forward rates at time t are a deterministic function of X_t. This is the whole point of the model, of course, but it so happens that this function is explicit:

$$f(t,T) = f(0,T) + \frac{\exp[-k(T-t)] - \exp[-2k(T-t)]}{k}\varphi(t) + \exp[-k(T-t)]X_t$$

2. It follows that all discount factors at time t are also a deterministic function of X_t:

$$DF(t,T) = DF(0,t,T)\exp\left[-\frac{1}{2}\left(\frac{1-e^{-k(T-t)}}{k}\right)^2\varphi(t) - \frac{1-e^{-k(T-t)}}{k}X_t\right]$$

3. Finally, forward Libors at time t are a deterministic function of X_t:

$$F(t,T_1,T_2) = \frac{[1+(T_2-T_1)F(0,T_1,T_2)] \times \exp\left(-\frac{1}{2}\frac{(e^{-k(T_2-t)}-e^{-k(T_1-t)})(e^{-k(T_2-t)}+e^{-k(T_1-t)}-2)}{k^2}\varphi(t) - \frac{e^{-k(T_2-t)}-e^{-k(T_1-t)}}{k}X_t\right) - 1}{T_2 - T_1}$$

This formula correctly incorporates Libor basis, but keeps it constant.

These formulas, called *reconstruction formulas*, are a fundamental part of the model, because they map a path of its state variable X to a scenario of market samples S. Contrary to the other models we introduced so far, the LGM dynamics is not directly written on market variables, so the scenario is not modeled directly, but implicitly, as its samples are deduced from X with the reconstruction formulas.

Remember that the purpose of this model, like any model, is to specify the joint distribution of the market observations $\{s_k\}$ in the scenario. In an interest rate context, these observations are the values on the event dates of forward discounts and Libors, as well as the numeraire on payment dates. What LGM models is its internal state variable X, which joint distribution on event dates is a multidimensional Gaussian distribution, with known expectations and a known covariance matrix (although we are not calculating them here). The discounts and Libors on the event date t are modeled implicitly as known functions of the state variable X_t on the same event date. But this is not the case for the risk-neutral numeraire β_t. From its definition:

$$\beta_t = \exp\left(\int_0^t r_u du\right)$$

it appears clearly that β_t is a functional of the continuous path of X over $[0, t]$ but not of X_t. Of more concern is that β_t is not a function of a *discrete path of* X. It would follow that, in order to fulfill its purpose and specify the distribution of the scenario, the model would need to track the numeraire, in addition to its state variable, on event dates, somewhat defeating the benefit of the one-dimensional Markov specification, and damaging the speed and accuracy of algorithms. This comment is not limited to LGM and concerns all interest rate models. The risk-neutral measure is generally *not* the wisest choice for their implementation.

The solution is to change the numeraire, of course, and use one numeraire N where N_t *is* a known function of X_t. We have seen that this is the case for zero-coupon bonds. In order to use the same numeraire for all cash-flows, we must use a bond of a longer maturity than the financial product we are working with. The preferred maturity is generally the payment date of the last coupon, and the related probability is known as the *terminal measure.*

When we demonstrated the Heath-Jarrow-Morton formula page 169, we proved that:

$$dW^T = dW^Q - \sigma_{DF}(t, T)dt$$

is a standard Brownian motion under Q^T. In LGM:

$$\sigma_{DF}(t, T) = -\int_t^T \sigma_f(t, u)du = -\sigma(t)\frac{1 - \exp(-k(T - t))}{k}$$

hence, the dynamics of the state variable X under the terminal measure is:

$$\begin{aligned} dX_t &= [\varphi(t) - kX_t]dt + \sigma(t)dW^Q \\ &= [\varphi(t) - kX_t + \sigma(t)\sigma_{DF}(t, T)]dt + \sigma(t)dW^T \\ &= \left[\varphi(t) - kX_t - \sigma(t)^2\frac{1 - \exp(-k(T - t))}{k}\right] dt + \sigma(t)dW^T \end{aligned}$$

the rest of the model, in particular, the reconstruction formulas, being unmodified by the change of measure.

Finally, an exact closed-form solution for coupon bond options was derived in this model by El Karoui and Rochet in 1989 [58]. The formula can be applied to price a swaption by turning the underlying swap into a coupon bond, although the exercise is less trivial with Libor basis. It also applies to caplets and floorlets, considered as single-period swaptions, hence to caps and floors.

LGM is a fundamental interest rate model, and, to a large extent, the Black (and Scholes) of the interest rate world. Like Black, LGM is simple,

Gaussian, and analytic. Like Black, LGM can be implemented in very fast algorithms, thanks to its one-factor Markov dimension and Gaussian transition probabilities. And like Black, LGM was one of the first operational models in its field, and provided a solid, if somewhat limited, basis for trading the first interest rate exotics.

The fundamental result, that HJM is Markov if and only if its volatility structure is separable, was independently found by multiple researchers in the mid-1990s, including Cheyette in 1992 [55] and Ritchken and Sankarasubramanian in 1995 [59]. Cheyette, in particular, derived the result in a very general context, not only in a multifactor framework, but also with a non-deterministic volatility. Cheyette demonstrated that reconstruction formulas hold with local and even stochastic volatility, as long as it remains separable, even though any other specification than Gaussian increases the Markov dimension of the model.

Modern implementation of interest rate models Research in interest rate models peaked in the decade 1998–2008 to cope with the rapidly growing size and complexity of rate exotic markets, the main focus being stochastic volatility and multi-factor extensions. LMM and Cheyette[40] roughly shared the scene and were both upgraded with multiple factors and stochastic volatility.

Cheyette implementations are more efficient due to a low Markov dimension, LMM being more flexible and better calibrated, although the calibration gap significantly narrows when the number of factors in Cheyette is increased.

Research on stochastic rate models continued after 2008–2011, although somewhat less actively, mainly with the incorporation of stochastic Libor basis and collateral discounting, and a focus on a fast implementation for the needs of xVA and regulations, therefore favoring the Cheyette family. We refer to [7] for an application of MFC to xVA.

A general formulation of dynamic models

The purpose of a model, dynamic or not, is to specify the distribution of the scenario $\Omega = \{s_k\}$, a collection of market observations on a collection of event dates, which definition is independent of the model. How models fulfill this purpose is down to their internal mechanics and depends on the specific model. Dynamic models are those that determine the distribution of the scenario, implicitly, with the specification of a stochastic dynamics.

---

[40]With the addition of local and/or stochastic volatility, the model is no longer Gaussian, although still linear and Markov, and is no longer called LGM, but Cheyette. Multifactor versions are often called multifactor Cheyette or MFC. A practical, efficient MFC implementation was published in 2005 by Andreasen [8].

Some models, like Black and Scholes, directly specify the dynamics of market variables. But this is not always the case. Other models, like LGM, specify the dynamics of some internal state variables, together with a mapping from state to market, which we called reconstruction formulas.

More generally, dynamic models define the stochastic dynamics of an internal state and specify a mapping from state to market variables. A model is the sum of the dynamics and the mapping. Both may be subject to parameters, like Black and Scholes's volatility, Dupire's bi-dimensional local volatility surface, or LGM's time-dependent volatility and mean-reversion.

More formally, a dynamic model defines Q^N by specifying the dynamics of an internal *state vector* (X_t) and a mapping:

$$\zeta : \{X_t, 0 \leq t \leq T\} \rightarrow \Omega = \{S_t, t \in TL\}$$

from a path of X to the market samples over a scenario's timeline, including the numeraire on payment dates. Note that we have a mapping from $\{X_t, 0 \leq t \leq T\}$ to Ω, not one from X_t to S_t for every event date t. This would be generally desirable, but not always possible. For example, we have seen that the risk-neutral numeraire β_t in LGM is a functional of $\{X_t, 0 \leq t \leq T\}$ but not a function of X_t. Particular models may have time-wise mappings, but in general, it is a path-wise mapping. This is a detail, but one that affects the architecture of simulation libraries.

We restrict the presentation to diffusive models, in which the state vector follows, under Q_N, a multidimensional diffusion of the type:

$$dX_t = \mu(X_t, t)dt + \sigma(X_t, t)dW$$

The size of the state vector is called the *Markov dimension* of the model. It may be as low as one, but we also introduced models where it may be large (LMM), or infinite and even continuous (HJM). We carefully distinguish the Markov dimension: the size n of the state vector X; and the number of factors: the size m of the Brownian Motion W. It is always the case that $n \geq m$, and, in many cases, n is vastly superior to m.

The drift μ is a vector-valued function of the state vector and time, which components are the risk-neutral drifts of the components of X (called *state variables*) *each as a function of the entire state vector* X. The volatility matrix σ is an $n \times m$ matrix that represents the (instantaneous) covariance of the state variables. Since W is a standard Brownian Motion of dimension m, its components are independent; hence, the covariance of X is $\sigma\sigma^t$, which also means that σ is a square root (in matrix terms) of the covariance matrix of X.

The specification of μ and σ, along with the parameters of the mapping ζ, constitute the parameter set of the model. The model parameters are either set in reference to initial values (today's spot and forward prices, discounts, or rates), estimated (correlation) or calibrated (typically, volatilities) to match the market prices of a number of active (generally, European) instruments. Derivatives of product values to model parameters are the risk sensitivities, which effective production is a topic of Part III.

Finally, the combination of the dynamics of X and the mapping ζ must guarantee that ratios of asset prices to the numeraire are martingales, and that initial values correspond to today's market prices. This guarantees that the model is arbitrage-free and respects today's market. Practically, it also ensures that the model correctly prices linear instruments. When, in addition, the model is correctly calibrated, it also correctly prices European instruments. Dynamic models also implicitly specify the joint distribution of the scenario, allowing them to price exotics, including hybrids and xVA, provided that the model correctly specifies the joint dynamics of all the relevant market variables.

Finite Difference Methods

We have seen that, with the exception of a small number of particularly simple cases, dynamic models cannot price exotics analytically. Numerical methods must be invoked. There exist essentially two families of generic numerical methods for pricing with dynamic models: finite difference methods (FDM) and Monte-Carlo simulations. Other methods exist, but they are generally niche and only work in specific contexts like Fourier transforms, or they are inefficient, like binomial or trinomial trees, which can be seen as less stable and less efficient versions of FDM.

We briefly introduce FDM now, and explore Monte-Carlo in detail in the next chapters. We shall not discuss FDM further thereafter, and refer to the dedicated chapters in Andersen and Piterbarg's book [6] or Andreasen's lecture notes [60].

For the purpose of this brief introduction to FDM, we disregard interest rates and work under the risk-neutral measure, where, in this case, all the discount factors, and the numeraire, are constant equal to one, and cash-flows and payoffs are the same thing. Let us start with a European transaction with a single cash-flow of maturity T with a one-factor, one-state-variable dynamic model, with a time-wise mapping $\zeta_t : x_t \rightarrow S_t$:

$$V_t = E_t[g(S_T)]$$

$$dx_t = \mu(x_t, t)dt + \sigma(x_t, t)dW$$

$$S_t = \zeta_t(x_t)$$

From the Markov property of x, we know that $V_t = v(t, x_t)$. From Feynman-Kac's theorem, we know that v is the solution of the partial differential equation (PDE):

$$v_t = -\mu v_x - \frac{\sigma^2}{2} v_{xx}$$

(where the subscripts denote partial derivatives) with the boundary condition:

$$v(x, T) = g[\zeta_T(x)]$$

(the value at maturity is the payoff).

The solution of this PDE may be approached as follows: consider a grid of N, say evenly spaced times $T_j = \frac{j}{N}T$ and N evenly spaced xs x_i ranging from, say, the expectation minus five standard deviations at maturity to the expectation plus five standard deviations at maturity.[41]

Now we have a grid where each node (i, j) represents a scenario where $x_{T_j} = x_i$. We are going to populate this grid with the corresponding future values $v_{ij} = v(x_i, T_j)$ of the transaction. First, we apply the boundary condition, which populates v_{iN}. Then, we approximate the PDE with finite differences over the nodes in the grid in place of the continuous derivatives, which permits to propagate values backwards in time and populate the grid entirely. The node $v_{i_0 0}$ (where i_0 is the position of x_0 in the grid) is today's value.

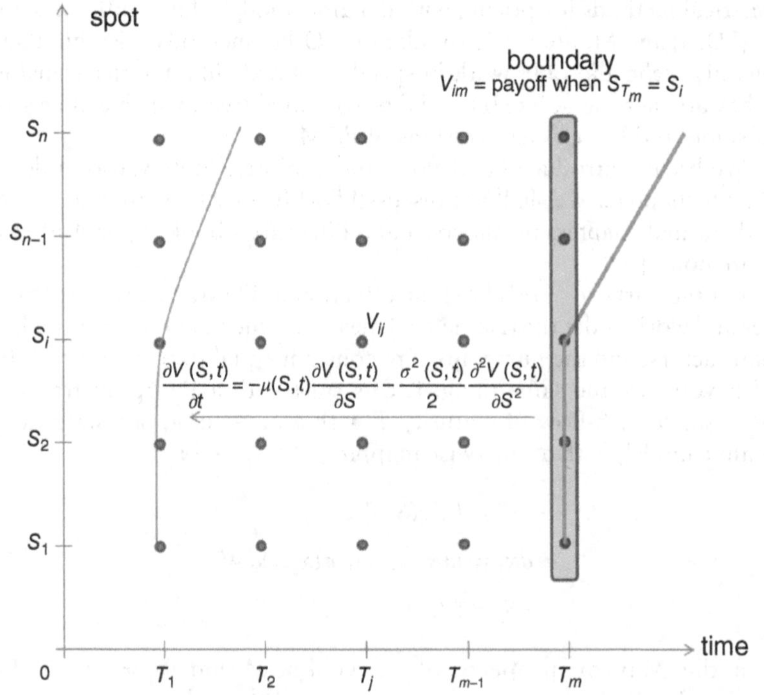

[41]In general, we do not have the same number of times and spots, but we go with that here in the interest of simplicity; in addition, the expectation and standard deviation of x_T may need to be estimated from the model parameters beforehand.

Unfortunately, it does not work that easily. We left out the details of the back-propagation process. What do we do exactly once we know the values at T_{j+1} to find the values at T_j? We could compute the derivatives in x over the known grid at T_{j+1}, and apply time differences to produce the grid at T_j. This is known as an *explicit* scheme and is notoriously unstable. An *implicit* scheme must be applied instead, where finite differences are computed over the *unknown* grid at T_j. This boils down to a tri-diagonal system of equations, something that is solvable efficiently in linear time. *Numerical Recipes* [20] provides the code of an efficient *tridag*() routine for this purpose. Even better, we can populate the grid at T_j with a mix (say, half–half) of the explicit and the implicit solutions, something known as the Crank-Nicolson (CN) scheme. CN not only guarantees the stability of the numerical solution (as long as it is at least 50% implicit), but also its convergence with order N^2.

The complexity of FDM is linear in both the size of the x grid and the size of the time grid. Linear complexity combined with quadratic convergence makes FDM *the most efficient valuation method ever invented.*

Of course, FDM is not only adequate for European profiles. The back-propagation equation is independent of the transaction. The cash flows only intervene through boundary conditions. We can incorporate many types of cash-flows by manipulating boundaries and without modification of back-propagation. For instance:

We can have multiple European cash flows $CF_j = g_j(S_{T_j})$ by applying the boundary:

$$v_{ij} += g_j[\zeta_{T_j}(x_i)]$$

(where we used the C style "+=" notation) *after* we solved the grid for T_j.

We can implement a barrier with the boundary condition that v is 0 above or below the barrier. The FDM will even correctly value a continuous barrier with a minor modification of the back-propagation scheme. For this reason, FDM is the method of choice for barrier options.

Since we are populating the different scenarios on the grid with the future values of the transaction, we can easily incorporate early exercise at time T_j with payoff $g_j(S_{T_j})$ with the boundary:

$$v_{ij} = \max\{v_{ij}, g_j[\zeta_{T_j}(x_i)]\}$$

again, to be applied after we solved the grid for T_j. We can also determine exercise regions:

$$R_{T_j} = \{x_i, v_{ij} < g_j[\zeta_{T_j}(x_i)]\}$$

and exercise boundaries. FDM can even correctly propagate the exercise between consecutive time steps as a continuous exercise; therefore it is adequate (actually, it is the method of choice) for American options.

What we *cannot* easily incorporate is general path dependence that cannot be interpreted as a boundary. For instance, the payoff of an Asian option (option on average) cannot be interpreted as a boundary. Because of that path-dependence, the future price of an Asian option is not a function of time and spot only, but also of the running average up to that time. Hence, the Markov dimension of the PDE is two in this case, although the Markov dimension of the model is still one. In general, FDM *can* handle path-dependency, but at the cost of an increase in dimension, something that FDM does not handle that well.

The major challenge with FDM is that its complexity grows exponentially with dimension for a fixed convergence. As the Markov dimension of the PDE grows to D, the number of nodes to process grows to N^{D+1} to maintain a numerical error in N^{-2}. If we have 100 steps in each direction, we have 10,000 nodes to process in dimension 1. A modern computer does that in a fraction of a millisecond. In dimension 2, we have 1M nodes, something a modern computer also handles easily. But we are no longer talking milliseconds. In dimension 3, this is 100M nodes. Now we are talking seconds. In dimension 4, it is 10BN nodes we have to process for a single valuation. We may be talking talking minutes. In addition to an exponentially growing complexity, dimension more than one creates difficulties in the treatment of the cross-derivatives, and the solutions (known as Craig-Sneyd's scheme and friends) further increase complexity.

We will see that Monte-Carlo simulations, which have a much lower convergence order of $\sqrt{N}$ in the number of simulations and a much higher complexity per scenario, still win because of their linear complexity in the Markov dimension of the model (not of the PDE, which dimension is at least equal to that of the model). FDM is not practical in dimension 4 or higher. That pretty much disqualifies it in the context of general path-dependent transactions and many models of practical relevance. Cheyette's Markov interest rate model with two factors, for example, is of dimension two with a deterministic volatility, but five otherwise. Hence, FDM is out for multi-factor interest rate models, too. More generally, the growing complexity of the exotic transactions in the decade 1998–2008, causing models to grow in sophistication to follow-up, increased the Markov dimension in most pricing contexts, disqualifying FDM and leaving the much slower and less accurate Monte-Carlo as the only solution. This is without even mentioning the hybrid models used for regulatory calculations, the dimension of which routinely stands in the hundreds.

One specific area in modern financial markets where FDM remains relevant is the so-called *first-generation foreign exchange exotics*. These are short-term (typically less than 6 months) barrier options (all sort of barriers, one touch, double no touch, knock-in, reverse knock-out, and so forth) written on currency exchange rates. These exotics trade in high volumes on specialized OTC[42] markets. The short maturity permits to neglect the impact of stochastic rates and keep dimension low, and FDM is ideal for barriers, as mentioned earlier.

Lipton's masterclass [61] provides an overview of forex exotics and analyses models with local volatility, stochastic volatility, and jumps in relation to those markets. One leading model is the so-called SLV (stochastic and local volatility) model defined by:

$$\frac{dS}{S} = (f(0,t) - y(t))dt + \sigma(S,t)\sqrt{v}dW^S$$

$$dv = -k(v-1)dt + \alpha\sqrt{v}dW^\sigma$$

$$\langle dW^S, dW^\sigma \rangle = \rho dt$$

with Heston's specification for the stochastic volatility, although other specifications are possible; see [51]. This is a two-factor, two-state-variable model suitable for FDM. The model is calibrated with Dupire's second formula (see [50]) implemented in the form of a forward FDM (see [44]), and it values barriers with backward FDM. Pricing takes a few 1/100s of a second, remarkably for a model of such sophistication.

It is, however, Monte-Carlo that accounts for the vast majority of valuation contexts of practical relevance in modern finance. But this is only because Monte-Carlo is the only viable method in such high dimension. Researchers *settled* for Monte-Carlo despite its slow convergence and CPU-demanding operation. Monte-Carlo is a *fallback* in a context where the superior FDM is no longer relevant.

For this reason, a computationally efficient implementation of Monte-Carlo simulations, including parallelism, and the differentiation of results, is a critical objective in modern finance. To give an idea of the time scales involved, it takes around one second to compute the full risk against 1,600 local volatilities of a barrier option in Dupire's local volatility model, with FDM, 100 time steps, 150 spot steps, by bumping, on a single modern CPU. To get a similar accuracy with Monte-Carlo takes at least 100,000 simulations over 200 time steps, and the full bump risk on a single core

[42] Over the counter, or directly between banks and institutional clients, as opposed to traded on an exchange.

takes about half an hour, around 2,000 times slower than FDM. With parallelism and AAD, FDM risk time is reduced under 1/100 of a second (on a quad core laptop). In relative terms, this is a fantastic 100+ speed-up. In absolute terms, 1 second or 1/100 of a second does not make much difference. With Monte-Carlo, parallelism and AAD reduce the time spent on the risk report *from half an hour to half a second* (on a quad core laptop or a quarter of a second on an octo-core workstation), without loss of accuracy. The difference this makes for investment banks and their trading desks is massive. This is the difference between a real-time risk report on a trader's workstation or an overnight computation in a data center.

The purpose of the techniques described in the rest of this publication is to bring the performance of FDM to Monte-Carlo simulations.

CHAPTER 5

Monte-Carlo

In this chapter, we introduce the Monte-Carlo (MC) algorithm as a practical solution for pricing financial products in dynamic models and lay the groundwork for the serial implementation of Chapter 6 and the parallel implementation of Chapter 7. We establish the theoretical foundations of our implementation, but we don't cover all the details and facets of Monte-Carlo simulations. Monte-Carlo is more extensively covered in dedicated publications, the established references in finance being Glasserman's [62] and Jaeckel's [63]. Glasserman covers the many facets of financial simulations in deep detail while Jaeckel offers a practitioner's perspective, focused on implementation and including the most complete available presentation of Sobol's sequence.

5.1 THE MONTE-CARLO ALGORITHM

Introduction

The Monte-Carlo algorithm consists in the computer simulation of a number of independent outcomes of a random experiment with known probability distribution. It is applied in many scientific fields. In finance, the Monte-Carlo algorithm estimates the value of a financial product in contexts where faster methods (analytic, numerical integration, FDM) cannot be applied. With the notations of the previous chapter, the value of a product is its expected (numeraire deflated) payoff g under an appropriate probability measure Q^N:

$$V_0 = E^N\{g[(S_t)_{t\in TL}]\}$$

where TL is the set of event dates, or timeline, and the scenario $\Omega = (S_t)_{t\in TL}$ is the set of corresponding market samples. Each sample S_j on the event date T_j is a collection of market observations s_j^k on the event date T_j, including the numeraire.

The payoff function g defines a financial product. We have seen that a model specifies a probability distribution for the scenario Ω, although *dynamic* models specify it *indirectly*. Dynamic models specify:

1. The multidimensional dynamics of a state vector X in dimension n, called the Markov dimension of the model.[1] We restrict our presentation to diffusion models. In this case, the dynamics of the state vector, under the pricing measure, is a stochastic differential equation (SDE):

$$dX_t = \mu(X_t, t)dt + \sigma(X_t, t)dW$$

where W is a standard Brownian motion in dimension $m \leq n$, called the number of factors in the model. In particular, the m components of W are independent, so the correlation structure of the model is contained in the matrix σ.

2. The mapping from state X to market S:

$$\zeta : \{X_t, 0 \leq t \leq T\} \rightarrow \Omega = \{S_t, t \in TL\}$$

where T is the last fixing date for the product, called maturity. We have seen that the mapping ζ can be *time-wise*:

$$\zeta_t : X_t \rightarrow S_t$$

but this is not always the case. We have seen on page 175 that LGM does not define a time-wise mapping under the risk-neutral measure because the numeraire at time t, β_t, is not a function of the state X_t on this date, but a functional of the past and present state $\{X_u, 0 \leq u \leq t\}$. We have also seen that we can rewrite LGM under a different measure Q^T where the numeraire $DF(t, T)$ is a function of X_t, so the mapping ζ *is* time-wise. This is a general property of many interest rate models: they typically don't admit a time-wise mapping under the risk-neutral measure, but they can be rewritten under a different measure, associated with a different numeraire, often a zero-coupon bond, where the mapping becomes time-wise, facilitating an efficient simulation.

The Monte-Carlo algorithm estimates the value V_0 by *brute force*:

1. Draw a large number N of realizations of the scenario Ω, also called paths Ω_i, from its probability distribution under Q^N, specified by the model. Draw the realizations independently of one another.

[1]We only consider models of finite dimension here. Infinite or continuous dimension is not suitable for a practical implementation. With infinite dimensional models like HJM, it is generally a finite dimensional version like LMM that is simulated.

2. Compute the payoff $g_i = g(\Omega_i)$ over each path, by evaluation of the product's payoff g.
3. Approximate the value $V_0 = E^N[g(\Omega)]$ by the average payoff $\overline{V_0} = \frac{1}{N}\sum_{i=1}^{N} g_i$ over the simulated paths.

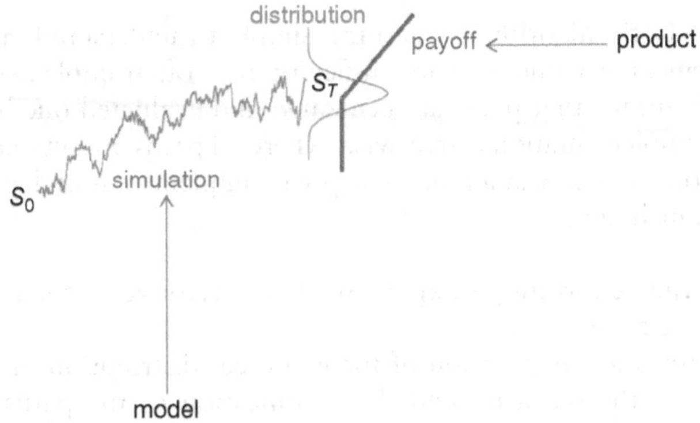

Convergence and complexity

By construction, the g_is are IID with distribution the distribution of $g[\Omega]$, therefore the MC estimate $\overline{V_0}$ converges *almost surely* to V_0 by the Strong Law of Large Numbers. Furthermore, the Central Limit Theorem states that $\overline{V_0}$ is asymptotically Gaussian with expectation V_0 and variance:

$$\frac{Var^N[g(\Omega)]}{N}$$

hence, a standard deviation inversely proportional to $\sqrt{N}$. This is a low convergence order. It means that in order to halve the error, we must quadruple the numbers of paths. To improve accuracy by a decimal takes a hundred times the number of paths, resulting in a hundred times slower simulation.

The complexity of MC is obviously linear in the number of paths. We will see that it is also linear in the number of time steps on the simulation timeline (number of time steps on the discrete path), the Markov dimension of the model (number of state variables updated on every time step), the dimension of the product (number of path-dependencies updated on event dates), and the dimension of the simulation (number of random variables drawn to simulate a path). Linear complexity in every direction, combined with an accuracy independent of dimension, are specific to Monte-Carlo and the reason why it is applicable in such a vast amount of situations.

This is why MC is by far the most widely used valuation algorithm in modern finance, despite its slow convergence. The complexity of other algorithms, like FDM, grows more than linearly with dimension for a given accuracy, restricting their application to low-dimension contexts.

Path-wise simulations

The Monte-Carlo algorithm is natural, simple to understand, and practically implemented without major difficulty. It is often implemented path-wise, which means that paths are generated and evaluated one by one. An alternative implementation is step-wise, where all paths are simulated simultaneously, one time step at a time. A step-wise implementation has a number of benefits, including:

- Offers the opportunity to apply SIMD to vectorize calculations across paths; see page 9.
- Facilitates the computation of the empirical distribution of the observations in the scenario, and the modification of the paths to match theoretical distributions, a technique known as *control variate*. This improves the ability of a limited number of paths to accurately represent the distribution of the scenario, improving the accuracy of results.

  In addition, we have seen that the probability distribution of an observation s_k on the event date T_k is equivalent to the prices of European options of all strikes with maturity T_k. It follows that a special application of control variate in a step-wise Monte-Carlo simulation is to calibrate a model to European options prices, either in a self-contained manner or in combination with the valuation of an exotic product. An application to extensions of Dupire's model is known as the *particle method* of [9].
- Permits the application of the *Latin hypercube* variance reduction optimization: to simulate the successive time steps for all paths, reuse the same set of random numbers, but in a different order. We will review the Latin hypercube method later in this chapter in the context of Sobol's quasi-random number generator.

Step-wise simulations also cause difficulties with path-dependent products, and prevent path-wise differentiation in the context of AAD. For these reasons, we only consider path-wise simulations here. Note that this doesn't prevent pre-simulating paths in a step-wise manner, storing them in memory, and picking them there, path by path, instead of drawing new random paths, in a path-wise simulation.

Application to pricing

Discrete path-dependence It follows from linear complexity and dimension-independent accuracy that MC can be applied to price a wide variety of products, in a vast number of models, in reasonable time. It is also practically applicable in a large number of contexts. In particular, a path-wise simulation naturally handles path-dependent products: the entire path is simulated before the evaluation of the payoff, so all the information necessary to deal with path-dependencies is available when the payoff is computed.

There are two features in a financial product that the Monte-Carlo algorithm is not equipped to process naturally: *continuous* path-dependence and *callable* features. In both cases, a prior transformation step may turn these features into (discrete) path-dependence before the application of the Monte-Carlo algorithm. We don't implement or discuss these steps in detail, but we introduce them quickly for the sake of completion and refer to Monte-Carlo literature for ample details.

Continuous path-dependence Monte-Carlo works with discrete scenarios, where the timeline TL is a finite set of event dates, so it cannot naturally handle continuous path-dependence. A classic example is the continuous barrier option, which dies if the barrier is breached *at any time* before expiry. The Monte-Carlo algorithm has no knowledge of what happens between event dates, so it cannot monitor the barrier continuously.

One solution is approximate the continuous barrier by one where the barrier is only active on a discrete set of event dates and remains unmonitored in between. This is called a *discretely monitored barrier*. It can be demonstrated (with the Brownian Bridge formula) that the value of the discretely monitored barrier converges to the continuous barrier when the time mesh ΔT between monitoring dates shrinks to 0, with error proportional to $\sqrt{\Delta T}$. This is therefore *not* a good approximation. The complexity of the simulation is inversely proportional to ΔT, so to reduce ΔT in the hope of approximating the continuous barrier may slow the computation down to an unacceptable extent.

A better solution is to find a discretely monitored barrier "equivalent" to a given continuous barrier. It should be intuitive that for a continuous up-and-out barrier B and a monitoring period ΔT, there exists an equivalent discrete barrier on a *lower* barrier $B - \epsilon$ such that the two have the same price in a reasonably sizeable region below $B - \epsilon$. A practical approximation was derived by [64], whereby ϵ is around 0.583 standard deviation on the barrier in between monitoring dates. We show an example in the next chapter. The approximation is very practical but unfortunately not precise enough to benchmark Monte-Carlo estimates.

Another solution is to deflate the payoffs of the paths where the barrier was not breached by the probability that the barrier was hit in between monitoring dates. This probability is given by the Brownian Bridge formula, under the approximation that the barrier index is Gaussian in between time steps,[2] and with an estimation of its volatility around the barrier.

These pre-calculations modify the product into an equivalent product[3] with discrete path-dependence, which is naturally handled with a subsequent Monte-Carlo valuation.

We don't implement the pre-processing of continuous path-dependent products in this publication. We only consider discrete path-dependence. Our implementation of Chapter 6 does not preclude continuous path-dependence, but it assumes that necessary pre-processing was performed in a prior step, so that our algorithms only deal with discrete path-dependence.

Callable products and xVA One exotic feature MC is not equipped to value is early exercise, as discussed in the previous chapter. With early exercise, cash-flows depend on the future values of the product, and when payoffs are computed, those future values are not available as part of the path. Regulatory calculations like CVA also subject cash-flows to future product values and are therefore vulnerable to the same problem.

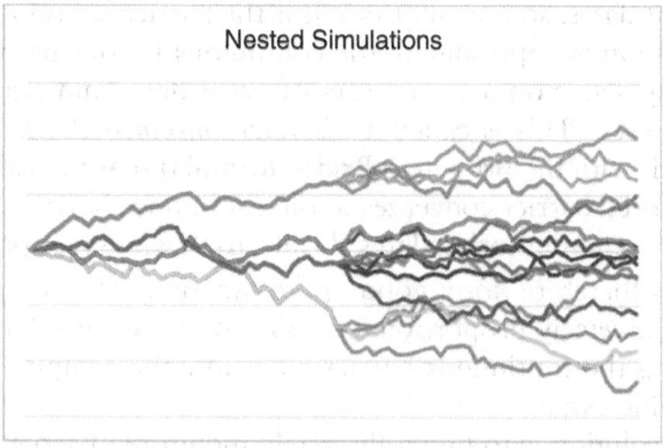

One solution could be to estimate future values with *nested MC simulations*, as illustrated above, from each exercise/exposure date in every path of the outer MC. This approach is obviously extremely expensive and inappropriate in a time-critical environment. The better solution, and the established

[2]An approximation MC also makes for path generation with the *Euler* scheme, as we will see shortly.
[3]At least approximately.

one in the industry, is the least squares method (LSM) algorithm of [14] and [13], illustrated below:

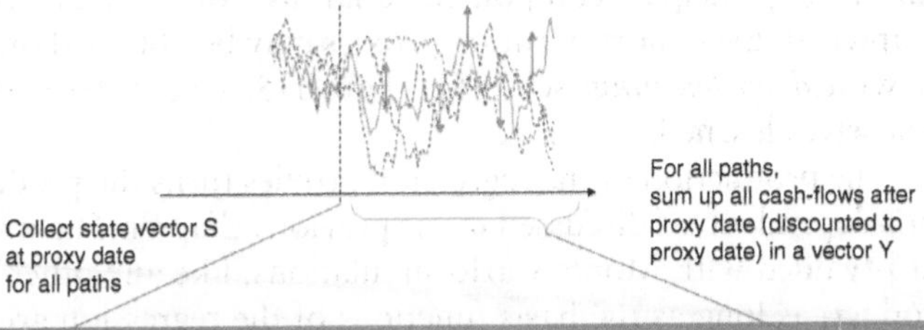

		basis functions : matrix ϕ					
path	state vector S	$\varphi_0 = 1$	φ_1	φ_2	...	φ_M	vector Y
1	S_1	1	$\varphi_1(S_1)$	$\varphi_2(S_1)$	...	$\varphi_M(S_1)$	Y_1
2	S_2	1	$\varphi_1(S_2)$	$\varphi_2(S_2)$	...	$\varphi_M(S_2)$	Y_2
...	...	...	...	...	...	...	...
N	S_N	1	$\varphi_1(S_N)$	$\varphi_2(S_N)$	...	$\varphi_M(S_N)$	Y_N

regression of Y over Φ: $\beta = (\Phi^t\Phi)^{-1}\,\Phi^t Y$
with Tikhonov regularization: $\beta = (\Phi^t\Phi + \lambda I)^{-1}\,\Phi^t Y$
proxy: $\tilde{V}_t = \beta \cdot \varphi(S_t)$

LSM approaches future values by regression over a set of basis functions of the state variables. The regression takes place *before* simulation. It uses simulations, too, which we call *pre-simulations* to distinguish them from the "main" Monte-Carlo valuation. Under these pre-simulations, we generate a *simulated learning set*: for every scenario, we store some basis functions of the state variables on the proxy date (exercise date) in a matrix ϕ, and the (numeraire deflated to the proxy date) sum of the cash flows *paid after the proxy date* in a vector Y. We use the simulated sample to *learn* the parameters of a functional dependence of the value V of the transaction on a proxy date, the conditional expectation of its (deflated) future cash-flows, to a linear combination of basis functions of the state variables, hence, a function of the state vector, by linear regression of Y over ϕ.

In case we have multiple call dates (or proxy dates), the regression must be repeated for every proxy date, where the sum of cash flows Y for a proxy date depends on the exercise strategy after that date, hence, on the proxies for the future proxy dates. Therefore, regressions must be performed recursively in reverse chronological order.

The linear regressions may be performed with the well-known *normal equation*:

$$\beta = (\Phi^t\Phi)^{-1}\Phi^t Y$$

although this is vulnerable to co-linearity and over-fitting, see [65], Chapter 3. For this reason, best practice applies more sophisticated regressions based on singular value decomposition (SVD, see [20], Chapters 2 and 15), principal component analysis, or *regularization* (see [65], Chapter 3). Even more accurate proxies may be obtained out of the learning set with *deep learning*; see [65], chapter 5, where the basis functions are themselves learned.

The production of the regression proxies turns the product into a classic path-dependent, as discussed in the previous chapter. The transformed product is valued with Monte-Carlo simulations, like any other path-dependent product, as long as the basis functions of the regression are included in the samples for their proxy date, so the proxies can be computed in the simulations.

LSM was initially designed to value callable exotics with Monte-Carlo, but it also became best practice for regulatory calculations after 2008–2011.

We do not cover LSM in this publication further than this brief introduction. We refer to the original papers of [14] and [13], chapter 8 of Glasserman [62] for details, and [31] for the modern implementation, application to xVA, and differentiation. Here, we only deal with (discrete) path-dependence, assuming that, if the product did have early exercise features, they were turned into path dependencies with a prior LSM step.

5.2 SIMULATION OF DYNAMIC MODELS

With the clarifications of the previous section, all that remains is to specify how exactly we "draw" N paths $\{\Omega_i, 1 \leq i \leq N\}$ from the distribution of the scenario Ω specified in the model. The path Ω_i is a collection S_j^i of samples of observations on a discrete set of event dates $\{T_j, 1 \leq j \leq J\}$. The definition of the timeline and of the samples on its event dates is part of the specification of the product. The responsibility to draw the paths belongs with the model, which specifies the probability distribution of Ω. We note, once again, the separation of responsibilities between the model and the product. It is this separation that allows the design of a generic, versatile simulation libraries, where models and products are *loosely coupled* and only communicate through simulated paths. The responsibility of the model is to simulate the paths in accordance with the definition of the scenario in the product. The product consumes a simulated path to compute a payoff. This neat separation is somewhat muddied in the context of continuously path-dependent or callable products, although it may be reestablished by a pre-processing step, which allows to implement the simulation library itself with a clean separation. We will see in the next chapter how exactly this is implemented in code.

Drawing samples from a distribution

To draw a random number from a probability distribution F, we can draw a *uniform* random number U in $(0, 1)$ and apply the transformation $X = F^{-1}(U)$. It is immediate that X follows the distribution F, since:

$$Pr(X \leq x) = Pr(U \leq F(x)) = F(x)$$

We can only do this when the distribution F is known. We can only do this *efficiently* when F^{-1} (or a good approximation of it) is known *and cheap to evaluate*.

To draw a standard Gaussian number, we apply the inverse cumulative Gaussian distribution N^{-1} to a uniform number. The function N^{-1} doesn't have a known analytic expression, but efficient approximations have been derived, like the one from Beasley, Springer, and Moro [66], widely implemented in financial systems, and which code is found in gaussians.h in our repository.[4] We can draw a Gaussian number with mean μ and variance v with a trivial transformation of a standard Gaussian number, as it is immediately visible that:

$$X = \mu + \sqrt{V} N^{-1}(U)$$

follows the desired distribution.

To draw a number from a chi-square distribution C_D with D degrees of freedom is more difficult. There exists, to our knowledge, no efficient, accurate approximation of the inverse chi-square distribution C_D^{-1}. It is generally best to draw D independent Gaussians and sum their squares. What is apparent in this example, where we need D independent uniform numbers to draw *one* chi-square distributed number, is that the dimension of the uniform sample does not always correspond to the dimension of the drawn distribution. This consideration affects the design of generic simulation libraries.

In general, it takes $D \geq 1$ independent uniform numbers (what we call a uniform *vector* of dimension D; we don't always ensure that the components are independent, but this is always assumed) to draw one random number from a given distribution F, the most efficient transformation of the D uniforms into one number with distribution F depending on the nature of the distribution. It follows that, to draw N independent random numbers from

[4]Moro's approximation is accurate and reasonably fast, but it contains control flow, and therefore, it cannot be vectorized; see our introduction to SIMD on page 9. Most known alternatives, like Acklam's interpolation cited by Jaeckel [63], also contain control flow. Intel's free IPP library offers a vectorized alternative *ippsErfcInv_64f_A26()* but this is not standard, portable C++, and therefore out of our scope.

some distribution F, we draw N independent uniform vectors U_i, each of dimension D, and apply the transformation $X_i = t_F(U_i)$ specific to F.

To draw a random number from a *multidimensional* distribution F_n, we follow the same steps. Depending on the nature of the distribution, an efficient transformation *may* exist that turns a uniform vector of dimension D into a vector of dimension n following the distribution F_n.

To draw a multivariate Gaussian distribution in dimension n, of mean μ and covariance matrix C, we note that there exists a squared matrix c of dimension n such that $cc^t = c^t c = C$. Denote N_n a vector of n independent standard Gaussian variables. Then:

$$\tilde{N}_n = \mu + cN_n$$

is a Gaussian vector with mean μ and covariance matrix C. A "square root" c of C can be found, for instance, by Cholesky's decomposition, of complexity cubic in n, covered in chapter 2 of *Numerical Recipes* [20] with code. It follows that it is expensive to draw one sample from a multidimensional Gaussian distribution, but it cheap to draw a large number N of independent, identically distributed samples, because the expensive Cholesky decomposition is performed only once. Also note that it takes n independent uniforms to draw an n-dimensional Gaussian; hence, in this particular case, the two dimensions coincide.

To draw a *path* in a model defined by SDE:

$$dX_t = \mu(X_t, t)dt + \sigma(X_t, t)dW$$

(where W is a standard m-dimensional Brownian motion with independent components and the state vector X is in the Markov dimension n) and the state to market mapping:

$$\zeta : \{X_t, 0 \leq t \leq T\} \rightarrow \Omega = \{S_j, 1 \leq j \leq J\}$$

also follows the same steps. Draw a uniform vector U in dimension D and apply a model-specific transformation to turn it into a path $\{S_j, 1 \leq j \leq J\}$. The dimension D of the uniform vector necessary to draw one path is called the *dimension* of the simulation.[5]

The transformation of U into $\{S_j, 1 \leq j \leq J\}$ is called a *simulation scheme*. Simulation schemes are model specific, and in particular, the means

[5]Not to be confused with the dimension of the scenario, the dimension of the samples, the Markov dimension of the model, or the number of its factors, all of those being different notions!

to efficiently turn U into a realization of Ω are dependent on the model, although general-purpose simulation schemes, applicable to many different models, exist in literature. We discuss the most common ones below.

Simulation schemes

A simulation scheme is a transformation that produces a path in a model out of a uniform vector. We discuss uniform number generators in the next section. In this section, we are given a uniform vector U, and we investigate its transformation into a path.

What complicates the simulation of paths is that dynamic models don't explicitly specify the multidimensional distribution of the observations in the scenario, but the SDE of a state vector X, and a mapping ζ that transforms a path of X into a collection of market samples S_j on the event dates $\{T_j, 1 \leq j \leq J\}$, and which may or may not be time-wise, as previously explained. It follows that simulation schemes involve two steps:[6]

1. Generate a path of X over a collection of time steps called *simulation timeline*, consuming the uniform vector U. The simulation timeline doesn't necessarily correspond to the collection of event dates in the product. Some simulation schemes, like Euler's (presented shortly in detail), are inaccurate over big steps and require additional time steps to accurately simulate X.
2. Apply the model specific state to market mapping ζ to turn the path of X over the simulation timeline into a collection of samples over the event dates. The transformation is conducted by iteration over the event dates T_j, and consumes either the state X_{T_j} in the case of a time-wise mapping or the path $\{X_t, 0 < t < T_j\}$ otherwise.

The number of uniform numbers required to generate a path of X, the dimension D of the simulation, depends on the simulation scheme.

Transition probability sampling

The simplest simulation scheme is the one that applies the transition probabilities $F(t_1, t_2, x)$ of the SDE of X, the distribution of X_{t_2} conditional to X_{t_1}:

$$F(t_1, t_2, x) = Pr(X_{t_2} < x | X_{t_1})$$

[6]The two steps are often conducted simultaneously in practice, but it helps to separate them for the purpose of the presentation.

with $t_1 < t_2$. X_0 is known, and X_{i+1} is sampled from $F(T_i, T_{i+1}, X_i)$ after X_i, sequentially building a path over the simulation timeline $\{T_0 = 0, T_1, ..., T_M\}$. Despite its simplicity, the transitional scheme is extremely efficient, because it draws increments from the transition distributions over big steps without loss of accuracy. This scheme can accurately simulate X over the event dates without the need for additional simulation time steps. It follows that the simulation timeline coincides with the product's timeline, and the simulation is conducted with the minimum possible number of time steps. Since its complexity is evidently linear in the number of time steps, the result is a particularly efficient simulation:

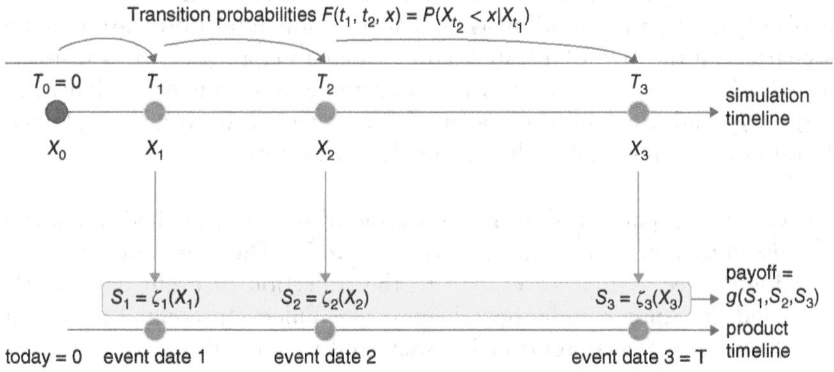

Of course, this scheme can only be applied with models where the transition probabilities are known and explicit. This not the case, for example, in Dupire's model, where volatility changes with spot and time in accordance with an arbitrary specification of the local volatility structure, and the resulting transition probabilities are unknown.

In addition, the transitional scheme can only be applied *efficiently* when the transition probabilities are cheap to sample. Gaussian sampling is fast and efficient, but other distributions may be expensive to sample and defeat the benefits of the scheme. For instance, in the constant elasticity volatility (CEV) extension of Black and Scholes:

$$ds_t = \sigma {s_t}^{\beta} dW$$

it was demonstrated in [67] that transition probabilities are displaced chi-square distributions with known parameters. To draw samples from a chi-square distribution is expensive, to the point where different simulation schemes, like Euler's, generally perform better, despite the additional time steps required on the simulation timeline.

In practice, the transitional scheme is mostly restricted to models with Gaussian transition probabilities, like Black-Scholes, LGM, or multidimensional extensions thereof. In addition, to simulate X over big steps is only possible with a time-wise mapping ζ. In this case, we can simulate X with big steps over the event dates and apply the mapping on each event date. When the model doesn't provide a time-wise mapping, the sample S_j on an event date T_j is a functional of $\{X_t, 0 < t < T_j\}$, and the path for X cannot be simulated over big steps without loss of accuracy. This being said, as seen with LGM on page 175, it is generally possible to rewrite a model under an appropriate measure so it accepts a time-wise mapping. This is desirable for performance, the alternative being to simulate X with small time steps over a dense simulation timeline, so that the mapping ζ can be applied accurately, at the cost of expensive additional time steps.

A textbook illustration of the transitional scheme is the Black and Scholes model, where the risk-neutral transition probabilities of the spot price are log-normal, and all market variables are deterministic functions of the spot price. We have seen on page 163 that the transition probability distributions in Black and Scholes, under the risk-neutral measure, are:

$$s_{t_2} = s_{t_1} \exp\left[f(t_1, t_2) - \frac{V(t_1, t_2)}{2} + \sqrt{V(t_1, t_2)}N\right]$$

where

$$V(t_1, t_2) = \int_{t_1}^{t_2} \sigma^2(t)dt, \quad f(t_1, t_2) = \int_{t_1}^{t_2} \mu(u)du, \quad \mu(u) = r(u) - y(u)$$

and N is a standard Gaussian variable. In addition, increments are independent, and it follows that a path for the spot $\{s_j\}$ over the event dates $\{T_j, 1 \leq j \leq J\}$, with the correct joint distribution, is produced with the simple recursion:

$$s_{j+1} = s_j \exp\left[f(t_j, t_{j+1}) - \frac{V(t_j, t_{j+1})}{2} + \sqrt{V(t_j, t_{j+1})}N^{-1}(U_{j+1})\right]$$

where s_0 is known, the Us are independent uniform numbers and hence the successive $N^{-1}(U_j)$ are independent, standard Gaussians. Rates and dividend yields are deterministic, and forwards $F(t, T)$ of all maturities T are deterministic, explicit functions of the spot price s_t:

$$F(t, T) = s_t e^{f(t,T)}$$

It follows that market samples on an event date T_j are all deterministic functions of s_{T_j}, therefore, there is no need for additional time steps on the simulation timeline: the joint distribution of the spot is correctly sampled over big steps, and the mapping to market samples on an event date only requires the spot on the event date. It follows that we need J independent uniform numbers to generate a path, the number of event dates in the product. The dimension of the simulation is therefore $D = J$, the lowest possible dimension permitted by the context and resulting in a particularly fast simulation.

The transitional scheme is applicable to Gaussian models like Black and Scholes or LGM, including multidimensional extensions, but with other models, like Dupire or non-Gaussian specifications of Cheyette, transition probabilities are either unknown or expensive to sample, and a different scheme must be invoked, one that does not require an explicit knowledge of transition probabilities. Euler's scheme offers a general-purpose fallback scheme, applicable to all kinds of diffusions, and always consuming Gaussian random numbers. The catch is that its accuracy deteriorates over big steps, so it requires additional time steps on the simulation timeline, with additional linear overhead.

Euler's scheme

Euler's scheme basically replaces the infinitesimal moves "d-something" in the SDE:

$$dX_t = \mu(X_t, t)dt + \sigma(X_t, t)dW$$

by small, but not infinitesimal, moves "Δ-something":

$$\Delta X_t = \mu(X_t, t)\Delta t + \sigma(X_t, t)\Delta W$$

More formally, and knowing that ΔW is a vector of m[7] independent Gaussians with mean 0 and variance Δt, Euler's scheme consists in the following recursion over the simulation timeline:

$$X_{i+1} = X_i + \mu(X_i, T_i)(T_{i+1} - T_i) + \sigma(X_i, T_i)\sqrt{(T_{i+1} - T_i)}N_{i+1}$$

where the N_i are a series of independent Gaussian vectors, each consisting in m independent Gaussian numbers. It follows that we need m random numbers for every step of Euler's scheme, a total simulation dimension

[7]We recall that m is the number of factors in the model, the dimension of the Brownian motion in its SDE.

of $D = Mm$ numbers for an entire path over a simulation timeline of M time steps.

Euler's scheme corresponds to a Gaussian approximation with constant diffusion coefficients between time steps; hence, it becomes inaccurate for large time steps. On the other hand, all diffusions are locally Gaussian, so Euler's scheme does converge.[8]

Therefore, when the event dates are widely spaced, we must insert additional time steps into the simulation timeline for the purpose of an accurate simulation. For instance, if a product has annual fixings, and we force a maximum space of three months on the timeline to limit inaccuracy, we evenly insert three additional simulation steps in between each event date. Those additional time steps are for the purpose of the simulation only. In particular, they don't produce market samples. The product's timeline, the one consumed in the computation of the payoff, is unchanged. It is important to distinguish the product's timeline, which is the collection of *event dates* where market samples are observed for the computation of the payoff, and depend on the product, and the simulation timeline, which is the collection of *time steps*, inserted for the purpose of an accurate simulation of the state vector X, depending on the model and its simulation scheme:

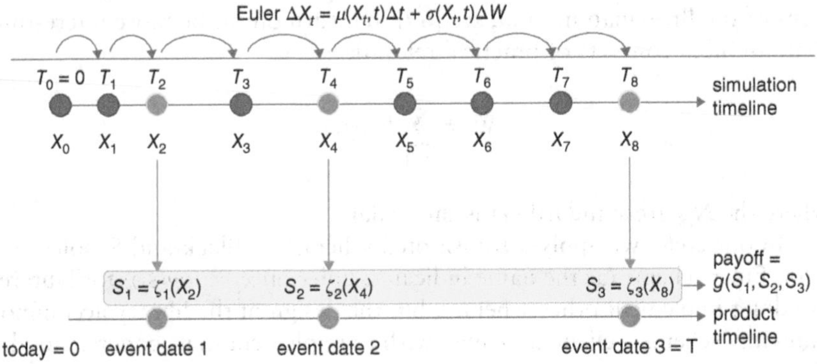

Alternative schemes

Simulation complexity is linear in the number of time steps, so it is desirable to minimize the number of the additional steps in the simulation timeline. To this end, among others, many alternative schemes were designed by researchers and practitioners, to provide a better approximation over big

[8]Its weak convergence is in ΔT, which means that in the limit of an infinite number of paths, where the only error comes from the discrete simulation timeline, that error is asymptotically proportional to ΔT.

steps in certain situations. We cite the most interesting ones, referring to [62], [63], and the original publications for details.

Euler's scheme is said to be of order one because it converges in ΔT. It was extended by Milstein to an order-two scheme, theoretically allowing longer time steps, but with smoothness constraints on the parameters of the SDE.

Andreasen and Huge's "random grids" [68] offer an original solution for the simulation of a wide class of local and stochastic volatility models, over big steps, with a scheme that combines the benefits of Monte-Carlo and FDM, but which implementation requires significant effort.

Special schemes for stochastic volatility models a la Heston were proposed by Andersen and Piterbarg in the Heston chapter of [6], and Andreasen in [8]. Heston models a stochastic variance, which cannot be correctly simulated over Euler's scheme: Euler's scheme is Gaussian in between time steps; therefore it allows negative samples over the discrete timeline even when the continuous time diffusion prohibits them. A variance obviously cannot be negative, so Euler's scheme is not suitable in this case. Heston, and stochastic volatility models in general, require specialized schemes.

As a curiosity, so-called *spectral* schemes produce *continuous* paths with a finite number of random numbers by sampling the harmonic basis functions of the Brownian motion, see [69], but turn out to be more interesting than useful in contexts of practical relevance:

$$W_t = \sum_{i=1}^{D} N_i \eta_i(t)$$

where the N_is are standard Gaussian scalars.

In our code, we apply a transitional scheme for Black and Scholes and a log-Euler scheme (as the name indicates, Euler on $x_t = log(s_t)$) for Dupire. We don't implement other schemes, but the design of the library accommodates any scheme a client code may wish to implement in its concrete model classes.

5.3 RANDOM NUMBERS

Simulation schemes consume a number D of uniform random numbers to generate a path. We complete our presentation of simulation algorithms with a discussion of random numbers. A program that produces (pseudo) random numbers is called a *random number generator* (RNG). Many programming languages like C++ offer RNGs in their standard libraries. The implementation is platform and compiler specific. Their statistical correctness is unknown: how well they span the uniform space $(0, 1)$ and how independent are the successive draws. Their implementation is hidden, which is a problem for a parallel application, as we will see in the next chapter.

For these reasons, a professional simulation library cannot rely on the standard library RNGs and must include a custom implementation of generators purposely designed for simulations with proven statistical correctness and convergence properties. The current industry standard is the combined recursive generator *mrg32k3a*, designed by L'Ecuyer in 1996 [70].

The design of random number generators is a discipline of its own. Interested readers will find a wealth of information on L'Ecuyer's page in the University of Montreal.[9] In particular, his synthetic note[10] offers a clear and useful overview in just 40 pages. L'Ecuyer also introduces efficient "skip-ahead" techniques introduced in the next chapter, and provides the C++ code for a number of generators.

We don't cover the design of random number generators, or investigate their statistical properties, which are discussed in deep detail in L'Ecuyer's documents. What we do is implement mrg32k3a. The algorithm is the following:

1. Initialize the first three values in two series of state variables X and Y with $X_{-2} = X_{-1} = X_0 = \alpha$ and $Y_{-2} = Y_{-1} = Y_0 = \beta$.

   The pair (α, β) is called *seed*. The RNG produces the exact same sequence of numbers when seeded identically. The seed must satisfy $0 < \alpha < m_1, 0 < \beta < m_2$ where $m_1 = 2^{32} - 209$ and $m_2 = 2^{32} - 22853$.
2. Successive values of X and Y are produced with the following recurrence:

$$X_k = {a_1}^t \begin{pmatrix} X_{k-1} \\ X_{k-2} \\ X_{k-3} \end{pmatrix} \bmod m_1, \; Y_k = {a_2}^t \begin{pmatrix} Y_{k-1} \\ Y_{k-2} \\ Y_{k-3} \end{pmatrix} \bmod m_2$$

   with:

$$a_1 = \begin{pmatrix} 0 \\ 1403580 \\ -810728 \end{pmatrix}, a_2 = \begin{pmatrix} 527612 \\ 0 \\ -1370589 \end{pmatrix}$$

3. The kth random number is

$$u_k = \frac{(X_k - Y_k) \bmod m_1}{m_1}$$

[9] https://www.iro.umontreal.ca/~lecuyer/.

[10] https://www.iro.umontreal.ca/~lecuyer/myftp/papers/handstat.pdf.

5.4 BETTER RANDOM NUMBERS

Variance reduction methods and antithetic sampling

Monte-Carlo is easy to understand and implement, and adequate in virtually any valuation context, but it is expensive in CPU time and its convergence in the number of paths is slow. A vast amount of research, collectively known as *variance reduction methods*, conducted in the past decades, attempted to accelerate its convergence.

Perhaps the simplest variance reduction technique is *antithetic sampling*. For every path generated with the uniform vector $U = (U_1, ..., U_D)$, generate another one with its antithetic:

$$(1 - U_1, ..., 1 - U_D)$$

Not only does antithetic sampling balance the paths with a desirable symmetry, it also generates two paths for the cost of one random vector.

When the scheme consumes Gaussian numbers, antithetic sampling is even more efficient: for every path generated with the Gaussian vector G, generate another one with $-G$, guaranteeing that the empirical mean of all components in the Gaussian vector is zero. This is a simple case of a family of variance reduction methods called *control variates*, where the simulation enforces that the empirical values of sample statistics such as means, standard deviations, or correlations match theoretical targets. In addition, we get two paths for the cost of one Gaussian vector, saving not only the generation of the uniforms, but also the not-so-cheap Gaussian transformations.

Antithetic sampling is a simple notion and translates into equally simple code. We implement antithetic sampling with mrg32k3a in the next chapter. We will mention other variance reduction methods, but we don't discuss them in detail, referring readers to chapter 4 of [62]. What we *do* cover is a special breed of "random" numbers particularly well suited to Monte-Carlo simulations.

Monte-Carlo as a numerical integration

We reiterate the sequence of steps involved in a Monte-Carlo valuation.[11] We generate and evaluate N paths, each path number i following the steps:

1. Pick a uniform random vector U_i in the *hypercube* in dimension D:

$$U_i = (u_{ik}) \in (0, 1)^D$$

[11]Assuming the simulation scheme consumes Gaussian numbers, as is almost always the case.

2. Turn U_i into a Gaussian vector with the application of the inverse cumulative Gaussian distribution N^{-1} to its components:

$$N_i = (n_{ik}) = G(U_i) = (N^{-1}(u_{ik}))$$

3. Feed the Gaussian vector N_i to the simulation scheme to produce a path X^i for the model's state vector X over the simulation timeline:

$$X^i = (X^i_{T_j})_{0 \le j \le M} = v(N_i)$$

4. Use the model's mapping to turn the path X^i into a collection of samples S^i over the event dates:

$$S^i = (S^i_{T_j})_{1 \le j \le J} = \zeta(X^i)$$

5. Compute the product's payoff g_i on this path:

$$g_i = g(S^i)$$

Hence, g_i is obtained from U_i through a series of successive transformations. Therefore:

$$g_i = h(U_i)$$

where $h \equiv g \circ \zeta \circ v \circ G$:

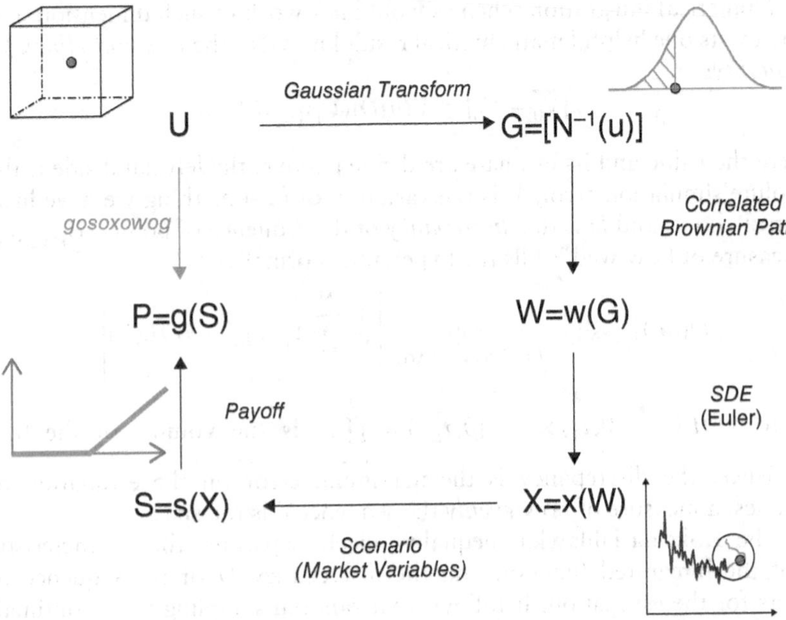

The MC estimate of the value is:

$$\overline{V_0} = \frac{1}{N}\sum_{i=1}^{N} g_i = \frac{1}{N}\sum_{i=1}^{N} h(U_i)$$

The *theoretical* value (modulo discrete time) is:

$$V_0 = E[h(U)] = \int_{(0,1)^D} h(u)du$$

Hence, Monte-Carlo is nothing more, nothing less, than the numerical approximation of the integral of a (complicated) function $h : (0,1)^D \to \mathbb{R}$ with a sequence $(U_i)_{1 \le i \le N}$ of random samples drawn in the hypercube.

Koksma-Hhlawka inequality and low-discrepancy sequences

A random sequence of points is not necessarily the optimal choice for numerical integration. Standard numerical integration schemes, covered, for example, in [20], work in low dimension. Dimension is typically high in finance. With Euler's scheme, it is the number of time steps times the number of factors. For a simple weekly monitored 3y barrier in Black and Scholes or Dupire's model, the dimension is 156. Typical dimension for exotics is in the hundreds. For xVA and other regulations, it its generally in the thousands or tens of thousands, due to the large number of factors required to accurately simulate the large number of market variables affecting a netting set.

Numerical integration schemes would not work in such dimension, but there exists one helpful mathematical result known as the *Koksma-Hhlawka inequality*:

$$|\overline{V_0} - V_0| \le V(h)D[(U_i)_{1 \le i \le N}]$$

where the value and its estimate are defined above, the left-hand side is the absolute simulation error, V is the variation of h, something we have little control about, and D is the *discrepancy* of the sequence of points $(U_i)_{1 \le i \le N}$, a measure of how well it fills the hypercube. Formally:

$$D[(u_i)_{1 \le i \le N}] = \sup_{E=[0,t_1)\times\ldots\times[0,t_D)} \left| \frac{1}{N}\sum_{i=1}^{N} 1_{\{u_i \in E\}} - VOL(E) \right|$$

where $VOL\{E = [0,t_1)\times \ldots \times [0,t_D)\} = \prod_{k=1}^{D} t_k$ is the volume of the box E. Hence, the discrepancy is the maximum error on the estimation of volumes, a measure of how *evenly* the sequence fills the space.

The Koksma-Hhlawka inequality nicely separates the characteristic V of the integrated function and the discrepancy D of the sequence of points for the estimation. It follows that random sampling is *not* optimal. Random sampling may cause clusters and holes in the hypercube, resulting in poor evenness and high discrepancy. It is best to use sequences of points purposely designed to minimize discrepancy. A number of such sequences were invented by Van Der Corput, Halton, or Faure, but the most successful one, by far, was designed by Sobol in 1967 USSR [71].

The following picture provides a visual intuition of the discrepancy of Sobol's sequence. It compares 512 points drawn randomly in $(0, 1)^2$ on the left to the 512 first Sobol points in dimension two on the right:

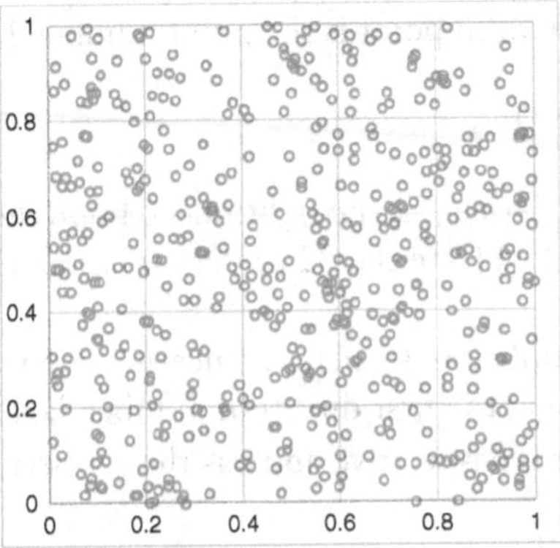

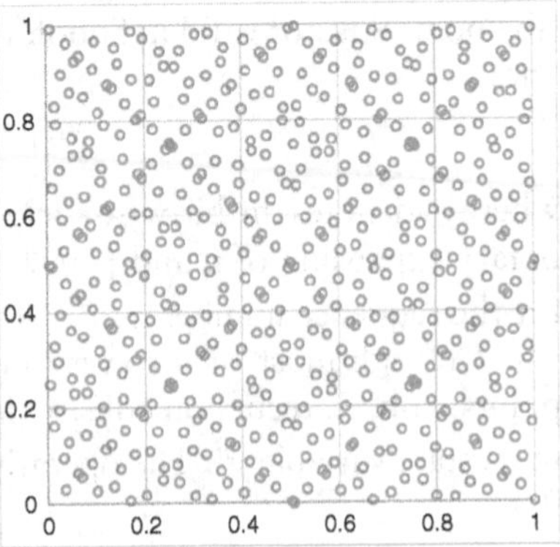

It is clear to the naked eye that Sobol's sequence effectively avoids the holes and clusters inevitable with a random sequence and fills the hypercube with a better "evenness." The mathematical notion of discrepancy formalizes the notion of "evenness," and the Koksma-Hhlawka inequality demonstrates the intuition that better evenness results in a more accurate integration.

Sobol's sequence

The construction speed of Sobol's sequence was massively improved in 1979 by Antonov and Saleev, whose algorithm generates successive Sobol points extremely fast, with just a few low-level bit-wise operations. It is this implementation that is typically presented in literature, including here. Over the past 20 years, Jaeckel [63] and Joe and Kuo [72], [73] performed considerable work on Sobol's *direction numbers*[12] so that the sequence could be practically applied in the very high dimension familiar to finance, achieving remarkable results. Sobol's sequence (with Antonov and Saleev's optimization and Jaeckel or Joe and Kuo's direction numbers) became a best practice in financial applications, which it remains to this day.

Sobol's sequence is not really a sequence of points in the hypercube $(0, 1)^D$, but a collection of D sequences of numbers in $(0, 1)$. The sequences of scalar numbers (x_i^d) on each axis d of the sequence is self-contained, and the first D coordinates of a sequence in dimension $D' > D$ exactly correspond to the sequence in dimension D.

[12] Explained shortly.

Sobol generates sequences of *integers* (y_i^d) between 0 and $2^{32} - 1$, so the ith number on the axis d is:

$$x_i^d = \frac{y_i^d}{2^{32}} \in (0, 1)$$

The integers (y_i^d) on a given axis d are produced by recursion: $y_0^d = 0$ (the point 0 is not valid in Sobol; the first valid point is the point number 1) and:

$$y_{i+1}^d = y_i^d \oplus DN_{J_i}^d$$

where $\oplus$ denotes the bit-wise exclusive or (xor), J_i is the rightmost 0 bit in the binary expansion of i, and $\{DN_j^d, 0 \leq j \leq 31\}$ are the 32 *direction numbers* for the sequence number d.

The rightmost bit of every even number is 0; hence, once every two points, Sobol's recursion consists in xor-ing the first direction number DN_0^d to its state variable y^d. The operation xor is associative and has the property that:

$$x \oplus x = 0$$

so DN_0^d flicks in and out of the state y^d every two points. For the same reason, DN_1^d flicks in and out every four points, DN_2^d flicks in and out every eight points, and, more generally, DN_k^d (with $0 \leq k \leq 31$) flicks in and out every 2^{k+1} points. Sobol's sequence is illustrated below:

i	DN0	DN1	DN2	DN3	y
0	0	0	0	0	0
1	1	0	0	0	DN0
2	1	1	0	0	DN0 + DN1
3	0	1	0	0	DN1
4	0	1	1	0	DN1 + DN2
5	1	1	1	0	DN0 + DN1 + DN2
6	1	0	1	0	DN0 + DN2
7	0	0	1	0	DN2
8	0	0	1	1	DN2 + DN3
9	1	0	1	1	DN0 + DN2 + DN3
10	1	1	1	1	DN0 + DN1 + DN2 + DN3
11	0	1	1	1	DN1 + DN2 + DN3
12	0	1	0	1	DN1 + DN3
13	1	1	0	1	DN0 + DN1 + DN3
14	1	0	0	1	DN0 + DN3
15	0	0	0	1	DN3

More generally, for $i < 2^k$, the state y_i^d is a combination of the k first direction numbers:

$$y_i^d = \sum_{j=0}^{k-1} a_j^i DN_j^d$$

where the weights a_j^i are either zero or 1: a_0 flicks every 2 points, a_1 every 4 points, a_2 every 8 points, a_j every 2^{j+1} points. It follows, importantly, that the 2^k first points in the sequence $y_i^d, 0 \leq i < 2^k$ span *all the* 2^k *possible combinations* of the k first direction numbers, each combination being represented exactly once.

It follows that the Sobol scalar x_i^d in (0, 1) is:

$$x_i^d = \frac{y_i^d}{2^{32}} = \sum_{j=0}^{k-1} a_j \frac{DN_j^d}{2^{32}} = \sum_{j=0}^{k-1} a_j D_j^d$$

where the $D_j^d \equiv \frac{DN_j^d}{2^{32}}$ are the normalized direction numbers.

Direction numbers on the first axis

On the first axis $d = 0$, the direction numbers are simply:

$$DN_k^0 = 2^{31-k}$$

and it follows that the normalized numbers are:

$$D_k^0 = \frac{1}{2^{k+1}}$$

so the sequence of the 32 normalized direction numbers D_k^0 is $\frac{1}{2}, \frac{1}{4}, \frac{1}{8}, \frac{1}{16}$, etc. Or, bit-wise, the kth direction number DN_k^0 has a bit of 1, k spaces from the right, and 0 everywhere else:

	kth direction number of the first axis							
space	31	30	29	...	31-k	..	1	0
bit	0	0	0	0	1	0	0	0
k	0	1	2	...	k		30	31
DN	2^31	2^30	2^29	...	2^(31-k)	...	2	1
D	1/2	1/4	1/8	...	2^(-k-1)	...	2^(-31)	2^(-32)

0 everywhere on the left — The kth direction number has bit 1 on space k — 0 everywhere on the right

It follows that the sequence of numbers on the first axis is:

$$\begin{aligned}
x_0^0 &= 0 \\
x_1^0 &= \frac{1}{2} \\
x_2^0 &= \frac{1}{2} + \frac{1}{4} = \frac{3}{4} \\
x_3^0 &= \frac{1}{4} \\
x_4^0 &= \frac{1}{4} + \frac{1}{8} = \frac{3}{8} \\
x_5^0 &= \frac{1}{2} + \frac{1}{4} + \frac{1}{8} = \frac{7}{8} \\
&\dots
\end{aligned}$$

or, graphically:

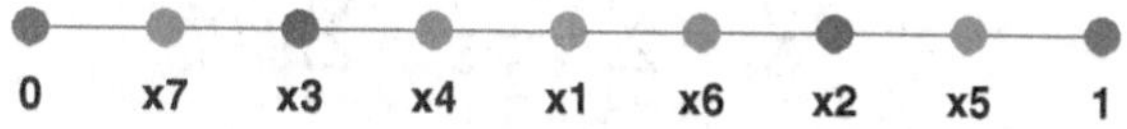

where we see that the normalized direction number $1/2^{k+1}$ does not kick in before all the combinations of the $1/2^j$ for $1 \le j \le k$ were exhausted. It is visible on the chart that x_1 goes in the middle of the axis, x_2 and x_3 go in the middle of the spaces on the left and right of x_1, and the following four numbers go straight in the middle of the spaces left in the previous sequence. The same pattern carries on indefinitely, progressively and incrementally filling the axis in the middle of the spaces left by the previous numbers.

It follows that the first 2^k Sobol scalars on the first axis ($2^k - 1$ excluding the first and invalid scalar $x_0 = 0$) evenly span the axis $(0, 1)$:

$$\{x_j^0, 1 \le j < 2^k\} = \left\{ \frac{j}{2^k}, 1 \le j < 2^k \right\}$$

and we begin to build an intuition for the performance of the sequence: the $2^k - 1$ first "random" numbers are even quantiles of the uniform distribution. When transformed into a Gaussian or another distribution, these numbers sample even quantiles on the target distribution.

Latin hypercube

This property, where the $2^k - 1$ first numbers are evenly spaced by $1/2^k$ on the axis $(0, 1)$, holds for all the axes. *Sobol samples the same numbers on all the axes, but in a different order.*

The direction numbers DN_k^d on all the axes have the following bit-wise property:

	kth direction number of *any* axis							
space	31	30	29	...	31-k	..	1	0
bit	1	0	1	0	1	0	0	0

Bits on the left are 0 or 1 depending on the axis d

The kth bit is 1

All the right bits are 0

The kth bit is still one, so 1/2 still flicks in and out every two numbers, 1/4 every four numbers, 1/8 every eight numbers and so forth. The bits on the right of k are still 0, so 1/16 will not kick in before all the combinations of 1/2, 1/4, and 1/8 are exhausted. It follows that the $2^k - 1$ first numbers are still the quantiles $\{j/2^k, 1 \leq j < 2^k\}$.

But the bits on the left of k are no longer all zero. Some are zero, some are one, depending on the axis, and, crucially, they are different on different axes. When the direction number DN_k^d flicks in or out every 2^{k+1} points in the sequence, it doesn't only flick $1/2^{k+1}$. Its bits on the left of k also flick some $1/2^j$ for $j \leq k$, shuffling their order of flickering.

It follows that the same numbers are sampled on all the axes, but in a different order. In addition, these numbers are even quantiles. The chart below shows the first 15 Sobol points in dimension 2, where it is visible that the points sample the 1/16 quantiles on both axes, but in a different order.

15 first Sobol points in dimension 2

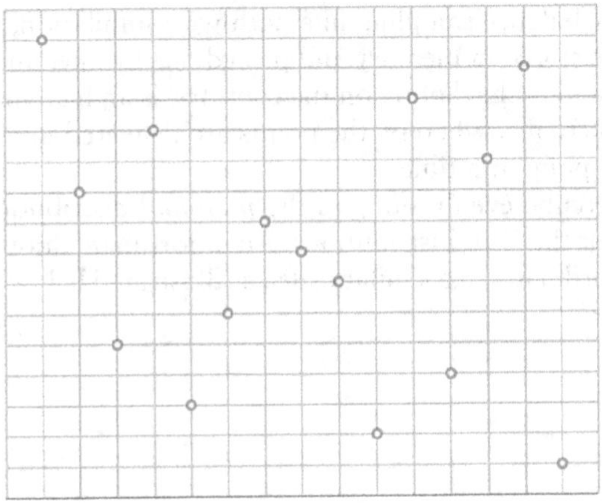

This property where N points x_i sample the hypercube $(0,1)^D$ in such a way that for every coordinate $d \in [0, D-1]$:

$$\{x_i^d, 0 \leq i \leq N-1\} = \left\{\frac{i+1}{N+1}, 0 \leq i \leq N-1\right\}$$

is not unique to Sobol and well known under the name *latin hypercube*. Latin hypercube samples the coordinates in a balanced manner. It is a form of control variate, since the N points correctly sample the $1/(N+1)$th quantiles of the marginal distributions by construction.

Latin hypercube may be implemented directly with a sampling strategy known as *balanced sampling*: to produce N points in the hypercube, sample the first axis with the sequence $x_i^0 = i/(N+1)$ for $1 \leq i \leq N$ and all the other axes by random permutations of the same values. Balanced sampling improves the convergence of Monte-Carlo simulations over random sampling, sometimes noticeably, and its construction is extremely efficient. Instead of randomly shuffling uniform samples before applying the Gaussian transformation, we can randomly shuffle the Gaussian quantiles $G_i^0 = N^{-1}(i/(N+1))$, so that the rather expensive Gaussian transformation is only applied on the first axis N times instead of the usual ND. Balanced sampling may be further optimized, both in speed and quality, in combination with antithetic sampling.

One catch is that balanced sampling is not an *incremental* sampling strategy. We cannot sample N points and then P additional points. All the points are generated together, coordinate by coordinate and not point by point. To apply balance sampling in a path-wise simulation, the N random vectors must be stored in memory and picked one by one, imposing a heavy load on RAM and cache. Sobol, on the contrary, is an incremental variation of the latin hypercube, whereby the points are delivered in a sequence, one D-dimensional point at a time.

Latin hypercube evenly samples the *marginal* distributions, but offers no guarantee for the *joint* distributions. An unfortunate draw may very well sample two coordinates in a similar order, as illustrated below with 15 points in dimension 2:

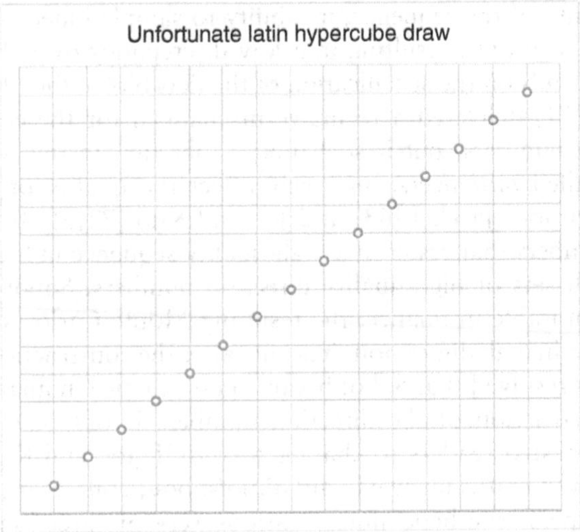

While the marginal distributions are still sampled evenly, it is clear that the empirical *joint* distribution is wrong: the two random numbers, supposedly independent, are 100% dependent in the sample. Seen from the discrepancy perspective, it is visibly poor, the points being clustered around the diagonal, leaving the rest of the space empty. Because the marginal distributions are correctly sampled, a simulation produced with the sample still correctly values all sets of European cash-flows, but the values of exotics would be heavily biased due to the poor sampling of the copula (see Section 4.3).

It therefore appears that, while the latin hypercube property of random numbers is a highly desirable one in the context of Monte-Carlo simulations, it is not in itself sufficient to guarantee a correct representation of joint distributions or a low discrepancy. Additional mechanisms should be in place so that the different axes are sampled in a dissimilar and independent *order*. Sobol's sequence achieves such "order independence" with the definition of its direction numbers.

Direction numbers on all axes

We did not, in our short presentation, specify the direction numbers applied in Sobol's sequence. We did define the 32 direction numbers of the first axis, and introduce a general property of all direction numbers, whereby the kth bit of the kth direction number is always one and the bits on the right of k are always zero. This is what guarantees the latin hypercube property. But the sampling order depends on the bits on the left of k. These bits are zero on the first axis, and specified in a different manner on all other axes, the specification on a given axis determining the sampling order on that axis.

The quality of the sequence, its ability to sample independent uniform numbers on each axis, resulting in a low discrepancy over the hypercube, therefore depends on the specification of the (left bits of the) direction numbers. Sobol [71] delivered a recursive mechanism for the construction of the direction numbers, but it so happens that the starting values for the recursion, called *initializers*, massively affect the quality of the sequence in high dimension. Jaeckel [63] and Joe and Kuo [72], [73] found sets of direction numbers that result in a high-quality sequence in high dimension. Without such sets of high-quality direction numbers, Sobol's sequence is practically unusable in finance, the resulting Monte-Carlo estimates being heavily biased in high dimension. It is only after the construction of direction numbers was resolved that Sobol became best practice in finance.

The construction of the direction numbers is out of our scope. We refer the interested readers to chapter 8 of [63]. Joe and Kuo's dedicated page, http://web.maths.unsw.edu.au/~fkuo/sobol/, collects many resources, including papers, synthetic notes, various sets of direction numbers in dimension up to 21,201, and demonstration code in C++ for the construction of the direction numbers and the points in Sobol's sequence. Our implementation of the next chapter uses their direction numbers in dimension up to 1,111 derived in their 2003 paper [72].

CHAPTER 6

Serial Implementation

In this chapter, we turn the ideas of the previous two chapters into a simulation library in C++. We introduce a library with a generic, modular architecture, capable of accommodating wide varieties of models and products. We also instantiate a small number of models and products, where the calculation code is mainly a translation in C++ of the mathematics in the previous chapters.

Other publications cover the design and development of financial libraries, the best to our knowledge being Joshi's [74] and Schlogl's [75]. Our purpose is not to build a complete financial library, but to introduce the generic design of valuation libraries and establish a baseline for our work on parallelism and differentiation in the following chapters. In particular, we only cover simulation. Calibration is introduced in Chapter 13. We don't implement interest rate markets, model hierarchies, or algorithms other than simulation.

6.1 THE TEMPLATE SIMULATION ALGORITHM

Object Oriented Design (OOD) reigned supreme for decades over the software development community and gave birth to popular programming languages like Java, C#, and C++. OOD lost traction in the past decade in favor of the more intellectually satisfying functional programming; see for instance the excellent [23], and the more efficient meta-programming, see for instance [76].

The main criticisms toward OOD are the run-time overhead of dynamic polymorphism, and the presence of mutable state causing race conditions in multi-threaded programs. While we don't disagree and generally prefer functional code and static polymorphism, we also like OOD architectures in relevant situations. We believe a simulation library is one of them. C++ is a multi-paradigm language, and OOD is one of its many facets. We believe that OOD is ideally suited to a modular architecture where models, products, and random number generators may be mixed and matched at run time.

Products, models, and random number generators are defined and discussed in detail in the previous two chapters. The previous chapter also articulated the detail of the steps involved in a simulation. We reiterate these steps here in software development words.

In the previous chapters, we assumed that the payoff g of a product was a scalar function of the scenario. We extend it here to a vector function, allowing a single product to represent P payoffs. Each payoff is not necessarily a separate cash-flow, but rather a separate aggregate of the cash-flows. This allows us to represent multiple products in one, like a barrier option together with its European counterpart, or a portfolio of many European options of different strikes and maturities. This allows us to evaluate the P payoffs together, in particular, with the same set of simulated scenarios, resulting in a performance improvement. The rest of the algorithm scrupulously follows the previous chapter.

1. Initialization phase
   (a) The product advertises its timeline $(T_j)_{1 \le j \le J}$ and the *definition* of the samples S_js for the event date T_js: the number of market observations on the event date, along with their natures and maturities.
   (b) The model initializes itself to simulate scenarios over the product timeline, knowing what market observations it is required to simulate. Among other initialization steps, the model establishes its own simulation timeline. After initialization, the model knows the dimension of the simulation D, how many random numbers it consumes to generate one path.
   (c) The model advertises D and the RNG initializes itself so as to generate D random numbers repeatedly during the simulation phase.
2. Simulation phase
   Simulate and evaluate N paths. For every path i:
   (a) The RNG generates a new Gaussian vector G in dimension D.[1]
   (b) The model consumes G to simulate the scenario $(S_j)_{1 \le j \le J}$.
   (c) The product runs its payoff function over the path to compute $g_i = g[(S_j)_{1 \le j \le J}]$, which is now a vector in dimension P.
3. Aggregation phase
   The algorithm returns the $N \times P$ matrix of all the g_is so client code may aggregate it to compute values (averages), standard errors (standard deviations over $\sqrt{N}$), quantiles, and so forth.

[1]The vast majority of simulations consume Gaussian numbers, so we encapsulate the inverse cumulative Gaussian transformation in the RNG for convenience. RNGs provide either uniform or Gaussian vectors on demand. Our simulations only use Gaussian numbers. A minor extension to the template code would be required to accommodate non-Gaussian schemes.

This summary is sufficient to design the base classes and the simulation algorithm. The code of this section is in mcBase.h in our repository.

Scenarios

We start with the scenarios, which are the objects that models and products communicate to one another. Scenarios are collections of samples:

```
template <class T>
using Scenario = vector<Sample<T>>;
```

Why the template? The full answer will be given in Part III. The short answer is that we template our code for the number type. We use a template type in place of doubles to represent real numbers. AAD is *almost* noninvasive, in the sense that it doesn't require any modification in the differentiated code, *except for the representation of its numbers*. AAD works with numbers represented with a custom type. It would be messy to code with that custom type in place of doubles. It is a much better practice to template code on the number representation type, so that the code may be instantiated with doubles for evaluation, or with the AAD number type for differentiation.

Our code is therefore templated in the number type, *but not all of it*. Once again, the full answer will be given in Chapter 12. The short answer is that *inactive* code, the code that doesn't depend on the parameters, is best left untemplated so it runs without unnecessary AAD overhead.

A sample is the collection of market observations on an event date for the evaluation of the payoff: the numeraire if the event date is a payment date, and collections of forwards, discounts, and libors, fixed on the event date:

```
template <class T>
struct Sample
{
    T           numeraire;
    vector<T>   forwards;
    vector<T>   discounts;
    vector<T>   libors;
};
```

This sample definition accommodates a wide class of single underlying price, rate, or price/rate hybrid models and products. In order to deal with other model or product types, like credit, the definition of the sample must be extended with additional observations, like credit spreads or credit events. It can also be extended to accommodate multi-underlying and multi-currency models and products with collections of forwards *for every underlying* and collections of rates (discounts and libors) *for every currency*, like this:

```
template <class T>
struct Sample
{
    T                       numeraire;
    vector<vector<T>>       forwards;
    vector<vector<T>>       discounts;
    vector<vector<T>>       libors;
};
```

Although we are not implementing these extensions here, it is worth noting that the range of admissible models and products can be easily extended with a straightforward modification of the definition of samples.

Together with the Sample, which stores the simulated market observations, we have the SampleDef, which specifies the definition of these observations:

```
// SampleDef = definition
//     of what data must be simulated
struct SampleDef
{
    // Need numeraire on this sample?
    //      true for payment dates
    //      false otherwise
    bool            numeraire = true;

    // Maturities of the forwards on this event date
    //      note the size = number of forwards
    vector<Time>    forwardMats;
    // Maturities of the discounts on this event date
    vector<Time>    discountMats;

    // Specification of a Libor rate: start date, end date, curve index
    struct RateDef
    {
        Time    start;
        Time    end;
        string  curve;

        RateDef(const Time s, const Time e, const string& c) :
            start(s), end(e), curve(c) {};
    };

    // Specifications of the Libors on this event date
    vector<RateDef> liborDefs;
};
```

Time is an alias for double, the number of years from today. A SampleDef contains the definition of a Sample, the number and nature of the market observations on a given event date: whether the sample requires a numeraire, the maturities of the forwards and discounts in the sample, and the specification (start date, end date and index) of the Libor observations. The sizes of the vectors forwardsMats, discountMats, and liborDefs also tell how many forwards, discounts, and Libors are simulated on this event date.

It is the responsibility of the product to determine the sample definition, and the responsibility of the model to simulate samples in accordance with their definition.

A Sample is initialized from its SampleDef. We split initialization into two different functions: one that allocates memory, and one that initializes values. Following the discussion of the Section 3.11 of our multi-threaded programming Chapter 3, we always separate allocation and initialization. Memory must be allocated initially, before simulations start and certainly before the execution of concurrent code, whereas, in some circumstances, it may be necessary to conduct initialization during the simulation phase.

```
template <class T>
struct Sample
{
    //  Data on the sample
    T            numeraire;
    vector<T>    forwards;
    vector<T>    discounts;
    vector<T>    libors;

    //  Allocate given corresponding SampleDef
    void allocate(const SampleDef& data)
    {
        forwards.resize(data.forwardMats.size());
        discounts.resize(data.discountMats.size());
        libors.resize(data.liborDefs.size());
    }

    //  Initialize defaults
    //  (unnecessary but convenient)
    void initialize()
    {
        numeraire = T(1.0);
        fill(forwards.begin(), forwards.end(), T(100.0));
        fill(discounts.begin(), discounts.end(), T(1.0));
        fill(libors.begin(), libors.end(), T(0.0));
    }
};
```

To give default values is not necessary but may be convenient. The standard library algorithm *fill*() is self-explanatory. Finally, we provide free functions to batch allocate and initialize a collection of samples:

```
template <class T>
inline void allocatePath(const vector<SampleDef>& defline, Scenario<T>& path)
{
    path.resize(defline.size());
    for (size_t i = 0; i < defline.size(); ++i)
    {
        path[i].allocate(defline[i]);
    }
}

```

```
template <class T>
inline void initializePath(Scenario<T>& path)
{
    for (auto& scen : path) scen.initialize();
}
```

The code is final and will not be modified in the rest of the publication.

Base classes

From the description of the steps involved in a simulation, we know exactly what is the responsibility of models, products, and RNGs. This is enough information to develop abstract base classes:

```
template <class T>
class Product
{
public:

    //  Access to the product timeline (event dates)
    //      along with the sample definitions (defline)
    virtual const vector<Time>& timeline() const = 0;
    virtual const vector<SampleDef>& defline() const = 0;

    //  Labels of all the payoffs in the product
    //      note vector size = number of payoffs in the product
    virtual const vector<string>& payoffLabels() const = 0;

    //  Compute payoffs given a path (on the product timeline)
    virtual void payoffs(
        //  path, one entry per event date (on the product timeline)
        const Scenario<T>&          path,
        //  pre-allocated space for resulting payoffs
        vector<T>&                  payoffs)
            const = 0;

    //  Virtual copy constructor
    virtual unique_ptr<Product<T>> clone() const = 0;

    //  Virtual destructor
    virtual ~Product() {}
};
```

The base Product class defines the product's *contract*, and directly follows a product's responsibilities in the sequential steps of a simulation: a product knows its timeline, and the definition of the samples on the event dates. It advertises this information with its accessors *timeline*() and *defline*(). We commented earlier that a product could encompass a number P of payoffs. Its accessor *payoffLabels*() advertises the names of the P payoffs, so client code knows which is which. Note that the accessors return information by const reference. This way, a product may store the information and make it available to client code, instead of having to regenerate it every time on demand.

The const method *payoffs*() is the main responsibility of a product in a simulation: compute the P payoffs from a given scenario Ω, what we denoted $g(\Omega)$ in the previous chapters. The P payoffs are returned in a preallocated vector to minimize allocations.

Finally, *clone*() implements the *virtual copy constructor* idiom. When an object is manipulated with a pointer or reference on its base class, C++ does not provide any means to make copies of the object and obtain *clones*, new objects *of the same concrete type*, with a copy of all the data members. Generic code, which manipulates objects with polymorphic base class pointers, cannot make copies, which substantially restricts its possibilities. In the absence of a native language solution, the development community worked out the virtual copy constructor solution, now a well-established idiom: a pure virtual method *clone*(), which contract is to return a copy of the invoking object through a (smart) pointer on its base class. The actual copy happens in the derived classes, where concrete types are known, and *clone*() is overridden to invoke the copy constructor of the concrete class. We will further clarify this point when we develop concrete products (and models and RNGs, where this idiom is also implemented).

The product class is final, but the model base class will need a minor upgrade for AAD in Part III. It is written in a similar manner:

```
template <class T>
class Model
{
public:

    //  Initialize with product timeline and defline
    //      separate allocation and initialization
    virtual void allocate(
        const vector<Time>&         prdTimeline,
        const vector<SampleDef>&    prdDefline)
        = 0;
    virtual void init(
        const vector<Time>&         prdTimeline,
        const vector<SampleDef>&    prdDefline)
        = 0;

    //  Access the MC dimension, after initialization
    virtual size_t simDim() const = 0;

    //  Simulate a path consuming a vector[simDim()]
    //      of independent Gaussians
    virtual void generatePath(
        const vector<double>&        gaussVec,
        //  return results in a pre-allocated scenario
        Scenario<T>&                 path)
            const = 0;

    //  Virtual copy constructor
    virtual unique_ptr<Model<T>> clone() const = 0;

```

```
    // Virtual destructor
    virtual ~Model() {}

    // Access to all the model parameters and what they mean
    //      size of vector = number of parameters
    virtual const vector<T*>& parameters() = 0;
    virtual const vector<string>& parameterLabels() const = 0;

    // Number of parameters
    size_t numParams() const
    {
        // parameters() is not const, must cast
        //      cast OK here, we know we don't modify anything
        return const_cast<Model*>(this)->parameters().size();
    }

    // Minor extensions for AAD
    // ...
}
```

A model initializes itself to perform simulations, knowing the timeline and scenario definition from the product. We separate memory allocation as usual, so we have two initialization methods: *allocate*() and *init*(). The first method only allocates working memory. The second method conducts the actual initialization. This is where a model initializes its simulation timeline, as explained on page 195, and pre-calculates some scenario independent results. After a model is initialized, it knows the dimension D of the simulation, how many random numbers it needs to simulate a scenario. It advertises D with its const method *simDim*() so that the RNG can initialize itself, too.

The const method *generatePath*() is the principal responsibility of a model in a simulation: to generate a scenario in accordance with the product's event dates and sample definitions, which the model knows from the initialization step. The process consumes a Gaussian vector in dimension D. The result is returned in a preallocated vector or preallocated samples.

Models implement the virtual copy constructor idiom in their const method *clone*().

The next three methods are something new. Different models have different parameters: in Black and Scholes, the parameters are the spot price, and time dependent rates, dividends, and volatilities (which could be represented with vectors, although we will implement a simplified version of Black and Scholes where all the parameters are scalar). In Dupire's model, it is a two-dimensional local volatility surface (which we represent with a bi-linearly interpolated matrix). In LGM (which we don't implement), it is a volatility curve and a mean reversion curve (which could be represented with two interpolated vectors). The parameters of different models are different in nature, dimension, and number.

Our generic algorithms manipulate models by base class pointers. They have no knowledge of the concrete model they are manipulating. Risk is the sensitivity to model parameters, so risk algorithms must access and manipulate the parameters. This is why it is a part of a model's contract to expose its parameters in a flat vector *of pointers* to its internal parameters, so client code not only can read, but also modify the parameters of a generic model with the *non const* accessor *parameters*(). Models also expose the names of their parameters, in the same way that products advertise the names of their payoffs, so client code knows what parameters in the vector correspond to what. Finally, the number of parameters in a model can be accessed with a call to *parameters*().*size*() but, since *parameters*() is not const, it would be considered a mutable call, which is of course not the case. This is why models offer a non-virtual method, written directly on the base class, to provide the number of parameters. The method is const and encapsulates ugly const casts so client code doesn't have to.

Finally, the base RNG class is as follows for now. We need nothing else for serial simulations, but parallel simulations will require a major extension to the base and concrete RNGs.

```
class RNG
{
public:

    //  Initialise with dimension simDim
    virtual void init(const size_t simDim) = 0;

    //  Compute the next vector[simDim] of independent Uniforms or Gaussians
    //  The vector is filled by the function and must be pre-allocated
    //  Note: non-const, modifies internal state
    virtual void nextU(vector<double>& uVec) = 0;
    virtual void nextG(vector<double>& gaussVec) = 0;

    //  Virtual copy constructor
    virtual unique_ptr<RNG> clone() const = 0;

    //  Virtual destructor
    virtual ~RNG() {}

    //  Access dimension
    virtual size_t simDim() const = 0;

    //  Major addition needed for parallel simulation
    //  ...
};
```

For now, the RNGs contract should be self-explanatory: an RNG is initialized with a simulation dimension D; it delivers either uniform or standard Gaussian vectors in dimension D in sequence, returning them in preallocated vectors. It also implements the virtual copy constructor idiom and advertises its dimension.

An important note is that the methods *nextU*() and *nextG*() are not *const*, since generators typically update their internal state when they compute a new set of numbers, as we have seen in the previous chapter with mrg32k3a and Sobol. This is in contrast to models and products, who respectively generate scenarios and compute payoffs without modification of internal state. It follows that models and products are thread safe after they are initialized, whereas RNGs are mutable objects. This comment affects the code of the next chapter.

Template algorithm

We have everything we need to code the Monte-Carlo algorithm, with a literal translation in C++ of the steps recalled in the beginning of this chapter.

```
//  MC simulator: free function that conducts simulations
//       and returns a matrix (as vector of vectors) of payoffs
//           (0..nPath-1 , 0..nPay-1)
inline vector<vector<double>> mcSimul(
    const Product<double>&          prd,
    const Model<double>&            mdl,
    const RNG&                      rng,
    const size_t                    nPath)
{
    //  Work with copies of the model and RNG
    //       which are modified when we set up the simulation
    //  Copies are OK at high level
    auto cMdl = mdl.clone();
    auto cRng = rng.clone();

    //  Allocate results
    const size_t nPay = prd.payoffLabels().size();
    vector<vector<double>> results(nPath, vector<double>(nPay));
    //  Init the simulation timeline
    cMdl->allocate(prd.timeline(), prd.defline());
    cMdl->init(prd.timeline(), prd.defline());
    //  Init the RNG
    cRng->init(cMdl->simDim());
    //  Allocate Gaussian vector
    vector<double> gaussVec(cMdl->simDim());
    //  Allocate and initialize path
    Scenario<double> path;
    allocatePath(prd.defline(), path);
    initializePath(path);

    //  Iterate over paths
    for (size_t i = 0; i<nPath; i++)
    {
        //  Next Gaussian vector, dimension D
        cRng->nextG(gaussVec);
        //  Generate path, consume Gaussian vector
        cMdl->generatePath(gaussVec, path);
        // Compute payoffs
        prd.payoffs(path, results[i]);
    }

    return results;    //    C++11: move
}
```

This generic Monte-Carlo simulation algorithm implements its successive steps with calls to the (pure virtual) methods of our product, model, and RNG base classes. We haven't yet implemented any model, product, or RNG, or written a concrete implementation of any of their methods. We wrote the skeleton of an algorithm in terms of generic steps, which concrete implementation belongs to the overridden methods of derived classes. Our generic algorithm never addresses concrete models, products, or RNGs; it only calls the *interface* defined in their base class. In GOF [25] terms,[2] this idiom is called a *template* pattern. Our generic simulation algorithm is a template algorithm.

The template code is remarkably short, and most of it deals with initialization and the preallocation of working memory. The actual simulations are implemented in the lines 31 to 40 and iterate over three lines of code 35, 37, and 39, excluding comments: generate a Gaussian vector, consume it to generate a scenario, compute the payoff over the scenario, repeat. But, of course, it is all the initializations that make the computation efficient: we will see that a lot of the work is performed there so that the repeated code only conducts fast calculations in the most efficient possible way.

In order to run this code, of course, we need concrete models, products, and RNGs. We start with the number generators.

6.2 RANDOM NUMBER GENERATORS

MRG32k3a

L'Ecuyer's random generator is coded in mrg32k3a.h in accordance with the algorithm on page 201. The file gaussians.h, referenced on the second line, contains an implementation of Moro's [66] inverse cumulative Gaussian *invNormalCdf*(). The implementation in mrg32k3a.h mainly translates in C++ the steps of the algorithm on page 201, with the incorporation of the important antithetic optimization, discussed on page 202.

```
1 #include "mcBase.h"
2 #include "gaussians.h"
3
```

[2]Erich Gamma, Richard Helm, Ralph Johnson, and John Vlissides, often called the "gang of four," or GOF, published in 1995 the book *Design Patterns: Elements of Reusable Object-oriented Software*, an extremely influential programming publication, describing a number of programming techniques, called *design patterns*, reusable in a wide number of situations, independently of the field of application, platform, or language. GOF's influence is still strong today. Modern languages offer native support for their patterns (for instance, the standard C++ library supports the *visitor* pattern in the *variant* class since C++17) and modern software typically implements their patterns in one way or another.

```
class mrg32k3a : public RNG
{

    //  Seed
    const double        myA, myB;

    //  Dimension
    size_t              myDim;

    //  State
    double              myXn, myXn1, myXn2, myYn, myYn1, myYn2;

    //  Antithetic
    bool                myAnti;
    //  false: generate new, true: negate cached
    vector<double>      myCachedUniforms;
    vector<double>      myCachedGaussians;

    //  Constants
    static constexpr  double    m1 = 4294967087;
    static constexpr  double    m2 = 4294944443;
    static constexpr  double    a12 = 1403580;
    static constexpr  double    a13 = 810728;
    static constexpr  double    a21 = 527612;
    static constexpr  double    a23 = 1370589;
    //  We divide the final uniform
    //      by m1 + 1 so we never hit 1
    static constexpr  double    m1p1 = 4294967088;

    //  Produce next number and update state
    double nextNumber()
    {
        //  Update X
        //  Recursion
        double x = a12 * myXn1 - a13 * myXn2;
        //  Modulus
        x -= long(x / m1) * m1;
        if (x < 0) x += m1;
        //  Update
        myXn2 = myXn1;
        myXn1 = myXn;
        myXn = x;

        //  Same for Y
        double y = a21 * myYn - a23 * myYn2;
        y -= long(y / m2) * m2;
        if (y < 0) y += m2;
        myYn2 = myYn1;
        myYn1 = myYn;
        myYn = y;

        //  Uniform
        const double u = x > y
            ? (x - y) / m1p1
            : (x - y + m1) / m1p1;
        return u;
    }
```

```

public:

    //  Constructor with seed
    mrg32k3a(const unsigned a = 12345, const unsigned b = 12346) :
        myA(a), myB(b)
    {
        reset();
    }

    //  Reset state to 0 (seed)
    void reset()
    {
        //  Reset state
        myXn = myXn1 = myXn2 = myA;
        myYn = myYn1 = myYn2 = myB;

        //  Anti = false: generate next
        myAnti = false;
    }

    //  Virtual copy constructor
    unique_ptr<RNG> clone() const override
    {
        return make_unique<mrg32k3a>(*this);
    }

    //  Initializer
    void init(const size_t simDim) override
    {
        myDim = simDim;
        myCachedUniforms.resize(myDim);
        myCachedGaussians.resize(myDim);
    }

    void nextU(vector<double>& uVec) override
    {
        if (myAnti)
        {
            //  Do not generate, negate cached
            transform(
                myCachedUniforms.begin(),
                myCachedUniforms.end(),
                uVec.begin(),
                [](const double d) { return 1.0 - d; });

            //  Generate next
            myAnti = false;
        }
        else
        {
            //  Generate and cache
            generate(
                myCachedUniforms.begin(),
                myCachedUniforms.end(),
                [this]() { return nextNumber(); });

```

```
            // Copy
            copy(
                myCachedUniforms.begin(),
                myCachedUniforms.end(),
                uVec.begin());

            // Do not generate next
            myAnti = true;
        }
    }

    void nextG(vector<double>& gaussVec) override
    {
        if (myAnti)
        {
            // Do not generate, negate cached
            // Note: we reuse the Gaussian numbers,
            //       we save not only generation, but also
            //       Gaussian transaformation
            transform(
                myCachedGaussians.begin(),
                myCachedGaussians.end(),
                gaussVec.begin(),
                [](const double n) { return -n; });

            // Generate next
            myAnti = false;
        }
        else
        {
            // Generate and cache
            generate(
                myCachedGaussians.begin(),
                myCachedGaussians.end(),
                [this]() { return invNormalCdf(nextNumber()); });

            // Copy
            copy(
                myCachedGaussians.begin(),
                myCachedGaussians.end(),
                gaussVec.begin());

            // Do not generate next
            myAnti = true;
        }
    }

    // Access dimension
    size_t simDim() const override
    {
        return myDim;
    }

    // Major extension required for parallel simulation
    // ...
};
```

The method *nextNumber*() implements the recursions on page 201. The constructor records the seeds, so the method *reset*() factory resets the state of the generator.

Note the virtual copy constructor in action. Generic client code calls the virtual *clone*() method to copy an RNG manipulated by base class pointer. The implementation of *clone*() is overridden in the concrete class. The implementation therefore knows that the RNG is an mrg32k3a, and easily makes a copy of itself with a call to its (default) copy constructor. The copy is stored on the heap and returned by base class (unique) pointer. The unique pointer on the concrete class implicitly and silently converts to a unique pointer of the base class when the method returns.

The override *init*() records the dimension, and allocates space for the caching of uniform and Gaussian numbers for the implementation of the antithetic logic. The antithetic logic itself is implemented in the overrides *nextU*() and *nextG*(), which return the next (respectively uniform or Gaussian) random vector in dimension *myDim*. The indicator *myAnti* starts *false* and flicks at every every call to *nextU*() or *nextG*(). When *myAnti* is *false*, a new random vector is generated by *myDim* repeated calls to *nextNumber*() for uniforms or *invNormalCdf*(*nextNumber*()) for Gaussians (encapsulated in the standard algorithm *generate*()). Before the random vector is returned, a copy is cached in the generator, so on the next call to *nextU*() or *nextG*(), with *myAnti* now *true*, no new numbers are generated; instead, a negation of the previously cached vector ($1 - U$ for uniforms, $-G$ for Gaussians) is returned, effectively producing two random vectors for the price of one, cutting in half the overhead of random number generation and Gaussian transformation.

Sobol

Our implementation of the Sobol sequence uses the 32×1111 direction numbers found by Joe and Kuo in 2003. The numbers are listed in sobol.cpp:

```
const unsigned jkDir1[] = {
    2147483648 , 2147483648 , 2147483648 , 2147483648 , 2147483648 ,
    2147483648 , 2147483648 , 2147483648 , 2147483648 , 2147483648 ,
    2147483648 , 2147483648 , 2147483648 , 2147483648 , 2147483648 ,
    2147483648 , 2147483648 , 2147483648 , 2147483648 , 2147483648 ,

//  ... 1111 repetitions of the same, the 1st number on all axes is 2^31

//  ... 32 sequences jkDirk ...

const unsigned jkDir32[] = {
    1 , 4294967295 , 1325465599 , 1342505107 , 806158221 ,
    3231291801 , 1076939793 , 2203259459 , 2233163655 , 2164226257 ,
```

```
    2221345463 , 2245004877 , 2218592819 , 894435425 , 2982991765 ,
    162569061 , 2918702895 , 3494266285 , 3691081979 , 2969567247 ,

//  ... 1111 numbers

const unsigned * const jkDir[32] =
{
    jkDir1, jkDir2, jkDir3, jkDir4, jkDir5, jkDir6, jkDir7, jkDir8,
    jkDir9, jkDir10, jkDir11, jkDir12, jkDir13, jkDir14, jkDir15, jkDir16,
    jkDir17, jkDir18, jkDir19, jkDir20, jkDir21, jkDir22, jkDir23, jkDir24,
    jkDir25, jkDir26, jkDir27, jkDir28, jkDir29, jkDir30, jkDir31, jkDir32
};

const unsigned * const * getjkDir()
{
    return jkDir;
}
```

It follows that our implementation is limited to dimension 1,111. The application crashes in a higher dimension. When a higher dimension is desired, for example, for the purpose of xVA or regulatory calculations, mrg32k3a may be used instead. The construction of mrg32k3a numbers *almost* matches the speed of Sobol, thanks to the antithetic optimization (not recommended with Sobol, since it would interfere with the naturally low discrepancy), but its convergence order is lower, together with accuracy and stability. Alternative sets of direction numbers in dimension up to 21,201 are found on Joe and Kuo's page. Jaeckel [63] delivers code that generates direction numbers on demand (with virtually no overhead) in dimension up to 8,129,334 (!)

The implementation of the sequence in sobol.h is a direct translation of the recursion on page 206:

```
#include "mcBase.h"
#include "gaussians.h"

#define ONEOVER2POW32 2.3283064365387E-10

const unsigned * const * getjkDir();

class Sobol : public RNG
{
    //  Dimension
    size_t                      myDim;

    //  State Y
    vector<unsigned>            myState;

    //  Current index in the sequence
    unsigned                    myIndex;

    //  The direction numbers listed in sobol.cpp
    //  Note jkDir[i][dim] gives the i-th (0 to 31)
    //      direction number of dimension dim
```

```
    const unsigned * const *    jkDir;

public:

    //  Virtual copy constructor
    unique_ptr<RNG> clone() const override
    {
        return make_unique<Sobol>(*this);
    }

    //  Initializer
    void init(const size_t simDim) override
    {
        //  Set pointer on direction numbers
        jkDir = getjkDir();

        //  Dimension
        myDim = simDim;
        myState.resize(myDim);

        //  Reset to 0
        reset();
    }

    void reset()
    {
        //  Set state to 0
        memset(myState.data(), 0, myDim * sizeof(unsigned));
        //  Set index to 0
        myIndex = 0;
    }

    //  Next point
    void next()
    {
        //  Gray code, find position j
        //      of rightmost zero bit of current index n
        unsigned n = myIndex, j = 0;
        while (n & 1)
        {
            n >>= 1;
            ++j;
        }

        //  Direction numbers
        const unsigned* dirNums = jkDir[j];

        //  XOR the appropriate direction number
        //      into each component of the integer sequence
        for (int i = 0; i<myDim; ++i)
        {
            myState[i] ^= dirNums[i];
        }

        //      Update count
        ++myIndex;
    }
```

```
    void nextU(vector<double>& uVec) override
    {
        next();
        transform(myState.begin(), myState.end(), uVec.begin(),
            [](const unsigned long i)
                {return ONEOVER2POW32 * i; });
    }

    void nextG(vector<double>& gaussVec) override
    {
        next();
        transform(myState.begin(), myState.end(), gaussVec.begin(),
            [](const unsigned long i)
                {return invNormalCdf(ONEOVER2POW32 * i); });
    }

    //  Major extension required for parallel simulation
    //  ...
};
```

6.3 CONCRETE PRODUCTS

We have introduced a generic, professional simulation environment with support for a wide variety of models and products. Professionally developed model hierarchies and product representations fit in this environment as long as they implement the Product and Model interfaces.

In the chapters that follow, we will extend this environment for parallel simulation and constant time differentiation, working on the template algorithm *without modification of models or products*. It follows that models and products interfaced with our environment automatically benefit from parallelism and AAD.[3]

In the rest of this chapter, we implement a few examples of concrete models and products so we can run concrete simulations. But those concrete implementations are meant for demonstration only; they do not constitute professional code suitable for production.

Concrete products, in particular, are sketchy, and for good reason. It is *not* best practice to represent products and cash-flows with dedicated C++ code in modern financial systems. The correct, generic, versatile representation of products and cash-flows is with a scripting library, where all cash-flows are represented in a consistent manner, products are created and customized at run time, and client code may inspect and manipulate cash-flows in many ways, including aggregation, compression, or decoration. Well-designed scripting libraries offer all these facilities, and much more,

[3]Provided, of course, that their design and code is correct, in particular, thread safe, lock-free, const correct, and properly templated for AAD.

without significant overhead. Scripting is a keystone of modern financial systems. It is also a demanding topic, which we address in detail in a dedicated publication [11]. Of course, scripted products implement the Product interface and seamlessly connect to our simulation library, including parallelism and AAD.

This being said, we need a few simple examples to effectively run simulations. All the demonstration products are listed in mcPrd.h in our repository. We start with a simple European call.

European call

All the concrete products implement the same pattern: they build their timeline and defline (definition of the samples on event dates, for lack of a better word) on construction, and evaluate payoffs against scenarios in their const *payoffs*() override, reading market observations on the scenario in accordance with the definition of the samples. They also expose the number of the name of their different payoffs, advertise their timeline and defline with const accessors, and implement the virtual copy constructor idiom in the same manner as RNGs.

```
template <class T>
class European : public Product<T>
{
    // Internal data
    double                myStrike;
    Time                  myExerciseDate;
    Time                  mySettlementDate;

    // Timeline, defline and payoff labels
    //      held internally
    //      for clients to accesses by const ref
    vector<Time>          myTimeline;
    vector<SampleDef>     myDefline;
    vector<string>        myLabels;

public:

    // Constructor: store data
    //      and build timeline, defline and labels
    European(const double    strike,
        const Time           exerciseDate,
        const Time           settlementDate) :
        myStrike(strike),
        myExerciseDate(exerciseDate),
        mySettlementDate(settlementDate),
        myLabels(1)
    {
        // Timeline = { exercise date }
        myTimeline.push_back(exerciseDate);

        // Defline

```

```
    // One sample on the exercise date
    myDefline.resize(1);

    // Payment of a cash flow ==> numeraire required
    myDefline[0].numeraire = true;

    // One forward to settlement date
    myDefline[0].forwardMats.push_back(settlementDate);

    // One discount to settlement date
    myDefline[0].discountMats.push_back(settlementDate);

    // Identify the product
    // One payoff = "call strike K exercise date settlement date"
    ostringstream ost;
    ost.precision(2);
    ost << fixed;
    if (settlementDate == exerciseDate)
    {
        ost << "call " << myStrike << " "
            << exerciseDate;
    }
    else
    {
        ost << "call " << myStrike << " "
            << exerciseDate << " " << settlementDate;
    }
    myLabels[0] = ost.str();
}

// Another constructor without a settlement date
//     implicitly settlement = exercise
European(const double   strike,
    const Time          exerciseDate) :
    European(strike, exerciseDate, exerciseDate)
{}

// Virtual copy constructor
unique_ptr<Product<T>> clone() const override
{
    return make_unique<European<T>>(*this);
}

// Accessors

// Timeline
const vector<Time>& timeline() const override
{
    return myTimeline;
}

// Defline
const vector<SampleDef>& defline() const override
{
    return myDefline;
}

```

```
90      // Labels
91      const vector<string>& payoffLabels() const override
92      {
93          return myLabels;
94      }
95
96      // Computation of the payoff on a path
97      void payoffs(
98          // path, one entry per event date
99          const Scenario<T>&          path,
100         // pre-allocated space for resulting payoffs
101         vector<T>&                  payoffs)
102             const override
103     {
104         // path[0] = exercise date
105         //      as specified in timeline
106
107         // forwards[0] = forward (exercise, settlement)
108         // discounts[0] = discount (exercise, settlement)
109         // numeraire at exercise
110         //      as specified in defline
111
112         payoffs[0] = max(path[0].forwards[0] - myStrike, 0.0)
113             * path[0].discounts[0]
114             / path[0].numeraire;
115     }
116 };
```

This class illustrates some key notions covered in Chapter 4. It represents a call option of strike K that pays $s_{T_s} - K$ on the settlement date T_s if exercised on the exercise date $T_e \leq T_s$. The holder exercises if $F(T_e, T_s) > K$, in which case the present value at T_e is:

$$CF_{T_e} = [F(T_e, T_s) - K]DF(T_e, T_s)$$

so that it may be modeled as a single cash-flow on the exercise date:

$$CF_{T_e} = [F(T_e, T_s) - K]^+DF(T_e, T_s)$$

The product holds its strike, exercise, and settlement date, as well as timeline and defline, and exposes them to client code. This is a single payoff product, so the vector of results filled by *payoffs*() and the vector of labels exposed in *payoffLabels*() are of dimension 1. The label, which identifies the product as "call K T_e T_s," is initialized on lines 45–60 in the constructor, along with the timeline and the defline.

The timeline $\{T_e\}$ is initialized on line 29. The defline is initialized on lines 31–43, with the definition of the sample on the exercise date: the forward and discount to the settlement date. We modeled the option as a single cash-flow on the exercise date, hence, T_e is considered a payment date and must include the numeraire.

The payoff, computed in *payoffs*(), was defined in Chapter 4 as the sum of the numeraire deflated cash-flows. In this case, the payoff is:

$$g(S_{T_e}) = g(F(T_e, T_s), DF(T_e, T_s), N_{T_e}) = \frac{CF_{T_e}}{N_{T_e}} = \frac{[F(T_e, T_s) - K]^{+} DF(T_e, T_s)}{N_{T_e}}$$

Note how the definition of the timeline and defline in the constructor synchronizes with the computation of the payoff in the *payoffs*() override. The timeline is set as vector with a unique entry myTimeline[0] = exerciseDate, so the model's contract is to generate paths with a single sample path[0], observed on the exercise date. It follows that the vector myDefline also has a unique entry myDefline[0] that specifies the market sample on the exercise date: what observations the model simulates. The numeraire property on the sample is set to true so the model simulates the numeraire on the exercise date in path[0].numeraire. The forwardMats and discountMats vectors on the sample definition are set to the unique entries forwardMats[0] = settlementDate and discountMats[0] = settlementDate. It follows that the model's contract is simulate $F(T_e, T_s)$ on path[0].forwards[0], and $DF(T_e, T_s)$ on path[0].discounts[0], in addition to the numeraire. The product doesn't require any other market observation to compute its payoff; in particular, the vector liborDefs on the sample definition is left empty, so the model doesn't need to simulate Libors. The product knows all this because it is in control of the timeline and defline and trusts the model to fulfill its contract. Therefore, in its *payoffs*() override, it knows that path[0].numeraire is N_{T-e}, paths[0].forwards[0] is $F(T_e, T_s)$, and path[0].discounts[0] is $DF(T_e, T_s)$, and confidently implements the payoff equation above on lines 112–114.

Barrier option

Next, we define a discretely monitored barrier option, more precisely an up-and-out call. We dropped the separate settlement date for simplicity. We define the up-and-out call and the corresponding European call together in the same product with two different payoffs, so we can evaluate them simultaneously against a shared set of scenarios.

```
#include "mcBase.h"

#define ONE_HOUR 0.000114469

//  Up and out call
template <class T>
class UOC : public Product<T>
{
    //  Internal data
    double                  myStrike;
    double                  myBarrier;
    Time                    myMaturity;
```

```

    //  Smoothing factor
    double              mySmooth;

    //  Timeline, defline and payoff labels
    vector<Time>        myTimeline;
    vector<SampleDef>   myDefline;
    vector<string>      myLabels;

public:

    //  Constructor: store data and build timeline, defline, labels
    UOC(const double    strike,
        const double    barrier,
        const Time      maturity,
        const Time      monitorFreq,
        const double    smooth)
        : myStrike(strike),
        myBarrier(barrier),
        myMaturity(maturity),
        mySmooth(smooth),
        myLabels(2)
    {
        //  Timeline
        //  Timeline = system date to maturity,
        //      with an event date every monitoring period

        //  Today
        myTimeline.push_back(systemTime);
        Time t = systemTime + monitorFreq;

        //  Barrier monitoring
        while (myMaturity - t > ONE_HOUR)
        {
            myTimeline.push_back(t);
            t += monitorFreq;
        }

        //  Maturity
        myTimeline.push_back(myMaturity);

        //

        //  Defline

        const size_t n = myTimeline.size();
        myDefline.resize(n);
        for (size_t i = 0; i < n; ++i)
        {
            //  Only last event date is a payment date
            //      numeraire only required on last event date
            myDefline[i].numeraire = false;

            //  spot(t) = forward (t, t) required on every event date
            //      for barrier monitoring and cash-flow determination
            myDefline[i].forwardMats.push_back(myTimeline[i]);
        }
```

```
        //  Numeraire only on last step
        myDefline.back().numeraire = true;

        //

        //  Identify the product
        //  label[0] = barrier = "call T K up and out B"
        //  label[1] = european = "call T K"
        ostringstream ost;
        ost.precision(2);
        ost << fixed;

        //  Second payoff = European
        ost << "call " << myMaturity << " " << myStrike ;
        myLabels[1] = ost.str();

        //  First payoff = barrier
        ost << " up and out "
            << myBarrier << " monitoring freq " << monitorFreq
            << " smooth " << mySmooth;
        myLabels[0] = ost.str();
    }

    //  Virtual copy constructor
    unique_ptr<Product<T>> clone() const override
    {
        return make_unique<UOC<T>>(*this);
    }

    //  Accessors

    //  Timeline
    const vector<Time>& timeline() const override
    {
        return myTimeline;
    }

    //  Defline
    const vector<SampleDef>& defline() const override
    {
        return myDefline;
    }

    //  Labels
    const vector<string>& payoffLabels() const override
    {
        return myLabels;
    }

    //  Payoffs
    void payoffs(
        //  path, one entry per event date
        const Scenario<T>&          path,
        //  pre-allocated space for resulting payoffs
        vector<T>&                  payoffs)
            const override
    {
```

```
127         //  Application of the smooth barrier technique to stabilize risks
128         //  See Savine's presentation on Fuzzy Logic, Global Derivatives 2016
129
130         //  Smoothing factor = x% of the spot both ways
131         //      untemplated
132         const double smooth = double(path[0].forwards[0] * mySmooth),
133             twoSmooth = 2 * smooth,
134             barSmooth = myBarrier + smooth;
135
136         //  Start alive
137         T alive(1.0);
138
139         //  Iterate through path, update alive status
140         for (const auto& sample: path)
141         {
142             //  Breached
143             if (sample.forwards[0] > barSmooth)
144             {
145                 alive = T(0.0);
146                 break;
147             }
148
149             //  Semi-breached: apply smoothing
150             if (sample.forwards[0] > myBarrier - smooth)
151             {
152                 alive *= (barSmooth - sample.forwards[0]) / twoSmooth;
153             }
154         }
155
156         //  Payoffs
157
158         //  European
159         payoffs[1] = max(path.back().forwards[0] - myStrike, 0.0)
160                      / path.back().numeraire;
161
162         //  Barrier
163         payoffs[0] = alive * payoffs[1];
164     }
165 };
```

This code follows the same pattern as the European option. The timeline is the set of barrier monitoring dates plus maturity, built on lines 36–54 in the constructor. All samples include the spot $S_t = F(t, t)$, and the sample on maturity includes the numeraire, too. We have two payoffs: the barrier and the European, with two corresponding labels.

The payoff implements the *smooth barrier* technique to mitigate the instability of Monte-Carlo risk sensitivities with discontinuous cash-flows. Smoothing is the approximation of a discontinuous cash-flow by a close continuous one. We apply a smoothing spread s (in percentage of s_0) so the barrier dies above $B + s$, lives below $B - s$, and in between loses a part of the notional interpolated between 1 at $B + s$ and 0 and $B - s$ and continues with the remaining notional. See our presentation [77] for an introduction to smoothing or Bergomi's [78] for an insightful discussion. We differentiate this code in Part III so we must apply smoothing to obtain correct, stable risks (as option traders *always* do). Note that smoothing is the responsibility of the product, not the model, because it consists in a modification of the payoff.

Like in the European option, the timeline and samples are defined in the constructor. The timeline is the union of the barrier monitoring schedule, today's date, and the maturity date. The sample on all event dates is defined as the spot, that is, the forward with maturity the same event date. So the product knows, when it accepts the path simulated by the model in its *payoffs*() override, that path[j].forwards[0] is the spot observed on T_j. In addition, the numeraire property on the sample definition is set to true on maturity only, so the product knows to read N_T on path.back().numeraire, where *back*() accesses the last entry of a STL vector, so path.back().forwards[0] is the spot at maturity. This is how the product confidently calculates the payoff of the barrier on line 163 and the corresponding European on line 159 . It knows that (provided the model fulfills its contract) the correct observations are in the right places on the scenario.

We shall not further comment on this correspondence between the definition of the timeline and defline on one side, and the computation of the payoff on the other side. This is a fundamental element of generic design in our simulation library. When a product sets the timeline and the sample definitions, it specifies what goes where on the path. To populate the path correctly, and in accordance with these instructions, is the model's responsibility, so the timeline and defline are communicated to the model on initialization. The product computes payoffs accordingly, reading the correct simulated observations in the right places on the scenario in accordance with its own specifications. The specifications in the product's constructor, and the computations in its *payoff*() override, must always remain synchronized, otherwise the product is incorrectly implemented and the simulations evaluate it incorrectly.

European portfolio

To demonstrate the simultaneous pricing of many different payoffs in a single product, we develop a product class for a portfolio of European options of different strikes and maturities.

```
template <class T>
class Europeans : public Product<T>
{
    //  Timeline
    vector<Time>                myMaturities;
    //  One vector of strikes per maturity
    vector<vector<double>>   myStrikes;
    //  One sample per maturity
```

```
    vector<SampleDef>        myDefline;

    //  Labels of all the European options in the portfolio
    vector<string>           myLabels;

public:

    //  Constructor: store data and build timeline
    //      input = maturity : vector of strikes
    Europeans(const map<Time, vector<double>>& options)
    {
        //  n: number of maturities
        const size_t n = options.size();

        //  Timeline = each maturity is an event date
        for (const pair<Time, vector<double>>& p : options)
        {
            myMaturities.push_back(p.first);
            myStrikes.push_back(p.second);
        }

        //  Defline = numeraire and spot(t) = forward(t,t) on every maturity
        myDefline.resize(n);
        for (size_t i = 0; i < n; ++i)
        {
            myDefline[i].numeraire = true;
            myDefline[i].forwardMats.push_back(myMaturities[i]);
        }

        //  Identify all the payoffs by maturity and strike
        for (const auto& option : options)
        {
            for (const auto& strike : option.second)
            {
                ostringstream ost;
                ost.precision(2);
                ost << fixed;
                ost << "call " << option.first << " " << strike;
                myLabels.push_back(ost.str());
            }
        }
    }

    //  Virtual copy constructor
    unique_ptr<Product<T>> clone() const override
    {
        return make_unique<Europeans<T>>(*this);
    }

    //  Accessors

    //  Access maturities and strikes
    const vector<Time>& maturities() const
    {
        return myMaturities;
    }
    const vector<vector<double>>& strikes() const
```

```
    {
        return myStrikes;
    }

    //  Timeline
    const vector<Time>& timeline() const override
    {
        return myMaturities;
    }

    //  Defline
    const vector<SampleDef>& defline() const override
    {
        return myDefline;
    }

    //  Labels
    const vector<string>& payoffLabels() const override
    {
        return myLabels;
    }

    //  Payoffs, maturity major
    void payoffs(
        //  path, one entry per event date
        const Scenario<T>&          path,
        //  pre-allocated space for resulting payoffs
        vector<T>&                  payoffs)
        const override
    {
        const size_t numT = myMaturities.size();

        auto payoffIt = payoffs.begin();
        for (size_t i = 0; i < numT; ++i)
        {
            transform(
                myStrikes[i].begin(),
                myStrikes[i].end(),
                payoffIt,
                [spot = path[i].forwards[0], num = path[i].numeraire]
                (const double& k)
                {
                    return max(spot - k, 0.0) / num;
                }
            );

            payoffIt += myStrikes[i].size();
        }
    }
};
```

It is very clumsy to code product portfolios as separate products. What we should code is a single portfolio class that aggregates a collection of products of any type into a single product.

This is possible, but it would take substantial effort. The portfolio's timeline is the union of its product's timelines, and the portfolio's samples are unions of its product's samples. But a product reads market variables on its own scenario to compute its payoff, not an aggregated scenario. Some clumsy, prone-to-error, inefficient boilerplate indexation code would have to be written so that timelines and samples may be aggregated in a way that keeps track of what belongs to whom. In contrast, aggregation is natural with scripted products.

The code above is therefore only for demonstration. We use it to check the calibration of Dupire's model in Chapter 13 by repricing a large number of European options, simultaneously and in a time virtually constant in the number of options. The simulations are shared, so only the (insignificant) computation of payoffs is linear in the number of options. In Chapter 14, we use this code to demonstrate the production of itemized risk reports (one risk report for every product in the book) with AAD, in *almost* constant time.

Contingent floater

Although our framework supports interest rate models, we are not implementing one in this publication, but we demonstrate the rate infrastructure with a basic hybrid product that pays Libor coupons, plus a spread, over a floating schedule, but only in those periods where the performance of some asset is positive.

The product is structured as a bond that also pays a redemption of 100% of the notional at maturity. The payoff of this contingent bond is:

$$g = \sum_{p=1}^{P} \frac{L(\delta(p-1), \delta p)\delta 1_{\{s_{\delta p} > s_{(\delta-1)p}\}}}{N_{\delta p}} + \frac{1}{N_{\delta P}}$$

where P is the number of coupons and δ is payment period. We smooth the digital indicators with a call spread, as is market practice, similarly to the smoothing we applied to the barrier, referring again to [77] for a quick introduction to smoothing. The resulting code is:

```
#define ONE_DAY 0.003773585

//  Payoff = sum { (libor(Ti, Ti+1) + cpn)
//        * coverage(Ti, Ti+1) only if Si+1 >= Si }
template <class T>
class ContingentBond : public Product<T>
{
    //  Data
    Time                myMaturity;
    double              myCpn;
    double              mySmooth;

    //  Timeline etc.
    vector<Time>        myTimeline;
```

```
    vector<SampleDef>   myDefline;
    vector<string>      myLabels;

    //  Pre-computed coverages
    vector<double>      myDt;

public:

    //  Constructor: store data and build timeline
    //  Timeline = system date to maturity,
    //      with event dates on every payment
    ContingentBond(
        const Time      maturity,
        const double    cpn,
        const Time      payFreq,
        const double    smooth) :
            myMaturity(maturity),
            myCpn(cpn),
            mySmooth(smooth),
            myLabels(1)
    {
        //  Timeline

        //  Today
        myTimeline.push_back(systemTime);
        Time t = systemTime + payFreq;

        //  Payment schedule
        while (myMaturity - t > ONE_DAY)
        {
            myDt.push_back(t - myTimeline.back());
            myTimeline.push_back(t);
            t += payFreq;
        }

        //  Maturity
        myDt.push_back(myMaturity - myTimeline.back());
        myTimeline.push_back(myMaturity);

        //

        //  Defline

        //  Payoff = sum { (libor(Ti, Ti+1) + cpn)
        //      * coverage(Ti, Ti+1) only if Si+1 >= Si }
        //  We need spot ( = forward (Ti, Ti) ) on every step,
        //      and libor (Ti, Ti+1) on on every step but the last
        //      (coverage is assumed act/365)

        const size_t n = myTimeline.size();
        myDefline.resize(n);
        for (size_t i = 0; i < n; ++i)
        {
            //  spot(Ti) = forward (Ti, Ti) on every step
            myDefline[i].forwardMats.push_back(myTimeline[i]);

            //  libor(Ti, Ti+1) and discount (Ti, Ti+1)
            //      on every step but last
```

```
            if (i < n - 1)
            {
                myDefline[i].liborDefs.push_back(
        SampleDef::RateDef(myTimeline[i], myTimeline[i + 1], "libor"));
            }

            //  Numeraire on every step but first
            //      first event date is not a payment date
            myDefline[i].numeraire = i > 0;
        }

        //  Identify the product
        //    single payoff = "contingent bond T cpn"
        ostringstream ost;
        ost.precision(2);
        ost << fixed;
        ost << "contingent bond " << myMaturity << " " << myCpn;
        myLabels[0] = ost.str();
    }

    //  Virtual copy constructor
    unique_ptr<Product<T>> clone() const override
    {
        return make_unique<ContingentBond<T>>(*this);
    }

    //  Accessors

    //  Timeline
    const vector<Time>& timeline() const override
    {
        return myTimeline;
    }

    //  Defline
    const vector<SampleDef>& defline() const override
    {
        return myDefline;
    }

    //  Labels
    const vector<string>& payoffLabels() const override
    {
        return myLabels;
    }

    //  Payoff
    void payoffs(
        //  path, one entry per event date
        const Scenario<T>&          path,
        //  pre-allocated space for resulting payoffs
        vector<T>&                  payoffs)
        const override
    {
        //  Application of the smooth digital technique to stabilize risks
        //  See Savine's presentation on Fuzzy Logic, Global Derivatives 2016

        //  Smoothing factor = x% of the spot both ways
```

```
        //      untemplated
        const double smooth = double(path[0].forwards[0] * mySmooth),
            twoSmooth = 2 * smooth;

        //  Period by period
        const size_t n = path.size() - 1;
        payoffs[0] = 0;
        for (size_t i = 0; i < n; ++i)
        {
            const auto& start = path[i];
            const auto& end = path[i + 1];

            const T s0 = start.forwards[0], s1 = end.forwards[0];

            //  Is asset performance positive?

            /*
                We apply smoothing here otherwise risks are
                unstable with bumps and wrong with AAD

                bool digital = end.forwards[0] >= start.forwards[0];
            */

            T digital;

            //  In
            if (s1 - s0 > smooth)
            {
                digital = T(1.0);
            }

            //  Out
            else if (s1 - s0 < - smooth)
            {
                digital = T(0.0);
            }

            // "Fuzzy" region = interpolate
            else
            {
                digital = (s1 - s0 + smooth) / twoSmooth;
            }

            //  ~smoothing

            payoffs[0] +=
                digital                 //  contingency
                * ( start.libors[0]     //  libor(Ti, Ti+1)
                + myCpn)                //  + coupon
                * myDt[i]               //  day count / 365
                / end.numeraire;        //  paid at Ti+1
        }
        payoffs[0] += 1.0 / path.back().numeraire;  // redemption at maturity
    }
};
```

6.4 CONCRETE MODELS

Finally, we implement two concrete models: Black-Scholes and Dupire. Contrary to the concrete products, which we only developed for demonstration, the model code is correct and professional, although somewhat simplified. We take the opportunity to demonstrate a few patterns for the efficient implementation of simulation models.

The implementations remain simplified compared to a production environment. We don't implement linear models or model hierarchies (although we do implement an IVS in Chapter 13) and store the initial market conditions as parameters in the dynamic models. The initial market conditions are simplified to the extreme: Black and Scholes is implemented with constant rate, dividend yield, and volatility, and Dupire is implemented with zero rates and dividends. We simplified the representation of the current market, which is not directly relevant for the purpose of this publication, and focused on the implementation of the simulated dynamics.

The main methods implemented in the models are overrides of *init*(), where the model accepts the definition of the scenario, so it knows what to simulate and initializes itself accordingly; and *generatePath*(), where the model simulates a scenario, in accordance with its definition, consuming a random vector in the process. Remember that initialization is split in two methods, memory allocations being gathered in a separate *allocate*() override.

Black and Scholes's model

Black and Scholes's model is implemented in mcMdlBs.h in our repository.

In order to illustrate the change of numeraire technique covered in length in Chapter 4, we implement the Black-Scholes model under the risk-neutral measure, or the spot measure, see page 163, at the client code's choice. Depending on the selected measure, the model implements different dynamics, and populates scenarios with the corresponding numeraire. The simulation under both measures results in the same price at convergence for all products, all the necessary logic remaining strictly encapsulated in the model, without any modification of the template algorithm or products.

```
template <class T>
class BlackScholes : public Model<T>
{
    //  Model parameters

    //  These would be picked on the linear market in a production system
    //  Today's spot
    T                    mySpot;
```

```
    //  Constant rate and dividend yield
    T                   myRate;
    T                   myDiv;

    //  Volatility
    T                   myVol;

    //  false = risk neutral measure
    //  true = spot measure
    const bool          mySpotMeasure;

    //  Similuation timeline = today + product timeline
    //  Black and Scholes implement big steps
    //      because its transition probabilities are known and cheap
    //      and mapping is time-wise
    vector<Time>        myTimeline;
    //  Is today on the product timeline?
    bool                myTodayOnTimeline;

    //  Reference on product's defline
    const vector<SampleDef>*    myDefline;

    //  Pre-calculated on initialization

    //  For the Gaussian transitional distributions

    //  standard deviations
    vector<T>           myStds;
    //  drifts
    vector<T>           myDrifts;

    //  For the mapping spot to sample

    //  forward factors exp((r - d) * (T - t))
    vector<vector<T>>   myForwardFactors;

    //  pre-calculated numeraires
    vector<T>           myNumeraires;
    //  pre-calculated discounts exp(-r * (T - t))
    vector<vector<T>>   myDiscounts;
    //  and rates = (exp(r * (T2 - T1)) - 1) / (T2 - T1)
    vector<vector<T>>   myLibors;

    //  Exported parameters
    vector<T*>          myParameters;
    vector<string>      myParameterLabels;

public:

    //  Constructor: store data
    template <class U>
    BlackScholes(
        const U             spot,
        const U             vol,
        const bool          spotMeasure = false,
        const U             rate = U(0.0),
        const U             div = U(0.0)) :
```

```
            mySpot(spot),
            myVol(vol),
            myRate(rate),
            myDiv(div),
            mySpotMeasure(spotMeasure),
            myParameters(4),
            myParameterLabels(4)
    {
        //  Set parameter labels once
        myParameterLabels[0] = "spot";
        myParameterLabels[1] = "vol";
        myParameterLabels[2] = "rate";
        myParameterLabels[3] = "div";

        //  Set pointers on parameters
        setParamPointers();
    }

private:

    //  Set pointers on parameters, reset on clone
    void setParamPointers()
    {
        myParameters[0] = &mySpot;
        myParameters[1] = &myVol;
        myParameters[2] = &myRate;
        myParameters[3] = &myDiv;
    }

public:

    //  Read access to parameters

    T spot() const
    {
        return mySpot;
    }

    const T vol() const
    {
        return myVol;
    }

    const T rate() const
    {
        return myRate;
    }

    const T div() const
    {
        return myDiv;
    }

    //  Access to all the model parameters
    const vector<T*>& parameters() override
    {
        return myParameters;
    }
```

```
    //  And their names
    const vector<string>& parameterLabels() const override
    {
        return myParameterLabels;
    }

    //  Virtual copy constructor
    unique_ptr<Model<T>> clone() const override
    {
        auto clone = make_unique<BlackScholes<T>>(*this);
        //  Must reset pointers, right now the clone's pointers
        //      point on the original model's parameters
        clone->setParamPointers();
        return clone;
    }

    //  Allocate simulation timeline
    void allocate(
        const vector<Time>&         productTimeline,
        const vector<SampleDef>&    defline)
            override
    {
        //  Simulation timeline = today + product timeline
        myTimeline.clear();
        myTimeline.push_back(systemTime);
        for (const auto& time : productTimeline)
        {
            if (time > systemTime) myTimeline.push_back(time);
        }

        //  Is today on the timeline?
        myTodayOnTimeline = (productTimeline[0] == systemTime);

        //  Take a reference on the product's defline
        myDefline = &defline;

        //  Allocate the standard devs and drifts
        //      over simulation timeline
        myStds.resize(myTimeline.size() - 1);
        myDrifts.resize(myTimeline.size() - 1);

        //  Allocate the numeraires, discount, forward factors and Libors
        //      over product timeline
        const size_t n = productTimeline.size();
        myNumeraires.resize(n);

        myDiscounts.resize(n);
        for (size_t j = 0; j < n; ++j)
        {
            myDiscounts[j].resize(defline[j].discountMats.size());
        }

        myForwardFactors.resize(n);
        for (size_t j = 0; j < n; ++j)
        {
            myForwardFactors[j].resize(defline[j].forwardMats.size());
        }
```

```
        myLibors.resize(n);
        for (size_t j = 0; j < n; ++j)
        {
            myLibors[j].resize(defline[j].liborDefs.size());
        }
    }

    // Initialize simulation timeline after allocation
    void init(
        const vector<Time>&          productTimeline,
        const vector<SampleDef>&     defline)
            override
    {
        // Pre-compute the standard devs and drifts over simulation timeline

        const T mu = myRate - myDiv;
        const size_t n = myTimeline.size() - 1;

        for (size_t i = 0; i < n; ++i)
        {
            const double dt = myTimeline[i + 1] - myTimeline[i];

            // Var[logST2 / ST1] = vol^2 * dt
            myStds[i] = myVol * sqrt(dt);

            //       under risk neutral measure,
            // E[logST2 / ST1] = logST1 + ( (r - d) - 0.5 * vol ^ 2 ) * dt
            //       under spot measure,
            // E[logST2 / ST1] = logST1 + ( (r - d) + 0.5 * vol ^ 2 ) * dt
            if (mySpotMeasure)
            {
                myDrifts[i] = (mu + 0.5*myVol*myVol)*dt;

            }
            else
            {
                myDrifts[i] = (mu - 0.5*myVol*myVol)*dt;
            }
        }

        // Pre-compute the numeraires, discount and forward factors
        // For mapping spot to sample, over event dates

        const size_t m = productTimeline.size();

        for (size_t i = 0; i < m; ++i)
        {
            // Numeraire
            if (defline[i].numeraire)
            {
                // Under the spot measure,
                //     the numeraire is the spot with reinvested dividends
                //     num(t) = spot(t) / spot(0) * exp(div * t)
                //     we precalculate exp(div * t) / spot(0)
                if (mySpotMeasure)
                {
```

```
                myNumeraires[i] = exp(myDiv * productTimeline[i]) / mySpot;
            }
            //  Under the risk neutral measure, in Black-Scholes
            //      numeraire is deterministic = exp(rate * t)
            else
            {
                myNumeraires[i] = exp(myRate * productTimeline[i]);
            }
        }

        //  Forward factors
        const size_t pFF = defline[i].forwardMats.size();
        for (size_t j = 0; j < pFF; ++j)
        {
            myForwardFactors[i][j] =
    exp(mu * (defline[i].forwardMats[j] - productTimeline[i]));
        }

        //  Discount factors
        const size_t pDF = defline[i].discountMats.size();
        for (size_t j = 0; j < pDF; ++j)
        {
            myDiscounts[i][j] =
    exp(-myRate * (defline[i].discountMats[j] - productTimeline[i]));
        }

        //  Libors
        const size_t pL = defline[i].liborDefs.size();
        for (size_t j = 0; j < pL; ++j)
        {
            const double dt
    = defline[i].liborDefs[j].end - defline[i].liborDefs[j].start;
            myLibors[i][j] = (exp(myRate*dt) - 1.0) / dt;
        }
    }   //  loop on event dates
}

//  MC Dimension
size_t simDim() const override
{
    return myTimeline.size() - 1;
}

private:

//  Mapping function, fills a Sample given the spot
inline void fillScen(
    const size_t        idx,    //  index on product timeline
    const T&            spot,   //  spot
    Sample<T>&          scen,   //  Sample to fill
    const SampleDef&    def)    //  and its definition
        const
{
    //  Fill numeraire
    if (def.numeraire)
    {
        scen.numeraire = myNumeraires[idx];
```

```
            if (mySpotMeasure) scen.numeraire *= spot;
        }

        //  Fill forwards
        transform(
            myForwardFactors[idx].begin(),
            myForwardFactors[idx].end(),
            scen.forwards.begin(),
            [&spot](const T& ff)
            {
                return spot * ff;
            }
        );

        //  Fill (deterministic) discounts and Libors

        copy(myDiscounts[idx].begin(), myDiscounts[idx].end(),
            scen.discounts.begin());
        copy(myLibors[idx].begin(), myLibors[idx].end(),
            scen.libors.begin());
    }
public:

    //  Generate one path, consume Gaussian vector
    //      path must be pre-allocated
    //      with the same size as the product timeline
    void generatePath(
        //  Gaussian vector, consumed to build the path
        const vector<double>&   gaussVec,
        //  Returned in a pre-allocated vector
        Scenario<T>&            path)
            const override
    {
        //  The starting spot
        //  We know that today is on the timeline
        T spot = mySpot;
        //  Next index to fill on the product timeline
        size_t idx = 0;
        //  Is today on the product timeline?
        if (myTodayOnTimeline)
        {
            fillScen(idx, spot, path[idx], (*myDefline)[idx]);
            ++idx;
        }

        //  Iterate through timeline, apply sampling scheme
        const size_t n = myTimeline.size() - 1;
        for (size_t i = 0; i < n; ++i)
        {
            //  Apply conditional distributions
            spot = spot * exp(myDrifts[i]
                + myStds[i] * gaussVec[i]);

            //  Store on the sample
            fillScen(idx, spot, path[idx], (*myDefline)[idx]);
```

```
352
353             ++idx;
354         }
355     }
356 };
```

In the constructor line 59, the model initializes its parameters: spot, volatility, rate, dividend. Note that the constructor is templated so a model that represent numbers with a type T can be initialized with numbers of another type U. The model also initializes a vector of four pointers on its four parameters, which it advertises in its method *parameters*(), line 119. It also initializes a vector of four strings in the same order "spot," "vol," "rate," "div" and advertises it with *parameterLabels*(), line 125, so client code knows what pointer refers to what parameter. The pointers are set in the private method *setParamPointers*() on line 86.

The virtual copy constructor idiom is implemented in *clone*() on line 130, where the model makes a copy of itself as usual. This copies all the data members, including the vector of pointers to parameters (myParameters), which are copied by value, so the pointers in the cloned model still point on the parameters of the original model. For this reason, the cloned model must reset its pointers to its own parameters to finalize cloning. This is why we made *setParamPointers*() a separate method.

A model's primary responsibility is to simulate scenarios and override *generatePath*() on line 320, which accepts a random vector and consumes it to produce a scenario in accordance with the timeline and sample definitions of the product. From the discussion on page 195, Black and Scholes is best simulated with the transitional scheme:

$$s_{j+1} = s_j \exp\left[f(t_j, t_{j+1}) - \frac{V(t_j, t_{j+1})}{2} + \sqrt{V(t_j, t_{j+1})}G_j\right]$$

under the risk-neutral measure, with large steps over the event dates.[4] With constant parameters, the forward factors are:

$$f(t_j, t_{j+1}) = (r - y)(t_{j+1} - t_j)$$

and $V(t_j, t_{j+1}) = \sigma^2(t_{j+1} - t_j)$. The G_js are the standard independent Gaussian numbers consumed in the simulation. We derived the dynamics under the spot measure on page 163, and the corresponding transitional scheme is:

$$s_{j+1} = s_j \exp\left[f(t_j, t_{j+1}) + \frac{V(t_j, t_{j+1})}{2} + \sqrt{V(t_j, t_{j+1})}G_j\right]$$

[4]With the addition of today's date, lines 146–155.

As the path of s over the timeline is generated, the model's mapping ζ is applied to produce the samples $S_j = \zeta_j(x_j)$ as discussed in depth in the previous chapter. The mapping is implemented in the private method *fillScen*(), line 283. The definition of samples on the event dates t_js was communicated along with the timeline on initialization, and the model took a reference myDefline (line 158) so it knows what to set on the samples. If timeline[j].numeraire is true, the model sets the correct numeraire N_{T_j} on path[j].numeraire: β_{t_j} under the risk-neutral measure or the spot with reinvested dividends, as explained on page 163, under the spot measure.[5] The mapping *fillScen*() also sets the forwards $F(j, T_k) = s_j exp(f(t_j, T_K))$ on paths[j].forwards[k], with the maturities T_k specified on (*myDefline)[j].forwardMats[k]. The *forward factors* $exp(f(t_1, t_2))$ are deterministic in this model, so they are not computed during the simulation. The are pre-calculated in *init*() and stored in the model's working memory, where they're picked when needed for the simulation of the scenario. The same applies to discounts and Libors, which are also deterministic in this model.

Note how this mechanics naturally articulates with that of products and permits a clear separation of models and products together with a synchronization through the timeline and the definition of samples.

The simulation is implemented in *generatePath*() line 320, the event date loop on lines 342–354 implementing the transitional scheme, including the mapping with a call to *fillScen*() on line 351. Lines 330–340 apply the same treatment to today's date, which is special, because it may or may not be an event date.

At this point, we state a key rule for the efficient implementation of simulation models:

Perform as much work as possible on initialization and as little as possible during simulations.

Initialization only occurs once. Simulations are repeated a large number of times. We must conduct all the *administrative* work, including allocation of working memory, on initialization. This is done in *allocate*(), line 140. But there is more. We can often pre-calculate parts of the computations occurring during simulation. The amount of computations moved to initialization time is a major determinant of the simulation's performance.

In the Black and Scholes model, a vast number of amounts in the simulation scheme and the mapping are deterministic, independent on random numbers, and may therefore be pre-calculated: the expectations, variances, and standard deviations for the Gaussian scheme, as well as the forward factors, and the deterministic discounts and libors for the mapping. For the

[5] We also always normalize numeraires with their values today, the framework assuming $N_0 = 1$.

numeraire, it is also deterministic under the risk-neutral measure, but not under the spot measure. In this case,

$$N_{T_j} = \frac{s_{T_j}}{s_0} exp(yT_j)$$

and we pre-calculate $exp(yT_j)/s_0$ so we only multiply by the spot s_{T_j} during simulations. The method *init*() on line 190 implements all these pre-calculations and stores them in the model's working memory. The memory is preallocated in *allocate*().

The code is not short, but the core functionality, the simulation in *generatePath*(), including mapping in *fillScen*(), takes around 50 lines, including many comments. Most computations are conducted in the initialization phase, in *allocate*() and *init*() (150 lines), as should be expected from an efficient implementation.

Dupire's model

To demonstrate the library with a more ambitious concrete model, we implement Dupire's local volatility model (although with zero rates and dividends) in mcMdlDupire.h:

$$\frac{ds_t}{s_t} = \sigma(s_t, t)dW$$

The local volatility function $\sigma(s_t, t)$ is meant to be calibrated – using Dupire's famous formula – to the market prices of European options. We will discuss, code, and differentiate Dupire's calibration in Chapter 13. For now, we focus on the simulation. We take the local volatility as a given matrix, with spots in rows and times in columns, and bi-linearly interpolate it in the simulation scheme. As discussed on page 195 of the previous chapter, a transitional scheme is not appropriate for Dupire's model, where transition probabilities are unknown, and we implement the log-Euler scheme instead:

$$logs_{j+1} = logs_j - \frac{\sigma(s_j, t_j)^2}{2}(t_{j+1} - t_j) + \sigma(s_j, t_j)\sqrt{t_{j+1} - t_j} G_j$$

where $\sigma(s_j, t_j)$ is bi-linearly interpolated spot and time over the local volatility matrix, keeping extrapolation flat. In addition, as discussed on page 195, we cannot accurately simulate over long time steps with this scheme. The simulation timeline cannot only consist of the product's event dates. We must insert additional time steps so that the space between them does not exceed a specified maximum ΔT.

It is clear that a lot of CPU time is spent in the interpolation. Bi-linear interpolation is expensive, and we have one in the innermost loop in the algorithm, in every simulation, on every time step. This is something we must optimize with pre-calculations. We therefore pre-interpolate volatilities *in time* on initialization, so we only conduct one-dimensional interpolations *in spot* in the simulation. We also pre-calculate the $t_{j+1} - t_j$s and their square roots.

Finally, we refer to our cache efficiency comments from Chapter 1. We will be interpolating in spot at simulation time, so our pre-interpolated in time volatilities must be stored in time major to be localized in memory in spot space, even though the matrix as a model parameter is spot major in our specification: $\sigma(s_i, t_j)$ not $\sigma(t_i, s_j)$.

Having discussed the challenges and the solutions specific to Dupire's model, we list the code below. The model mainly follows the same pattern as Black and Scholes, somewhat simplified in the absence of rates and dividends, the only differences coming for the local volatility matrix and its interpolation. We reuse our matrix class from Chapter 1 in the file matrix.h. We also need a few utility functions. One is for linear interpolation (with hard-coded flat extrapolation):

$$i(x_0) = Y_1 1_{\{x_0<X_1\}} + Y_n 1_{\{x_0\geq X_n\}} + \frac{(x_0 - X_i)(Y_{i+1} - Y_i)}{X_{i+1} - X_i} 1_{\{X_i\leq x_0<X_{i+1}\}}$$

The code is in interp.h file in our repository and listed below.

```
#include <algorithm>
using namespace std;

//  Utility for interpolation
//  Interpolates the vector y against knots x in value x0
//  Interpolation is linear, extrapolation is flat
template <class ITX, class ITY, class T>
inline auto interp(
    //  sorted xs
    ITX                         xBegin,
    ITX                         xEnd,
    //  corresponding ys
    ITY                         yBegin,
    ITY                         yEnd,
    //  interpolate for point x0
    const T&                    x0)
    ->remove_reference_t<decltype(*yBegin)>
{

    //  STL binary search, returns iterator on 1st element no less than x0
    //  upper_bound guarantees logarithmic search
    auto it = upper_bound(xBegin, xEnd, x0);

    //  Extrapolation?
```

```
    if (it == xEnd) return *(yEnd - 1);
    if (it == xBegin) return *yBegin;

    // Interpolation
    size_t n = distance(xBegin, it) - 1;
    auto x1 = xBegin[n];
    auto y1 = yBegin[n];
    auto x2 = xBegin[n + 1];
    auto y2 = yBegin[n + 1];
    return y1 + (y2 - y1) / (x2 - x1) * (x0 - x1);
}
```

We coded the function generically, STL style. The *xs* and the *ys* are passed through iterators, the type T of x_0 is templated, and the return type is deduced on instantiation. To code this algorithm generically is not only best practice, it is also convenient for our client code, where we will be using it with different types and different kinds of containers.

The return type is auto, with a *trailing syntax* that specifies it:

$$remove\_reference\_t < decltype(*\, yBegin) >$$

all of which is specialized C++11 syntax meaning "raw type of the elements referenced by the iterator yBegin, with references removed," or more simply the "type of the *ys*." The reason for the complications is that the *ys* are passed by iterator.

We need another utility function to fill a schedule with additional time steps so the spacing would not exceed a given amount. That is how we produce a simulation timeline out of the product timeline, as explained on page 195. The following code is in utility.h:

```
#define EPS 1.0e-08
//  Utility for filling data
template<class CONT, class T, class IT = T*>
//  Returns filled container
//      has all original points
//      plus additional ones if requested
//      plus additional ones so maxDx is not exceeded
//  Original container and addPoints must be sorted
//  Returned container is sorted
inline CONT
fillData(
    //  The original data, sorted
    const CONT&                         original,
    //  The maximum spacing allowed
    const T&                            maxDx,
    //  Minimum distance for equality
    const T&                            minDx = T(0.0),
    //  Specific points to add, by iterator, sorted
    IT                                  addBegin = nullptr,
    IT                                  addEnd = nullptr)
{
    //  Results
```

```
    CONT filled;

    //  Add points?
    CONT added;
    const size_t addPoints = addBegin || addEnd
        ? distance(addBegin, addEnd)
        : 0;

    if (addPoints > 0)
    {
        set_union(
            original.begin(),
            original.end(),
            addBegin,
            addEnd,
            back_inserter(added),
            [minDx](const T x, const T y) { return x < y - minDx; });
    }
    const CONT& sequence = addPoints > 0 ? added : original;

    //  Position on the start, add it
    auto it = sequence.begin();
    filled.push_back(*it);
    ++it;

    while (it != sequence.end())
    {
        auto current = filled.back();
        auto next = *it;
        //  Must supplement?
        if (next - current > maxDx)
        {
            //  Number of points to add
            int addPoints = int((next - current) / maxDx - EPS) + 1;
            //  Spacing between supplementary points
            auto spacing = (next - current) / addPoints;
            //  Add the steps
            auto t = current + spacing;
            while (t < next - minDx)
            {
                filled.push_back(t);
                t += spacing;
            }
        }
        //  Push the next step on the product timeline and advance
        filled.push_back(*it);
        ++it;
    }

    return filled;
}
```

The routine takes the product timeline, adds some specified points that we may want on the timeline, fills the schedule so that maxDx is not exceeded, and does not add points distant by less than minDx from existing. It is coded in a generic manner, in the sense that it doesn't specify the type of the containers or the type of the elements, although the code is mostly boilerplate.

The code for the model itself is listed below. It is very similar to the Black and Scholes code, the differences being extensively described above.

```
#include "matrix.h"
#include "interp.h"
#include "utility.h"

#define HALF_DAY 0.00136986301369863

template <class T>
class Dupire : public Model<T>
{
    //  Model parameters

    //  Today's spot
    //  That would be today's linear market in a production system
    T                       mySpot;

    //  Local volatility structure
    const vector<double>    mySpots;
    //  We store log spots to interpolate in log space
    vector<double>          myLogSpots;
    const vector<Time>      myTimes;
    //  Local vols
    //  Spot major: sigma(spot i, time j) = myVols[i][j]
    matrix<T>               myVols;

    //  Maximum space allowed between time steps
    const Time              myMaxDt;

    //  Similuation timeline
    vector<Time>            myTimeline;
    //  true (1) if the time step is on the product timeline
    //  false (0) if it is an additional simulation step
    vector<bool>            myCommonSteps;

    //  The pruduct's defline byref
    const vector<SampleDef>*    myDefline;

    //  Pre-calculated on initialization

    //  volatilities pre-interpolated in time for each time step
    //      here time major: iv(time i, spot j) = myInterpVols[i][j]
    matrix<T>               myInterpVols;
    //      volatilities are stored multiplied by sqrt(dt)

    //  Exported parameters
    vector<T*>              myParameters;
    vector<string>          myParameterLabels;

public:

    //  Constructor: store data

```

```
    template <class U>
    Dupire(const U                  spot,
        const vector<double>    spots,
        const vector<Time>      times,
        const matrix<U>         vols,
        const Time maxDt =      0.25)
        : mySpot(spot),
        mySpots(spots),
        myLogSpots(mySpots.size()),
        myTimes(times),
        myVols(vols),
        myMaxDt(maxDt),
        myParameters(myVols.rows() * myVols.cols() + 1),
        myParameterLabels(myVols.rows() * myVols.cols() + 1)
    {
        //  Compute log spots
        transform(mySpots.begin(), mySpots.end(), myLogSpots.begin(),
            [](const double s) {return log(s); });

        //  Set parameter labels once

        //  First parameter is s0
        myParameterLabels[0] = "spot";

        //  Labels for all the local volatilities = "lvol Si Tj"
        size_t p = 0;
        for (size_t i = 0; i < myVols.rows(); ++i)
        {
            for (size_t j = 0; j < myVols.cols(); ++j)
            {
                ostringstream ost;
                ost << setprecision(2) << fixed;
                ost << "lvol " << mySpots[i] << " " << myTimes[j];

                myParameterLabels[++p] = ost.str();
            }
        }

        setParamPointers();
    }

private:

    void setParamPointers()
    {
        //  First parameter is s0
        myParameters[0] = &mySpot;
        //  Subsequent parameters are local vols
        transform(myVols.begin(), myVols.end(), next(myParameters.begin()),
            [](auto& vol) {return &vol; });
    }

public:

    //  Read access to parameters

    T spot() const
```

```
    {
        return mySpot;
    }

    const vector<double>& spots() const
    {
        return mySpots;
    }

    const vector<Time>& times() const
    {
        return myTimes;
    }

    const vector<T>& vols() const
    {
        return myVols;
    }

    //  Access to all the model parameters and labels
    const vector<T*>& parameters() override
    {
        return myParameters;
    }
    const vector<string>& parameterLabels() const override
    {
        return myParameterLabels;
    }

    //  Virtual copy constructor
    unique_ptr<Model<T>> clone() const override
    {
        auto clone = make_unique<Dupire<T>>(*this);
        clone->setParamPointers();
        return clone;
    }

    //  Allocate and initialize timeline

    void allocate(
        const vector<Time>&         productTimeline,
        const vector<SampleDef>&    defline)
            override
    {
        //  Fill simulation timeline

        //  Do the fill
        myTimeline = fillData(
            productTimeline,    // Event dates
            myMaxDt,            // Maximum space allowed
            HALF_DAY,           // Minimum distance = half day
            &systemTime, &systemTime + 1);  //  Hack to include system time

        //  Which steps are event dates?
        myCommonSteps.resize(myTimeline.size());
        transform(myTimeline.begin(), myTimeline.end(), myCommonSteps.begin(),
            [&](const Time t)
```

```
        {
            return binary_search(
                productTimeline.begin(),
                productTimeline.end(),
                t);
        });

        //  Take a reference on the product's defline
        myDefline = &defline;

        //  Allocate the local volatilities
        //      pre-interpolated in time over simulation timeline
        myInterpVols.resize(myTimeline.size() - 1, mySpots.size());
    }

    void init(
        const vector<Time>&         productTimeline,
        const vector<SampleDef>&    defline)
            override
    {
        //  Compute the local volatilities
        //      pre-interpolated in time and multiplied by sqrt(dt)
        const size_t n = myTimeline.size() - 1;
        for (size_t i = 0; i < n; ++i)
        {
            const double sqrtdt = sqrt(myTimeline[i + 1] - myTimeline[i]);
            const size_t m = myLogSpots.size();
            for (size_t j = 0; j < m; ++j)
            {
                myInterpVols[i][j] = sqrtdt * interp(
                    myTimes.begin(),
                    myTimes.end(),
                    myVols[j],
                    myVols[j] + myTimes.size(),
                    myTimeline[i]);
            }
        }
    }

    //  MC Dimension
    size_t simDim() const override
    {
        return myTimeline.size() - 1;
    }

private:

    //  Mapping function, fills a sample given the spot
    inline static void fillScen(const T& spot, Sample<T>& scen)
    {
        //  Only fill forwards, the rest stays on default (zero rates)
        //      With zero rates and divs, all forwards = spot
        fill(scen.forwards.begin(), scen.forwards.end(), spot);
    }

public:

```

```
    //  Generate one path, consume Gaussian vector
    //      path must be pre-allocated
    //      with the same size as the product timeline
    void generatePath(
        const vector<double>& gaussVec,
        Scenario<T>& path)
            const override
    {
        //  The starting spot
        //  We know that today is on the timeline
        T logspot = log(mySpot);
        Time current = systemTime;
        //  Next index to fill on the product timeline
        size_t idx = 0;
        //  Is today on the product timeline?
        if (myCommonSteps[idx])
        {
            fillScen(exp(logspot), path[idx]);
            ++idx;
        }

        //  Iterate through timeline
        const size_t n = myTimeline.size() - 1;
        const size_t m = myLogSpots.size();
        for (size_t i = 0; i < n; ++i)
        {
            //  Interpolate local volatility in log spot
            T vol = interp(
                myLogSpots.begin(),
                myLogSpots.end(),
                myInterpVols[i],
                myInterpVols[i] + m,
                logspot);
            //  vol comes out * sqrt(dt)

            //  Apply Euler's scheme
            logspot += vol * (- 0.5 * vol + gaussVec[i]);

            //  Store on the path if event date
            if (myCommonSteps[i + 1])
            {
                fillScen(exp(logspot), path[idx]);
                ++idx;
            }
        }
    }
};
```

Note that the local volatilities are stored in the templated number type, whereas their labels (spots for rows and times for columns) are stored in native types (doubles and times, which is another name for a double in our implementation). Similarly, all the computations related to timelines use native types. We will see in Part III that those amounts are *inactive*; they don't contribute to the production of differentials. It is a strong optimization to leave them out of differentiation logic. A practical means of doing this is store them as native number types. The details will be given in Chapter 12. The RNGs are untemplated for the same reason.

The simulation timeline is computed in *init*(), where we build the simulation timeline from the product's event dates, and insert additional steps, keeping track of which time steps on the simulation are original event dates and which are not, so we only apply the mapping *fillScen*() on the event dates, as explained on page 195. Local volatilities are interpolated in time over the simulation timeline on initialization, and in spot during simulations in *generatePath*().

Dupire has a large number of parameters: all the cells in the volatility matrix, making it an ideal model for the demonstration of AAD. The constructor carefully labels all the parameters so client code can identify every local volatility with the corresponding spot and time.

6.5 USER INTERFACE

We have everything we need to run the library and price concrete products with concrete models. One of the benefits of a generic architecture is that we can mix and match models and products at run time. A particularly convenient user interface is one that stores models and products in memory, by name, and exports a function that takes the name of a model, the name of a product, and executes the simulation.

The code to do this is in store.h in our repository. It uses hash maps, a classic, efficient data structure for storing and retrieving labeled data. Hash maps are standard C++11. The data structure is called unordered_map, it is defined in the header <unordered_map> and its interface is virtually identical to the classic C++ map data structure.[6]

```
#include "mcBase.h"
#include "mcMdl.h"
#include "mcPrd.h"
#include <unordered_map>
#include <memory>
using namespace std;

using ModelStore =
unordered_map<string, unique_ptr<Model<double>>>;
using ProductStore =
unordered_map<string, unique_ptr<Product<double>>>;

ModelStore modelStore;
ProductStore productStore;
```

[6] We could have more properly implemented the store with GOF's singleton template, like we did for the thread pool in Section 3.18. In the case of such a simple data structure, however, it felt like overkill.

We have a global model store that stores models by unique pointer under string labels, and another global product store for products. We implement functions to create and store models, one function per model type, and a unique function to retrieve a model from the map:

```
void putBlackScholes(
    const double            spot,
    const double            vol,
    const bool              qSpot,
    const double            rate,
    const double            div,
    const string&           store)
{
    unique_ptr<Model<double>> mdl = make_unique<BlackScholes<double>>(
        spot, vol, qSpot, rate, div);

    modelStore[store] = move(mdl);
}

void putDupire(
    const double            spot,
    const vector<double>&   spots,
    const vector<Time>&     times,
    //  spot major
    const matrix<double>&   vols,
    const double            maxDt,
    const string&           store)
{
    unique_ptr<Model<double>> mdl = make_unique<Dupire<double>>(
        spot, spots, times, vols, maxDt);

    modelStore[store] = move(mdl);
}

const Model<double>* getModel(const string& store)
{
    auto it = modelStore.find(store);
    if (it == modelStore.end()) return nullptr;
    else return it->second.first.get();
}
```

When a model is stored under the same name as a model previously stored, the store destroys the previous model and removes it from memory. We have the exact same pattern for products:

```
void putEuropean(
    const double            strike,
    const Time              exerciseDate,
    const Time              settlementDate,
    const string&           store)
{
    unique_ptr<Product<double>> prd = make_unique<European<double>>(
        strike, exerciseDate, settlementDate);

    productStore[store] = move(prd);
}
```

```
void putBarrier(
    const double        strike,
    const double        barrier,
    const Time          maturity,
    const double        monitorFreq,
    const double        smooth,
    const string&       store)
{
    const double smoothFactor = smooth <= 0 ? EPS : smooth;

    unique_ptr<Product<double>> prd = make_unique<UOC<double>>(
        strike, barrier, maturity, monitorFreq, smoothFactor);

    productStore[store] = move(prd);
}

void putContingent(
    const double        coupon,
    const Time          maturity,
    const double        payFreq,
    const double        smooth,
    const string&       store)
{
    const double smoothFactor = smooth <= 0 ? 0.0 : smooth;

    unique_ptr<Product<double>> prd = make_unique<ContingentBond<double>>(
        maturity, coupon, payFreq, smoothFactor);

    productStore[store] = move(prd);
}

void putEuropeans(
    //  maturities must be given in increasing order
    const vector<double>&   maturities,
    const vector<double>&   strikes,
    const string&           store)
{
    //  Create map
    map<Time, vector<double>> options;
    for (size_t i = 0; i < maturities.size(); ++i)
    {
        options[maturities[i]].push_back(strikes[i]);
    }

    unique_ptr<Product<double>> prd = make_unique<Europeans<double>>(
        options);

    productStore[store] = move(prd);
}

const Product<double>* getProduct(const string& store)
{
    auto it = productStore.find(store);
    if (it == productStore.end()) return nullptr;
    else return it->second.first.get();
}
```

Even though this is not implemented to keep the code short, this is the ideal place to conduct some sanity checks: are spots and times sorted in the specification of local volatility? Is volatility strictly positive? A violation could cause the program to crash or return irrelevant results.

Finally, in main.h, we have the higher level functions in the library:

```
#include "mcBase.h"
#include "mcMdl.h"
#include "mcPrd.h"
#include "mrg32k3a.h"
#include "sobol.h"
#include <numeric>
#include <fstream>
using namespace std;

#include "store.h"

struct NumericalParam
{
    bool                parallel;
    bool                useSobol;
    int                 numPath;
    int                 seed1 = 12345;
    int                 seed2 = 1234;
};

inline auto value(
    const Model<double>&    model,
    const Product<double>&  product,
    //  numerical parameters
    const NumericalParam&   num)
{
    //  Random Number Generator
    unique_ptr<RNG> rng;
    if (num.useSobol) rng = make_unique<Sobol>();
    else rng = make_unique<mrg32k3a>(num.seed1, num.seed2);

    //  Simulate
    const auto resultMat = num.parallel
        ? mcParallelSimul(product, model, *rng, num.numPath)
        : mcSimul(product, model, *rng, num.numPath);

    //  Return 2 vectors : the payoff identifiers and their values
    struct
    {
        vector<string> identifiers;
        vector<double> values;
    } results;

    //  Average over paths
    const size_t nPayoffs = product.payoffLabels().size();
    results.identifiers = product.payoffLabels();
    results.values.resize(nPayoffs);
    for (size_t i = 0; i < nPayoffs; ++i)
    {
        results.values[i] = accumulate(
```

```
            resultMat.begin(), resultMat.end(), 0.0,
            [i](const double acc, const vector<double>& v)
            { return acc + v[i]; }
        ) / num.numPath;
    }

    return results;
}

//  Generic valuation
inline auto value(
    const string&          modelId,
    const string&          productId,
    //  numerical parameters
    const NumericalParam&   num)
{
    //  Get model and product
    const Model<double>* model = getModel<double>(modelId);
    const Product<double>* product = getProduct<double>(productId);

    if (!model || !product)
    {
        throw runtime_error("value() : Could not retrieve model and product");
    }

    return value(*model, *product, num);
}
```

The first overload of *value*() accepts a model, a product, and some numerical parameters as arguments, runs the simulations, averages the results for all payoffs in the product, and returns them along with the payoff labels.[7]

The second overload picks a model and a product in the store by name and runs the first overload. It is the most convenient entry point into the simulation library.

Note that those functions have a boolean parameter "parallel" for calling *mcParallelSimul*() in place of *mcSimul*(). The template algorithm *mcParallelSimul*() is the parallel version built in the next chapter.

The functions can be executed on the console with some test data, although it is most convenient to export them to Excel. We have included in our online repository a folder xlCpp that contains a tutorial and some necessary files to export C++ functions to Excel. The tutorial teaches this technology within a couple of hours, skipping many details. A detailed discussion can be found in Dalton's dedicated [79].

The code in our repository is set up to build an xll. Open the xll in Excel, like any other file, and the exported functions appear along Excel's native functions. In particular, xPutBlackScholes, xPutDupire, xPutEuropean,

[7]Remember that we can have multiple payoffs in a product, and that the product advertises their names as part of a product's contract.

xPutBarrier, xPutContingent, xPutEuropeans, and xValue are the exported counterparts of the functions in store.h and main.h listed above.

Readers can easily build the xll or use the prebuilt one from our repository (they may need to install the included redistributables on a machine without a Visual Studio installation) and price all the products in all the models with various parameters. We detail below some results for the barrier option in Dupire. Barriers in Dupire are best priced with FDM, very accurately and in a fraction of a millisecond. However, many contexts don't allow FDM and the purpose of this book is to accelerate Monte-Carlo as much as possible. The test below is therefore our baseline, which we accelerate with parallelism and AAD in the rest of the book.

Finally, the spreadsheet xlTest.xlsx in the repository is set up to work with the functions of the library in a convenient Excel environment, and, among other things, reproduce all the results and charts in the publication. The xll must be loaded in Excel for the spreadsheet to work.

6.6 RESULTS

We set Dupire's model with a local volatility matrix of 60 times and 30 spots, monthly in time up to 5y, and every 5 points in spot (currently 100), from 50 to 200. We set local volatility to 15% everywhere for now[8] so our Dupire is really a Black-Scholes. We value a 3y 120 European call, with and without a knock-out barrier 150, simulating and monitoring the barrier over 156 weekly time steps. The theoretical value of the European option is known from the Black and Scholes formula, around 4.04. For the discrete barrier, we use the result of a simulation with 5M paths as a benchmark, where Sobol and mrg32k3a agree to the second decimal at 1.20.[9] We

[8]We calibrate the model in Chapter 13.

[9]To be compared to the continuous barrier price (with a closed-form formula available in Black and Scholes) around 1.07. As expected, the difference is substantial, even with weekly observations. A result derived by Glasserman, Broadie, and Kou [64] shows that the price of a continuous barrier is approximately equal to that of a discretely monitored one with the barrier pulled closer by 0.583 standard deviations at the barrier between monitoring dates. This result confirms the convergence of the discrete barrier in $\sqrt{\Delta_T}$ and provides a practical measure of the error: in our example, the standard deviation at the barrier is $\sigma\sqrt{\Delta T} \approx 0.0208$, $0.583 * 0.0208 \approx 0.0121$. It follows that our 150 discrete barrier is roughly equivalent to a $150 exp(0.0121) \approx 151.83$ continuous barrier, for which the analytic price is effectively around 1.207. The approximation is very useful in practice, but not precise enough to benchmark our tests. With 10M paths, simulation results are around 1.200.

simulate with Sobol and mrg32k3a, resetting mrg32k3a seeds in between simulations to produce a visual glimpse of the standard error. Results are shown in the charts below.

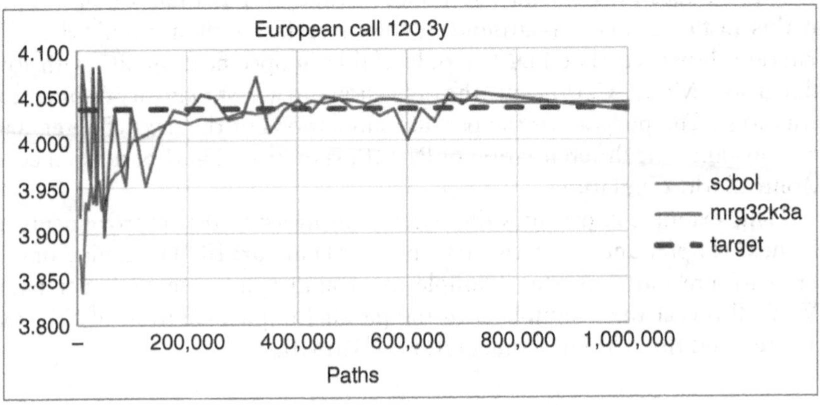

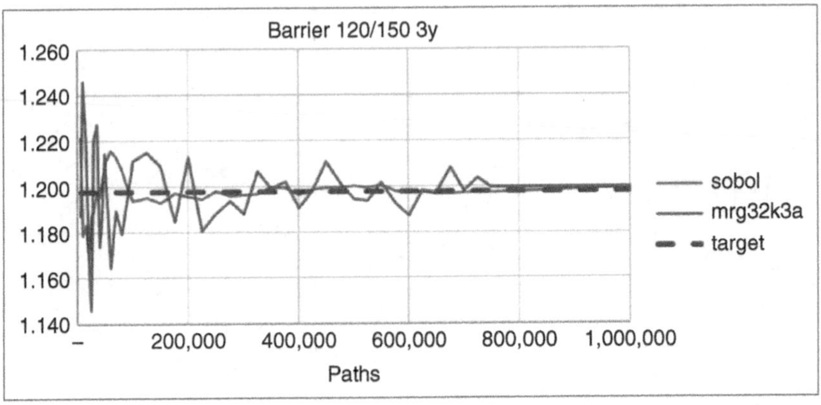

These confirm that convergence is slow with Monte-Carlo: it takes a large number of paths, at least 100,000, to obtain results accurate within the second decimal. Sobol converges faster and in a more reliable manner, with estimates within 2×10^{-2} of the benchmark with 100,000 paths or more. Although Sobol errors may be as large with a small number of paths, its convergence speed is visibly superior.

On our iMac Pro, it takes around 3 seconds with 500,000 paths to obtain an accurate price (for the European and the barrier simultaneously, as discussed on page 234) with either mrg32k3a or Sobol. Both return 1.20 for the barrier, correct to the second decimal. Sobol returns 4.04 for the

European, correct to the second decimal. Results obtained with mrg32k3a depend on the seeds; with (12345, 12346) we got 4.00.

In the context of a European or barrier option in Dupire's model, results of a comparable accuracy could be obtained with one-dimensional FDM, with 150 nodes in spot and 100 nodes in time, in less than a millisecond. In this instance, FDM is around 3,000 times faster than Sobol MC. We reiterate, however, that FDM is only viable in specific contexts with low dimension, Monte-Carlo being the only choice in a vast majority of practical situations. The purpose of the parallel simulations of the next chapter, and the constant time differentiation of Part III, is to bring FDM performance to Monte-Carlo simulations.

This example constitutes the baseline to measure the improvements of the next chapter and the performance of AAD in Part III. The timing of our serial Monte-Carlo, in this example, is around 6 microseconds per path. We shall revisit this example with the parallel implementation of the next chapter, and the AAD instrumentation of Part III.

CHAPTER 7

Parallel Implementation

In this chapter, we extend the code from the previous chapter with parallel simulations, applying everything we learned in Part I. It is generally not an easy task to parallelize an existing engine, because odds are that various parts were developed without thread safety in mind. We developed our engine from scratch in the previous chapter, knowing that we would eventually extend it with parallelism, so our work here is substantially less challenging than in real life.

It follows that we will not modify base or concrete models or products in any way, but we will need a substantial extension in the base and concrete RNGs.

The parallel algorithm is implemented in a parallel version of the template algorithm of the previous chapter, *mcParallelSimul*(), also in mcBase.h in our repository.

7.1 PARALLEL CODE AND SKIP AHEAD

Parallel simulation design

With serial path-wise MC, we processed the paths sequentially in a loop, the processing of every path consisting in the three following steps:

1. Generate a random Gaussian vector in dimension D with a call to $RNG :: nextG()$.
2. Consume the random vector to generate a scenario for the market observations on the event dates with a call to $Model :: generatePath()$.
3. Compute the payoffs over that path with a call to $Product :: payoffs()$.

We now process different paths in parallel. We know how to do this with the thread pool of Section 3.18 and the instructions on page 108. We spawn the paths for parallel processing into the thread pool, with the syntax:[1]

$$ThreadPool :: getInstance()- > spawnTask()$$

The thread pool schedules the parallel execution of the task and immediately returns a future that the main thread uses to monitor the status of the task. After the main thread finished dispatching the *nTasks* tasks, it waits for the tasks to complete and helps the worker threads executing the tasks in the meantime, with the syntax:

```
for (size_t task = 0; task < nTasks; ++task)
{
    ThreadPool::getInstance()->activeWait(futures[task]);
}
```

where *futures* is the vector of the *nTasks* futures associated to the *nTasks* tasks.

It follows that the parallel code is very similar to *mcSimul*() from the previous chapter, with the exception that the three lines which constitute the processing of a path are executed in parallel in the thread pool instead of sequentially on the main thread.

In reality, each task will consist not of a single path, but a batch of paths. We refer to the discussion on page 112, where we concluded that it is best to:

- spawn a large number of parallel tasks compared to the number of available cores, to take advantage of load balancing; and
- make tasks large enough so they take at least 0.1 ms to 1 ms of CPU time, so that the admin cost of sending the task to the pool remains negligible compared to its execution cost.

We estimated that a path takes around 6 microseconds of CPU time, and therefore default to batches of 64 paths each.

[1]We must start the thread pool with a call to:

$$ThreadPool :: getInstance()- > start()$$

when the application launches, and stop it with a call to:

$$ThreadPool :: getInstance()- > stop()$$

when the application closes. In an Excel-based application, we start the pool in *xlAutoOpen*(), right after the registration of our exported functions. In a console application, we would do that in the first lines of *main*(). The thread pool always remains available in the background, and it was designed so that it doesn't consume any resources when unused.

Thread safe design

If we are going to process different paths from different threads, the processing must be thread safe. *RNG* :: *nextG*(), *Model* :: *generatePath*() and *Product* :: *payoffs*() must be safe to call concurrently without interference. We knew that when we designed the simulation library, so we carefully coded the methods *Model* :: *generatePath*() and *Product* :: *payoffs*() (correctly) const. This is an example of thread safe design, since we did not, and will not, apply any lock in our simulation code. Those two methods can be safely called on the same model and product from different threads.

RNG :: *nextG*(), however, is not a const method: it modifies the state of the generator when it produces numbers. We cannot use the same RNG from different threads concurrently. The RNG is a *mutable object*, as defined on page 109, together with detailed instructions for making copies of such objects for every thread, and work with the executing thread's own copy from concurrent code. Our code scrupulously follows these instructions.

In addition to the RNG, we have two mutable objects on the workspace: a preallocated vector to hold the Gaussian numbers, and a preallocated vector of samples, or scenario, filled by *Model* :: *generatePath*() and consumed by *Product* :: *payoffs*(). We cope with these mutable objects in the exact same manner.

Skip ahead

All that precedes is a direct application of the parallel computing idioms from Part I to the embarrassingly parallel code in *mcSimul*() from the previous chapter. Readers are encouraged to review Section 3.19.

The only real difficulty is the following: we need the concurrent code to process in parallel the *same* paths that the serial code processes sequentially. With random sampling, this is nice to have. The multi-threaded algorithm produces the exact same results as the single-threaded version for a finite number of paths, and not only on convergence. With Sobol, this is *compulsory*. The low discrepancy of a Sobol sequence is only guaranteed when the points are sampled in a sequence.

The problem is that RNGs draw random numbers or vectors in a sequence, updating internal state with every new draw. The state Y_n that produces the random vector number n is the result of the prior $n - 1$ draws. To produce the same results in parallel, RNGs must acquire the ability to *skip ahead* to the state Y_n efficiently.

One obvious way to implement skip ahead is draw n samples and throw them away. This is obviously an inefficient process, linear in n. Many concrete RNGs can be skipped ahead much more efficiently, but this is entirely dependent on the RNG. In the case of both mrg32k3a and Sobol, a skip ahead of *logarithmic* complexity can be implemented.

For now, we add a virtual skip ahead method *RNG* : : *skipTo*() to our base RNG class, with a default implementation that generates and discards the first *n* draws. Concrete RNGs may override it with a more efficient skip ahead, when available:

```
// Class RNG

    //  Skip ahead
    virtual void skipTo(const unsigned b)
    {
        vector<double> dummy(simDim());
        for (int i = 0; i < b; ++i) nextG(dummy);
    }
```

We can proceed with the development and testing of the parallel template algorithm, and override *skipTo*() with fast skip ahead routines in the concrete RNGs thereafter.

The parallel template algorithm

The parallel template algorithm consists in rather minor modifications to the serial version, following to the letter the instructions of Section 3.19.

```
#include "threadPool.h"

#define BATCHSIZE 64
//  Parallel equivalent of mcSimul()
//  Same inputs and outputs
inline vector<vector<double>> mcParallelSimul(
    const Product<double>&          prd,
    const Model<double>&            mdl,
    const RNG&                      rng,
    const size_t                    nPath)
{
    auto cMdl = mdl.clone();

    const size_t nPay = prd.payoffLabels().size();
    vector<vector<double>> results(nPath, vector<double>(nPay));

    cMdl->allocate(prd.timeline(), prd.defline());
    cMdl->init(prd.timeline(), prd.defline());

    //  Allocate space for Gaussian vectors and paths,
    //      one for each thread
    ThreadPool *pool = ThreadPool::getInstance();
    const size_t nThread = pool->numThreads();
    vector<vector<double>> gaussVecs(nThread+1);     //  +1 for main
    vector<Scenario<double>> paths(nThread+1);
    for (auto& vec : gaussVecs) vec.resize(cMdl->simDim());
    for (auto& path : paths)
    {
```

```
        allocatePath(prd.defline(), path);
        initializePath(path);
    }

    //  Build one RNG per thread
    vector<unique_ptr<RNG>> rngs(nThread + 1);
    for (auto& random : rngs)
    {
        random = rng.clone();
        random->init(cMdl->simDim());
    }

    //  Reserve memory for futures
    vector<TaskHandle> futures;
    futures.reserve(nPath / BATCHSIZE + 1);

    //  Start
    //  Same as mcSimul() except we send tasks to the pool
    //      instead of executing them

    size_t firstPath = 0;
    size_t pathsLeft = nPath;
    while (pathsLeft > 0)
    {
        size_t pathsInTask = min<size_t>(pathsLeft, BATCHSIZE);

        //  Spawn here
        futures.push_back( pool->spawnTask ( [&, firstPath, pathsInTask] ()
        {
            //  Concurrent code

            //  Index of the executing thread
            const size_t threadNum = pool->threadNum();

            //  Pick the correct pre-allocated workspace
            //      for the executing thread
            vector<double>& gaussVec = gaussVecs[threadNum];
            Scenario<double>& path = paths[threadNum];

            //  Pick this thread's RNG and skip it
            auto& random = rngs[threadNum];
            random->skipTo(firstPath);

            //  Conduct the simulations, exactly the same as sequential
            for (size_t i = 0; i < pathsInTask; i++)
            {
                //  Gaussian vector
                random->nextG(gaussVec);
                //  Path
                cMdl->generatePath(gaussVec, path);
                //  Payoffs
                prd.payoffs(path, results[firstPath + i]);
            }

            //  Remember tasks must return bool
            return true;

```

```
            //  End of concurrent code
        }));

        pathsLeft -= pathsInTask;
        firstPath += pathsInTask;
    }

    //  Wait and help
    for (auto& future : futures) pool->activeWait(future);

    return results; //  C++11: move
}
```

We can execute the code straightaway, having already coded the interface in the entry point function *value*() in Section 6.6. The code works nicely, to execute it occupies 100% of the hardware threads, and it produces the *exact same* results as its single-threaded counterpart, approximately 10 to 20 times *slower*. The default skip ahead is so expensive that it exceeds by far the benefits of parallelism. We must now implement fast skip ahead algorithms for our concrete RNGs.

7.2 SKIP AHEAD WITH MRG32K3A

The transition equations on page 201 for mrg32k3a may be written in the compact matrix form:

$$\begin{pmatrix} X_k \\ X_{k-1} \\ X_{k-2} \end{pmatrix} = A \begin{pmatrix} X_{k-1} \\ X_{k-2} \\ X_{k-3} \end{pmatrix} \text{ mod } m1 \quad , \quad \begin{pmatrix} Y_k \\ Y_{k-1} \\ Y_{k-2} \end{pmatrix} = B \begin{pmatrix} Y_{k-1} \\ Y_{k-2} \\ Y_{k-3} \end{pmatrix} \text{ mod } m2$$

where:

$$A = \begin{pmatrix} 0 & 1403580 & -810728 \\ 1 & 0 & 0 \\ 0 & 1 & 0 \end{pmatrix} \quad , \quad B = \begin{pmatrix} 527612 & 0 & -1370589 \\ 1 & 0 & 0 \\ 0 & 1 & 0 \end{pmatrix}$$

and:

$$\begin{pmatrix} X_0 \\ X_{-1} \\ X_{-2} \end{pmatrix} = \begin{pmatrix} \alpha \\ \alpha \\ \alpha \end{pmatrix} \quad , \quad \begin{pmatrix} Y_0 \\ Y_{-1} \\ Y_{-2} \end{pmatrix} = B \begin{pmatrix} \beta \\ \beta \\ \beta \end{pmatrix}$$

Since moduli and matrix products are commutative, it immediately follows that the state $Z_k \equiv (Z_k^X, Z_k^Y)$ obtained after the production of k numbers:[2]

$$Z_k^X \equiv \begin{pmatrix} X_k \\ X_{k-1} \\ X_{k-2} \end{pmatrix}, \quad Z_k^Y \equiv \begin{pmatrix} Y_k \\ Y_{k-1} \\ Y_{k-2} \end{pmatrix}$$

is:

$$Z_k^X = A^k Z_0^X \quad \mathrm{mod}\ m1 = A^k \begin{pmatrix} \alpha \\ \alpha \\ \alpha \end{pmatrix} \quad \mathrm{mod}\ m1 ,$$

$$Z_k^Y = B^k Z_0^Y \quad \mathrm{mod}\ m2 = B^k \begin{pmatrix} \beta \\ \beta \\ \beta \end{pmatrix} \quad \mathrm{mod}\ m2$$

It follows that mrg32k3a is skipped ahead to state k by the left multiplication of the initial state vectors $(\alpha, \alpha, \alpha)^t$ and $(\beta, \beta, \beta)^t$ by the 3×3 matrices A^k and B^k (and the application of the moduli).

A classical and well-known logarithmic algorithm exists to compute the power matrix A^k. Consider the binary decomposition of k, that is the unique set of $a_i \in \{0, 1\}$ such that

$$k = \sum_i a_i 2^i$$

Then:

$$A^k = \prod_i (A^{2^i})^{a_i} = \prod_i A_i^{a_i} = \prod_{a_i=1} A_i$$

where we denoted $A_i \equiv A^{2^i}$. The A_is trivially satisfy:

$$\begin{cases} A_0 = A \\ A_{i+1} = A_i^2 \end{cases}$$

Hence, all the A_is (and hence, A^k) are computed in a number of matrix products corresponding to the number of bits in the binary decomposition of k, that is $log_2(k)$, a maximum of 32 for a 32-bit integer. The skip ahead completes in logarithmic time (17 matrix products in dimension 3×3 skipping 100,000 numbers).

[2]Careful that mrg32k3a works with numbers, not vectors.

A practical implementation must overcome the additional difficulty of overflow consecutive to the repeated multiplication of big numbers. We note that the product of two 32-bit integers cannot exceed 64 bits. Therefore, we store results in 64-bit unsigned long longs, and nest moduli after every scalar operation. This is slower than the double-based implementation of the generator in the previous chapter, but it doesn't matter for a logarithmic algorithm. First, we develop custom matrix products as private static methods of mrg32k3a, and a private method to skip k numbers with the algorithm we just described:

```
//  mrg323ka class

private:

    //  Matrix product with modulus
    static void mPrd(
        const unsigned long long     lhs[3][3],
        const unsigned long long     rhs[3][3],
        const unsigned long long     mod,
        unsigned long long           result[3][3])
    {
        //  Result go to temp, in case result points to lhs or rhs
        unsigned long long temp[3][3];

        for (size_t j = 0; j<3; j++)
        {
            for (size_t k = 0; k<3; k++)
            {
                unsigned long long s = 0;
                for (size_t l = 0; l<3; l++)
                {
                    // Apply modulus to innermost product
                    unsigned long long tmpNum = lhs[j][l] * rhs[l][k];
                    // Apply mod
                    tmpNum %= mod;
                    //  Result
                    s += tmpNum;
                    //  Reapply mod
                    s %= mod;
                }
                //  Store result in temp
                temp[j][k] = s;
            }
        }

        //  Now product is done, copy temp to result
        for (int j = 0; j < 3; j++)
        {
            for (int k = 0; k < 3; k++)
            {
                result[j][k] = temp[j][k];
            }
        }
    }
```

```

    //  Matrix by vector, exact same logic
    //  Except we don't implement temp,
    //      we never point result to lhs or rhs
    static void vPrd(
        const unsigned long long     lhs[3][3],
        const unsigned long long     rhs[3],
        const unsigned long long     mod,
        unsigned long long           result[3])
    {
        for (size_t j = 0; j<3; j++)
        {
            unsigned long long s = 0;
            for (size_t l = 0; l<3; l++)
            {
                unsigned long long tmpNum = lhs[j][l] * rhs[l];
                tmpNum %= mod;
                s += tmpNum;
                s %= mod;
            }
            result[j] = s;
        }
    }

    void skipNumbers(const unsigned b)
    {
        if ( b <= 0) return;
        unsigned skip = b;

        static constexpr unsigned long long
            m1l = unsigned long long(m1);
        static constexpr unsigned long long
            m2l = unsigned long long(m2);

        unsigned long long Ab[3][3] = {
            { 1, 0 ,0 },
            { 0, 1, 0 },
            { 0, 0, 1 }
        },
            Bb[3][3] = {
                { 1, 0 ,0 },
                { 0, 1, 0 },
                { 0, 0, 1 }
        },
            Ai[3][3] = {         //  A0 = A
                {
                    0,
                    unsigned long long (a12) ,
                    unsigned long long (m1 - a13)
                    //  m1 - a13 instead of -a13
                    //  so results are always positive
                    //  and we can use unsigned long longs
                    //  after modulus, we get the same results
                },
                { 1, 0, 0 },
                { 0, 1, 0 }
        },
```

```
            Bi[3][3] = {          //  B0 = B
                {
                    unsigned long long (a21),
                    0 ,
                    unsigned long long (m2 - a23)
                    //  same logic: m2 - a32
                },
                { 1, 0, 0 },
                { 0, 1, 0 }
        };

        while (skip>0)
        {
            if (skip & 1)    //  i.e. ai == 1
            {
                //  accumulate Ab and Bb
                mPrd(Ab, Ai, m11, Ab);
                mPrd(Bb, Bi, m21, Bb);
            }

            //  Recursion on Ai and Bi
            mPrd(Ai, Ai, m11, Ai);
            mPrd(Bi, Bi, m21, Bi);

            skip >>= 1;
        }

        //  Final result
        unsigned long long X0[3] =
        {
            unsigned long long (myXn),
            unsigned long long (myXn1),
            unsigned long long (myXn2)
        },
            Y0[3] =
        {
            unsigned long long (myYn),
            unsigned long long (myYn1),
            unsigned long long (myYn2)
        },
            temp[3];

        //  From initial to final state
        vPrd(Ab, X0, m11, temp);

        //  Convert back to doubles
        myXn = double(temp[0]);
        myXn1 = double(temp[1]);
        myXn2 = double(temp[2]);

        //  Same for Y
        vPrd(Bb, Y0, m21, temp);
        myYn = double(temp[0]);
        myYn1 = double(temp[1]);
        myYn2 = double(temp[2]);
    }
```

Next, we implement the public override *skipTo*(), taking into account the vector and antithetic logics:

```
//  mrg323ka class

public:

    void skipTo(const unsigned b) override
    {
        //  First reset to 0
        reset();

        //  How many numbers to skip
        unsigned skipnums = b * myDim;
        bool odd = false;

        //  Antithetic: skip only half
        if (skipnums & 1)
        {
            //  Odd
            odd = true;
            skipnums = (skipnums - 1) / 2;
        }
        else
        {
            //  Even
            skipnums /= 2;
        }

        //  Skip state
        skipNumbers(skipnums);

        //  If odd, pre-generate for antithetic
        if (odd)
        {
            myAnti = true;

            //  Uniforms
            generate(
                myCachedUniforms.begin(),
                myCachedUniforms.end(),
                [this]() { return nextNumber(); });

            //  Gaussians
            generate(
                myCachedGaussians.begin(),
                myCachedGaussians.end(),
                [this]() { return invNormalCdf(nextNumber()); });
        }
        else
        {
            myAnti = false;
        }
    }
```

7.3 SKIP AHEAD WITH SOBOL

Recall from the equation on page 207 that Sobol's state is a (xor) combination of direction numbers with coefficients 0 or 1:

$$y_k^d = \sum_{j=0}^{[log_2(k)]} a_j^k DN_j^d$$

and it immediately follows that Sobol is skipped to state y_k in logarithmic time.

In addition, we have seen that DN_0 flicks in and out every two numbers, starting with number 1. It follows that $a_0^k = 1$ when:

$$\left[\frac{k+1}{2}\right]$$

is odd, zero otherwise. DN_1 flicks every four numbers, starting with number 2, so $a_1^k = 1$ when:

$$\left[\frac{k+2}{4}\right]$$

is odd, zero otherwise. DN_2 flicks every eight numbers, starting with number 4, so $a_2^k = 1$ when:

$$\left[\frac{k+4}{8}\right]$$

is odd, zero otherwise. DN_j flicks every 2^{j+1} numbers, starting with number 2^j, so $a_j^k = 1$ when:

$$\left[\frac{k+2^j}{2^{j+1}}\right]$$

is odd, zero otherwise. And it follows that:

$$y_k^d = xor \sum_{j \in J_k} DN_j^d$$

where the sum means xor-sum and:

$$J_k = \left\{ j \leq \log_2 k, \left[\frac{k+2^j}{2^{j+1}}\right] \text{ is odd} \right\}$$

The code is listed below:

```
// sobol class

void skipTo(const unsigned b) override
{
    // Check skip
    if (!b) return;

    // Reset Sobol to 0
    reset();

    // The actual Sobol skipping algo
    unsigned im = b;
    unsigned two_i = 1, two_i_plus_one = 2;

    unsigned i = 0;
    while (two_i <= im)
    {
        if (((im + two_i) / two_i_plus_one) & 1)
        {
            for (unsigned k = 0; k<myDim; ++k)
            {
                myState[k] ^= jkDir[i][k];
            }
        }

        two_i <<= 1;
        two_i_plus_one <<= 1;
        ++i;
    }

    // End of skipping algo

    // Update next entry
    myIndex = unsigned(b);
}
```

7.4 RESULTS

The parallel code in this chapter brings together everything we learned so far. It produces the exact same results as the sequential code, faster by the number of available CPU cores. On our iMac Pro (8 cores), we exactly reproduce the examples of Section 6.6 in 350 ms with either mrg32k3a or Sobol, with a perfect parallel efficiency.[3] It is remarkable that we can run 500,000 simulations over 156 time steps in something like a quarter of a second.

[3]Acceleration is superior to the number eight of physical cores thanks to hyper-threading.

Timing is very satisfactory, as is parallel efficiency. Still, with 350 ms pricing, it would take several minutes to compute a risk report of all the sensitivities to the 1,800 (30 × 60) local volatilities with traditional finite differences. In the final and most important part of this publication, we will learn techniques to make this happen in *half a second* without loss of accuracy.

PART III

Constant Time Differentiation

Introduction

OPENING REMARKS

The main goal of quantitative research and derivatives libraries is not valuation but risk: what amounts of what assets we must trade to hedge the market risk or the regulatory adjustment of a derivatives portfolio. It is a major achievement of financial theory to have demonstrated that hedge ratios correspond to differential sensitivities. The goal is therefore to produce fast and accurate differentials of the valuation function. The determination and implementation of the valuation function is merely a step on the way, since, obviously, a function must be defined before it can be differentiated. We dealt with valuation in the previous part. This part is dedicated to differentiation.

Differentials are traditionally computed in finance by finite differences – or *bumping*: bump market inputs one by one and repeat valuation every time. This procedure may take unreasonable time, since valuation is repeated as many times as we have market inputs. *The complexity of the differentiation is linear in the number of differentials.* This number is typically large, several thousands for a derivatives book or the xVA of a netting set. It is not viable when the valuation is expensive, as is the case with regulatory calculations, and Monte-Carlo simulations in general. In recent years, the financial industry adopted the much more efficient algorithmic adjoint differentiation (AAD), which computes all sensitivities in constant time.

Despite being vastly superior and widely adopted, AAD is still largely misunderstood. This part explains AAD and its implementation, both in a general context and for the differentiation of Monte-Carlo simulations.

As promised, the publication comes with a professional implementation in C++, which can be found in our online repository. The files AAD.cpp, AAD.h, AADExpr.h, AADNode.h, AADNumber.h, and AADTape.h form a self-contained implementation of AAD,[1] including advanced constructs like memory management and expression template acceleration. The code is explained and commented in detail in the following chapters.

AAD is arguably the strongest addition to finance of the past decade. It changed the way we design and develop derivatives systems and gave us the means to compute the full risk of the xVA on a large netting set within

[1]With a dependency on gaussians.h for the analytic differentiation of Gaussian functions and blocklist.h for memory management.

minutes on a laptop. Without AAD, such calculations had to be conducted overnight in data centers.

For example, the final tests of our implementation of simulations in Dupire's model from Chapter 7 took 1,801 inputs, 1,800 local volatilities and one spot, and returned the price of a 3y barrier option. With our parallel implementation, it took 350 ms with 500,000 paths and 156 steps with Sobol numbers. If we differentiated this calculation with finite differences, we would compute the sensitivity to the spot (delta) and the 1,080 active volatilities[2] (what Dupire calls "microbucket" vega) by repeating valuation 1,081 times, each time bumping one input by a small amount and keeping the other inputs constant. Differentiation time is *linear* in the number of inputs. It would take around 380s or six and a half minutes to complete.[3]

With AAD, we obtain the same results in *half a second.*

In the case of a large xVA, valuation may take minutes, even with all the theoretical, numerical, and computational improvements explained in [10], [80], [7], [30], [81], [31] and the optimized simulations of Part II. In addition to credit, xVA typically depends on many thousands of market variables: all the underlying asset prices, as well as rate, spread, and other curves and all the implied volatility surfaces for all the currencies and assets in the netting set. The computation of the risk could take days.

With AAD, it takes minutes on a laptop, without loss of accuracy. On the contrary, if anything, AAD differentials are *more* accurate. AAD achieves this by computing all sensitivities analytically and *in constant time.*

---

[2]We have 1,800 local volatilities, out of which 1,080 are between now and 3y, hence active for the 3y barrier.

[3]In general, the goal of derivatives risk systems is to compute sensitivities to hedge instruments, not model parameters. Model parameters are computed from hedge instruments, generally by calibration. Sensitivities to hedge instruments can be produced in two ways: directly, by bumping the value of hedge instruments one by one and repeating the whole valuation process, including calibration; or, by computing sensitivities to model parameters first, and using a variety of techniques explained in Chapter 13 to turn them into sensitivities to hedge instruments. The second option is a much better one: it avoids a potentially unstable differentiation through calibration, provides valuable information in terms of sensitivity to model parameters, and offers the possibility to choose different instruments for calibration and risk. But, with a large number of model parameters, it may be prohibitively expensive with bumping; therefore, many risk systems based on bumping implement the first option. We will see that AAD offers a faster alternative to bumping, one that is constant in the number of sensitivities. Hence, an additional benefit of AAD is to permit the implementation of the better option for the production of risk without additional cost (actually, orders of magnitude faster). With AAD, it is best practice to compute sensitivities to model parameters first, and turn them into sensitivities to hedge instruments next. This is exactly the procedure we follow: Chapter 12 computes sensitivities to model parameters and Chapter 13 explains how to turn them into sensitivities to hedge instruments.

That differentiation is analytic is often put forward, but it makes little difference in practice. AAD is all about the constant time. In theory, the constant time is about that of three evaluations. Empirically, the number varies, depending on the valuation algorithm, the quality of its instrumentation, and the performance of the AAD library. Such speed is unprecedented since FDM. Its impact on the derivatives industry is substantial, profound, and irreversible. In addition to the obvious application to risk sensitivities, AAD is applied, for example, to speed up calibrations or compute regulatory amounts like FRTB; see [82].

Like finite differences, AAD is noninvasive. This means that developers only write the calculation code, and the algorithm produces its differentials automatically. Automatic differentiation is usually presented in two flavors: forward and backward. Forward auto-differentiation computes differentials analytically but in linear time, like bumping, and usually marginally slower. It is therefore mostly an interesting curiosity without practical relevance in our field. We coded forward AD in production systems in the early 2010s and never used it. In this book, we skip it altogether and focus on the automatic *backward* auto-differentiation, which is the one that produces differentials in constant time.

AAD is also sometimes advocated as a means to compute *better* differentials: AAD differentials are analytic, hence, deemed more accurate. While this may be correct in a small number of specific situations, in most cases of practical relevance, differentials produced with AAD or bumping are almost indiscernible.[4] In most cases, AAD produces similar differentials but it produces them faster. AAD therefore doesn't resolve all the problems with differentiation; it only resolves the problem of *speed*. AAD cannot, for instance, compute the differentials of discontinuous functions any better than bumping. In particular, AAD cannot differentiate code through control flow ("if this then that" statements). It cannot compute the sensitivities of discontinuous payoffs, like digitals or barriers. Evidently, such functions are not differentiable and AAD should not be expected to perform an impossible task. Bump risk over discontinuous functions is known to be extremely unstable. AAD risk over such functions ignores control flow. Both are wrong. Risk over discontinuous functions is not resolved with AAD; it is resolved by the approximation of those discontinuous functions by close continuous ones, a technique known as *smoothing*. We investigate smoothing in a general context in our talk [77] and our publication [11], where we abstract smoothing into *Fuzzy Logic* and produce general smoothing algorithms for scripted payoffs. For an enlightening presentation of digital and barrier smoothing with convolution kernels, see Bergomi's [78].

[4]Although we will discuss an important counter-example in Chapter 13, where AAD produces a more stable "superbucket" risk.

The origins of AAD can be traced back to 1964 [83]. Adjoint differentiation (AD), the main principle behind AAD, has been applied for a long time in the field of machine learning, under the name "backpropagation." "Backprop," as it is called now, contributed to the recent and spectacular success of deep learning models. It is the constant time calculation of the many derivatives of the loss functions that allows deep artificial neural networks to learn their parameters in a reasonable time. Machine learning, neural networks, and backpropagation are discussed in a vast amount of literature. We are partial to Bishop's excellent [65].

AAD was introduced to finance in 2006 by Giles and Glasserman's award-winning "Smoking Adjoints" [5], followed by pioneering work like Luca Capriotti's [84], [85], [86] or [87]. In the early 2010s, Danske Bank delivered, and publically demonstrated, a large-scale implementation throughout its production and regulatory systems. Today, AAD is widely considered the only acceptable alternative for the computation of risk sensitivities for xVA and other bank-wide regulatory amounts.

SUMMARY

AAD is the subject of a number of publications like [88], and the author was personally involved in lectures, professional training, workshops, and presentations following the implementation of AAD in Danske Bank's systems. The theory and mathematics of AAD are simple, yet people generally find it hard to wrap their minds around it at first. We hope our introductory Chapters 8 and 9 help with that.

A *basic* implementation is not hard, either. It is based on operator overloading and can be written in languages that support it, like C++.

A *professional* implementation must also provide efficient memory management; otherwise cache and memory load may defeat the benefits of AAD and result in code *slower* than bumping. Chapter 10 delivers an efficient, general-purpose AAD library, including memory management facilities.

Chapter 11 discusses the application and limits of AAD for the differentiation of arbitrary and financial calculation code.

Chapter 12 differentiates the parallel simulation library of Part II.

Chapter 13 introduces the fundamental notion of checkpointing, a crucial idiom for the practical implementation of AAD, and applies it to the problem of risk propagation from local to implied volatilities in the context of Dupire's model.

Chapter 14 extends the AAD library to differentiate not one, but multiple results of a calculation with respect to its inputs, and applies it to produce an itemized risk report for a portfolio of financial products, in a time virtually constant in the number of risk reports. In reality, AAD is always linear in the dimension of the differentiated result. What we achieve is a small marginal time per additional result in situations of practical relevance, hence the title of the chapter: "multiple differentiation in *almost* constant time."

Chapter 15 introduces a different breed of AAD implementation based on *expression templates*, something initially proposed in [89]. Expression templates may accelerate the production of AAD differentials by a significant factor, depending on the differentiated code, and approach the performance of the manual adjoint differentiation covered in Chapter 8. The differentiation of our simulation library is around twice as fast with expression templates.

A FIRST APPROACH TO ADJOINT DIFFERENTIATION

For a first introduction to adjoint differentiation (AD), we borrow Huge's analogy [90] to the more familiar field of matrix algebra. Consider a calculation represented by the function $F : \mathbb{R}^n \to \mathbb{R}$ that produces a scalar result z out of an input X in dimension n. Assume F may be broken down into a sequence of sub-calculations $G : \mathbb{R}^n \to \mathbb{R}^m$, $K : \mathbb{R}^m \to \mathbb{R}^p$ and $H : \mathbb{R}^p \to \mathbb{R}$, such that:

$$F(X) = H\{K[G(X)]\}$$

where G, K, and H are simple enough that their Jacobians are known analytically and computed without difficulty.

All calculations can be broken down in this manner, although, in general, they consist in not three, but thousands or millions of elementary calculations with derivatives known in closed form. In programming languages with support for *operator overloading*, all calculations can be *automatically* decomposed into elementary operations: additions, multiplications, logarithms, square roots, and so forth, which derivatives are obviously known. Operator overloading allows developers to overload what happens when operators like + or $*$ or functions like *log*() or *sqrt*() are invoked. With AAD, operators are overloaded to *record* the sequence of operations.

To evaluate F means evaluate G, K, and H:

$$X \in \mathbb{R}^n \overset{G}{\to} Y \in \mathbb{R}^m \overset{K}{\to} Z \in \mathbb{R}^p \overset{H}{\to} z \in \mathbb{R}$$

in the sequence:

$$\begin{aligned} 1\ Y_m &= G(X_n) \\ 2\ Z_p &= K(Y_m) \\ 3\ z\ &= H(Z_p) \end{aligned}$$

where lowercase letters are scalars and capital letters are vectors with dimension in subscript. It follows immediately from the chain rule that the Jacobian matrices of F, G, K, and H are related by matrix algebra:

$$\left(\frac{\partial F}{\partial X}\right)^t_{n\times 1} = \left(\frac{\partial G}{\partial X}\right)^t_{n\times m} \left(\frac{\partial K}{\partial Y}\right)^t_{m\times p} \left(\frac{\partial H}{\partial Z}\right)^t_{p\times 1}$$

where the subscripts denote the dimensions of the Jacobian matrices. It is our working assumption that G, K, and H are simple enough that their Jacobians are known and explicit. To find the Jacobian of F, conventional algorithms multiply the Jacobian matrices, in the order of evaluation, left to right:

$$\left(\frac{\partial F}{\partial X}\right)^t_{n\times 1} = \left[\left(\frac{\partial G}{\partial X}\right)^t_{n\times m} \left(\frac{\partial K}{\partial Y}\right)^t_{m\times p}\right]_{n\times p} \left(\frac{\partial H}{\partial Z}\right)^t_{p\times 1}$$

The complexity of the matrix product between brackets is nmp, and its product by the right-hand side vector has complexity np. The complexity of the differentiation is therefore $(m + 1)np$.

Because matrix products are associative, we can also accumulate the differentials of F in the *reverse* order:

$$\left(\frac{\partial F}{\partial X}\right)^t_{n\times 1} = \left(\frac{\partial G}{\partial X}\right)^t_{n\times m} \left[\left(\frac{\partial K}{\partial Y}\right)^t_{m\times p} \left(\frac{\partial H}{\partial Z}\right)^t_{p\times 1}\right]_{m\times 1}$$

Now, the right-hand-side term between brackets has complexity mp and results in a vector of dimension m, which product by the left-hand-side matrix is of complexity nm. The complexity of the differentiation of F is therefore $m(n + p)$, one order of magnitude lower.

Differentials accumulate one magnitude faster in reverse order.

It must be so, because F is a scalar function. For this reason, the right-most Jacobian of the decomposition must be a vector. When we multiply Jacobians right to left, the successive results always collapse into a vector, so all the matrix products are of quadratic complexity instead of cubic.

What traditional differentiation does is expand the *Jacobians* to X from the innermost function G to the outermost function F:

1 $\dfrac{\partial G}{\partial X}$

2 $\dfrac{\partial K}{\partial X} = \dfrac{\partial K}{\partial Y}\dfrac{\partial G}{\partial X}$

3 $\dfrac{\partial H}{\partial X} = \dfrac{\partial H}{\partial Z}\dfrac{\partial K}{\partial X} = \dfrac{\partial F}{\partial X}$

where it is clear that the differentiation occurs in the order G, K, H of the evaluation, or *forward*.

Reverse, or *backward*, differentiation propagates the *adjoints*, the derivatives of F to Z, Y, and X, in this (reverse) order. It starts with the adjoint of Z:

$$\frac{\partial F}{\partial Z} = \frac{\partial H}{\partial Z}$$

Next, it computes the adjoint of Y:

$$\frac{\partial F}{\partial Y} = \frac{\partial F}{\partial Z}\frac{\partial K}{\partial Y}$$

Finally, it produces the adjoint of X:

$$\frac{\partial F}{\partial X} = \frac{\partial F}{\partial Y}\frac{\partial G}{\partial X}$$

So Jacobians expand forwards, whereas adjoints propagate backwards. The adjoints of scalar calculations accumulate one order of magnitude faster than their Jacobians. To compute differentials in this manner is called *adjoint differentiation* or AD. When adjoints accumulate over a sequence of operations *automatically* extracted from a calculation, it is called *automatic adjoint differentiation* (AAD).

AAD reduces differentiation complexity by an order of magnitude, but only for scalar functions. If F (and H) were multidimensional functions valued in $\mathbb{R}^q$, AD complexity would be linear in q, and slower than conventional differentiation when $q > n$.

In addition, to compute the Jacobians $\partial H/\partial Z$ and $\partial K/\partial Y$, Z and Y must be known. So not only the entire sequence of operations in a calculation must be stored in memory, but also all the intermediate results. It is not a problem with three functions, but real-life calculations are decomposed into thousands or millions of elementary operations, imposing a heavy load on RAM and cache.

The practical performance is therefore entirely dependent on how fast operations and results are recorded. AAD is more dependent on implementation than most algorithms. A naive implementation could result in slow differentiation and RAM exhaustion. Efficient memory management is crucial.

The differentiation of multidimensional functions is covered in Chapter 14. Memory management is discussed in deep detail in Chapter 10, where we manage to record operations and results with very little overhead. The techniques of Chapter 13 split differentiations to alleviate RAM and cache footprint. It is, however, impossible to completely eliminate recording overhead, and the number of operations in a calculation is typically large. The techniques of Chapter 15 reduce the number of records by recording entire *expressions* in place of elementary operations.

The next chapter explains adjoint differentiation from a different angle, where it is shown that AD finds the common factors in the expression of differentials, reducing the complexity of their computation. Chapter 9 explains in detail how calculations are automatically split and their operations recorded to accumulate adjoints in the reverse order. Chapters 10 and following deal with the many aspects of a practical implementation.

CHAPTER 8

Manual Adjoint Differentiation

In this first introductory chapter to AAD, we introduce adjoint differentiation (AD) through simple examples. We explain how AD may compute many derivative sensitivities in constant time and show how this can be achieved by coding AD by hand.

In the next chapter, we will generalize and formalize adjoint mathematics, and show how adjoint calculations can be executed automatically through autogenerated calculation graphs without the need to code any form of differentiation manually.

In Chapter 10, we will turn these ideas into a general-purpose AAD library in C++, and, in subsequent chapters, we will apply this AAD library to differentiate serial and parallel Monte-Carlo simulations in constant time.

8.1 INTRODUCTION TO ADJOINT DIFFERENTIATION

We start with a simple example where we differentiate a well-known analytic function with respect to all its inputs and introduce adjoint calculations as a means to compute all those derivatives from one another in constant time. From the example, we abstract and illustrate a general methodology for the manual computation of adjoints and the production of corresponding code.

Example: The Black and Scholes formula

The well known Black and Scholes formula, derived in 1973 in [22], gives the price of a call option in the absence of arbitrage, under the assumption that the volatility of the underlying asset is a known constant, as a function of a small number of observable variables and the constant volatility. Under this simple model, the value of the option may be expressed in closed-form, as demonstrated on page 165:

$$C(S_0, r, y, \sigma, K, T) = DF[FN(d_1) - KN(d_2)]$$

The function has 6 inputs:

- The current spot price S_0
- The interest rate r
- The dividend yield y
- The volatility σ
- The strike K and maturity T of the option

and carries the following calculations:

- The discount factor $DF = \exp(-rT)$
- The forward $F = S_0 \exp[(r - y)T]$
- The standard deviation $std = \sigma\sqrt{T}$ of the return to maturity
- The log-moneyness $d = \frac{\log\left(\frac{F}{K}\right)}{std}$
- $d_1 = d + \frac{std}{2}, d_2 = d - \frac{std}{2}$
- The probabilities to end in the money under the spot measure $N(d_1)$ and under the risk-neutral measure $N(d_2)$ where N is the cumulative normal distribution
- Finally the call option price $C = DF[FN(d_1) - KN(d_2)]$

Note $N()$ has no analytic form but precise (if somewhat expensive) approximations exist, such as the function *normalCdf*() in the file gaussians.h in our repository.

To implement Black and Scholes' formula in C++ code is trivial:

```
double blackScholes(
    const double S0,
    const double r,
    const double y,
    const double sig,
    const double K,
    const double T)
{
    // 1
    const double df = exp(-r*T);
    // 2
    const double F = S0 * exp((r-y)*T);
    // 3
    const double std = sig * sqrt(T);
    // 4
    const double d = log(F/K) / std;
    // 5, 6
    const double d1 = d + 0.5 * std, d2 = d1 - std;
    // 7, 8
    const double nd1 = normalCdf(d1), nd2 = normalCdf(d2);
```

```
    // 9
    const double c = df * (F * nd1 - K * nd2);

    return c;
}
```

where the function was purposely written in a form that highlights and stores the intermediate calculations.

It follows that, despite its closed form, the evaluation of Black-Scholes is not especially cheap: it takes two evaluations of the normal distribution, two exponentials, a square root, and 10 multiplications and divisions.

In what follows, we call "operation" a piece of the calculation that produces an intermediate result out of previously available numbers and for which partial derivatives are known analytically. Our function has nine such operations, labeled accordingly in the code.

Differentiation

We now investigate different means of computing the six differentials of the formula to its inputs as efficiently as possible.

One solution is finite difference:

$$\frac{\partial C}{\partial S_0} \approx \frac{C(S_0 + \varepsilon, r, y, \sigma, K, T) - C(S_0, r, y, \sigma, K, T)}{\varepsilon}$$

$$\frac{\partial C}{\partial \sigma} \approx \frac{C(S_0, r, y, \sigma + \varepsilon, K, T) - C(S_0, r, y, \sigma, K, T)}{\varepsilon}$$

etc.

Besides the lack of accuracy, such differentiation through one-sided finite differences performs six additional evaluations of the function, once per input. The benefit of finite differences is that they are easily automated. The drawback is that the computation of differentials is expensive, linear in the number of desired sensitivities.

In our simple example, the differentials of Black and Scholes' formula are derived in closed form without difficulty, and are actually well known:

$$\frac{\partial C}{\partial S_0} = \frac{DF \cdot F}{S_0} \cdot N(d_1)$$

$$\frac{\partial C}{\partial r} = DF \cdot KT \cdot N(d_2)$$

$$\frac{\partial C}{\partial y} = -DF \cdot FT \cdot N(d_1)$$

$$\frac{\partial C}{\partial \sigma} = DF \cdot F \cdot n(d_1) \cdot \sqrt{T}$$
$$\frac{\partial C}{\partial K} = -DF \cdot N(d_2)$$
$$\frac{\partial C}{\partial T} = DF \cdot F \cdot n(d_1) \cdot \frac{\sigma}{2\sqrt{T}} + rK \cdot DF \cdot N(d_2) - y \cdot DF \cdot F \cdot N(d_1)$$

where n is the normal density: $n(x) = \frac{\partial N(x)}{\partial x} = \frac{1}{\sqrt{2\pi}} \exp\left(-\frac{x^2}{2}\right)$.

Perhaps surprisingly, the closed-form derivatives don't help performance: to compute the six sensitivities, we must still conduct six additional calculations, each of a complexity similar to the Black and Scholes formula.

But we note that all those derivatives share many common factors, so if we compute all the sensitivities *together*, we can calculate those common factors once and achieve a much better performance. In particular, the expensive $N(d_1)$ and $N(d_2)$ may be calculated once and reused six times. Besides, these quantities are known from the initial evaluation of the formula. We have other common factors, like $n(d_1)$, which is shared between theta and vega.[1]

It turns out that this is not a coincidence: the different sensitivities of a given calculation always share common factors as a consequence of the chain rule for derivatives:

$$\frac{\partial f[g(x)]}{\partial x} = \frac{\partial f}{\partial g}\frac{\partial g}{\partial x}$$

from which it follows that, when a calculation with inputs x and y computes an intermediate result $z = g(x, y)$ and uses it to produce a final result $f(z)$, we have:

$$\frac{\partial f}{\partial x} = \frac{\partial f}{\partial z}\frac{\partial z}{\partial x}, \frac{\partial f}{\partial y} = \frac{\partial f}{\partial z}\frac{\partial z}{\partial y}$$

so the two sensitivities share the common factor $\frac{\partial f}{\partial z}$. In Black and Scholes, C is a function of F and F is a function of S_0 and y (and r and T). S_0 and y don't contribute to the result besides affecting the forward, so it follows from the chain rule that:

$$\frac{\partial C}{\partial S_0} = \frac{\partial C}{\partial F}\frac{\partial F}{\partial S_0}, \frac{\partial C}{\partial y} = \frac{\partial C}{\partial F}\frac{\partial F}{\partial y}$$

[1]Derivatives practitioners give Greek names to the sensitivities of Black-Scholes: delta for the spot, vega for the volatility, theta for the time to expiry, etc.

The two sensitivities share the common factor $\frac{\partial C}{\partial F} = N(d_1)$, as seen in their analytic expression.[2] We can produce them together cheaply by computing $\frac{\partial C}{\partial F}$ first, and then multiply it respectively by $\frac{\partial F}{\partial S_0}$ and $\frac{\partial F}{\partial y}$. We note that in the *evaluation* of the function, F is computed first out of S_0, and C is computed next as a function of F. We *reversed* that order to compute sensitivities: we computed the common factor $\frac{\partial C}{\partial F}$ first, then $\frac{\partial C}{\partial S_0} = \frac{\partial C}{\partial F}\frac{\partial F}{\partial S_0}$ and $\frac{\partial C}{\partial y} = \frac{\partial C}{\partial F}\frac{\partial F}{\partial y}$ as a function of the common factor.

In order to compute all the sensitivities with maximum performance, we must identify all the factors common to multiple derivatives and compute them once. In general, we don't know the derivatives analytically and we can't inspect their formula to identify the common factors by the eye. We must find a systematic way to identify all those common factors and compute them first. This is where adjoint differentiation comes into play. AD builds on the chain rule to find an order of computation for the derivatives that maximizes factoring and guarantees that all derivatives are computed in a time proportional to the number of operations, but independent of the number of inputs. We will see that this is achieved by systematically reversing the calculation order for the computation of differentials.

Adjoint differentiation

We identify the computation order for differentials that optimally reuses common factors with the help of two fundamental notions: the *adjoints* and the *calculation graph*. We illustrate these notions here and formalize them in the next chapter.

We call "adjoint of x" and denote $\overline{x}$ the derivative of the final result to x. In Black and Scholes, for instance:

$$\overline{\sigma} \equiv \frac{\partial C}{\partial \sigma} = DF \cdot F \cdot n(d_1) \cdot \sqrt{T}$$

To compute all the derivative sensitivities of a calculation means to calculate the adjoints of all its inputs. We will see that we can compute the adjoints of all the *operations*, including those of inputs, in constant time.

[2]The sensitivities to r and T are also dependent on this factor, *in part*, for instance,

$$\frac{\partial C}{\partial T} = \frac{\partial C}{\partial F}\frac{\partial F}{\partial T} + \ldots$$

where the rest ... is due to contributions of T to C besides affecting the forward. This will be clarified in what follows.

It helps to draw a visual representation of all the operations involved and their dependencies. This is something we can easily infer from the code:

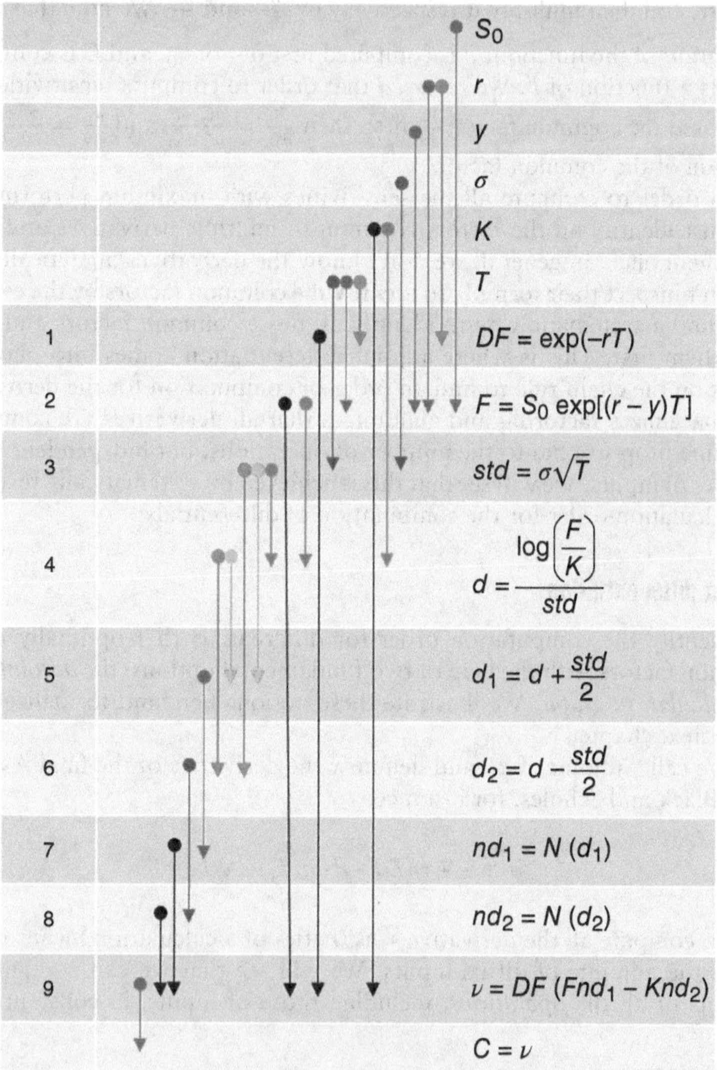

The figure shows the sequence of the nine operations involved in Black and Scholes, top to bottom. The arrows represent the dependencies between them. This is called a *calculation graph*. In the next chapter, we will learn to generate such graphs automatically from the code. For now, we note that every calculation defines a graph, which can be drawn by hand from an inspection of the code, without major difficulty. The graph offers a visual glimpse of the dependencies between the operations, which, in conjunction with the chain rule, provides the means to compute sensitivities in a way that reuses all the common factors at best.

Starting with S_0, we have only one branch out of S_0 that connects to F; hence, by an immediate application of the chain rule:

$$\frac{\partial C}{\partial S_0} = \frac{\partial C}{\partial F}\frac{\partial F}{\partial S_0}$$

or, using adjoint notations:

$$\overline{S_0} = \frac{\partial F}{\partial S_0}\overline{F} = \frac{F}{S_0}\overline{F}$$

Hence, we can compute $\overline{S_0}$ *after* we know $\overline{F}$. We note once again that the adjoint relationship is reversed: F is a function of S_0, but it is $\overline{S_0}$ that is a function of $\overline{F}$.

Two branches connect from F, respectively to d and v, hence (again, from an immediate application of the chain rule):

$$\begin{cases} d = \frac{\log\left(\frac{F}{K}\right)}{std} \\ v = DF[Fnd_1 - Knd_2] \end{cases} \Rightarrow \overline{F} = \frac{\overline{d}}{F \cdot std} + DF \cdot nd_1 \cdot \overline{v} = \frac{\overline{d}}{F \cdot std} + DF \cdot nd_1 \cdot 1$$

That the adjoint of v is 1 is evident from the definition: the adjoint of the final result is the sensitivity of the result to itself. The adjoint of the final result of any calculation is always 1. We can compute $\overline{F}$ as a function of $\overline{d}$, once again, in the reverse order from the evaluation of the function where d is a function of F. Further, we see two branches out of d connecting to d_1 and d_2, hence:

$$\begin{cases} d_1 = d + \frac{std}{2} \\ d_2 = d - \frac{std}{2} \end{cases} \Rightarrow \overline{d} = \overline{d_1} + \overline{d_2}$$

Further, nd_1 and nd_2 depend on d_1 and d_2; hence, $\overline{d_1}$ and $\overline{d_2}$ depend on $\overline{nd_1}$ and $\overline{nd_2}$:

$$nd_1 = N(d_1) \Rightarrow \overline{d_1} = n(d_1)\overline{nd_1}$$

$$nd_2 = N(d_2) \Rightarrow \overline{d_2} = n(d_2)\overline{nd_2}$$

and $\overline{nd_1}$ and $\overline{nd_2}$ both depend on $\overline{v} = 1$:

$$v = DF[Fnd_1 - Knd_2] \Rightarrow \begin{cases} \overline{nd_1} = DF \cdot F \cdot \overline{v} = DF \cdot F \\ \overline{nd_2} = -DF \cdot K \cdot \overline{v} = -DF \cdot K \end{cases}$$

With those equations, we compute $\overline{nd_1}$ and $\overline{nd_2}$, then $\overline{d_1}$ and $\overline{d_2}$, then $\overline{d}$, then $\overline{F}$, and eventually $\overline{S_0}$, in the exact reverse order from the evaluation of the function. This is a general rule: adjoints are always computed from one another, in the reverse order from the calculation, starting with the adjoint of the final result, which is always 1:

$$\begin{aligned}
\overline{S_0} &= \frac{F}{S_0}\overline{F} \\
&= \frac{F}{S_0}\left(\frac{\overline{d}}{F \cdot std} + DF \cdot nd_1\right) \\
&= \frac{F}{S_0}\left(\frac{\overline{d_1} + \overline{d_2}}{F \cdot std} + DF \cdot nd_1\right) \\
&= \frac{F}{S_0}\left(\frac{n(d_1)\overline{nd_1} + n(d_2)\overline{nd_1}}{F \cdot std} + DF \cdot nd_1\right) \\
&= \frac{F}{S_0}\left(\frac{n(d_1)DF \cdot F - n(d_2)DF \cdot K}{F \cdot std} + DF \cdot nd_1\right) \\
&= DF\frac{F}{S_0}\left[N(d_1) + \frac{n(d_1) - \frac{K}{F}n(d_2)}{std}\right] \\
&= \frac{DF \cdot F}{S_0}N(d_1)
\end{aligned}$$

as expected, since a simple calculation shows that $n(d_1) - \frac{K}{F}n(d_2) = 0$.

Moving on to the adjoint of K, we can compute it in the exact same manner. From the graph, K affects d and v, hence:

$$\overline{K} = \frac{\partial d}{\partial K}\overline{d} + \frac{\partial v}{\partial K}\overline{v} = -\frac{1}{K \cdot std}\overline{d} - DF \cdot nd_2{}^{3}$$

[3]One confusion that must be avoided is that in this equation, $\frac{\partial v}{\partial K}$ is not the sensitivity $\overline{K}$ of v to K, but the partial derivative of the *expression* of $v = DF[Fnd_1 - Knd_2]$ to K, hence, $-DF \cdot nd_2$.

since $\overline{v} = 1$. So, we can also compute $\overline{K}$ as a function of $\overline{d}$, which we already know from our decomposition of $\overline{S_0}$, and the result follows. The details are left as an exercise.

Moving on to $\overline{\sigma}$, the graph shows one connection to *std*, hence:

$$std = \sigma\sqrt{T} \Rightarrow \overline{\sigma} = \sqrt{T}\overline{std}$$

where *std* affects d, d_1, and d_2, hence:

$$\begin{cases} d = \frac{\log\left(\frac{F}{K}\right)}{std} \\ d_1 = d + \frac{std}{2} \\ d_2 = d - \frac{std}{2} \end{cases} \Rightarrow \overline{std} = -\frac{d}{std}\overline{d} + \frac{\overline{d_1}}{2} - \frac{\overline{d_2}}{2}$$

where we already know that $\overline{d} = \overline{d_1} + \overline{d_2}$, hence:

$$\overline{std} = \left(\frac{1}{2} - \frac{d}{std}\right)\overline{d_1} - \left(\frac{1}{2} + \frac{d}{std}\right)\overline{d_2}$$

and $\overline{d_1}$ and $\overline{d_2}$ are known from the previous computation of $\overline{S_0}$, so we can compute $\overline{\sigma}$ with a few multiplications (which details are also left as an exercise).

The equations just above illustrate two general properties of adjoints:

1. $\overline{std}$ is the sum of the three terms corresponding to the three operations that directly depend on *std* on the graph. More generally, an adjoint $\overline{x}$ is always the sum, *over all the subsequent operations that use x as an argument*, of the adjoints of those operations, weighted by their partial derivatives to x. We will formalize this in the next chapter; for now, we note that this is an immediate consequence of the chain rule.
2. $\overline{std}$, in addition to its dependency on $\overline{d_1}$ and $\overline{d_2}$, also depends on $\frac{d}{std}$. The partial derivatives that intervene in adjoint equations of the form:

$$\overline{std} = \left(\frac{1}{2} - \frac{d}{std}\right)\overline{d_1} - \left(\frac{1}{2} + \frac{d}{std}\right)\overline{d_2}$$

generally depend on the values of the arguments. Therefore, the values of all the intermediate results, computed in the direct evaluation order, must be accessible when we compute the adjoints in the reverse order.

We compute $\overline{T}$, $\overline{r}$ and $\overline{y}$ in the same manner. We noted that $\overline{K}$ and $\overline{\sigma}$ are computed very quickly once the common factors $\overline{d}$, $\overline{d_1}$, and $\overline{d_2}$ are known, and the same applies to $\overline{T}$, $\overline{r}$, and $\overline{y}$. We already dealt with $\overline{y}$ earlier and noted that:

$$\overline{y} = \frac{\partial F}{\partial y}\overline{F} = -FT\overline{F}$$

since the only contribution of y to C is through the forward; hence, $\overline{y}$ is immediately deduced from $\overline{F}$, which we already know from our computation of $\overline{S_0}$. Similar reasoning applies to $\overline{T}$ and $\overline{r}$, left as an exercise.

Back-propagation

It follows from the rules we extracted from our example that:

1. Adjoints are computed from one another in the reverse order of the calculation.
2. The adjoint $\overline{x}$ of an operation resulting in x is the sum of the adjoints of the future operations that use x, weighted by the partial derivatives of the future operations to x.
3. The adjoint of the final result is 1.

It follows that we can always calculate all adjoints with the following algorithm, called *adjoint differentiation*:

1. Evaluate the calculation, that is, the sequence of all the operations of the form:

   $$b = f(a_1, ..., a_n)$$

   keeping track of all the intermediate as and bs so that we can later compute the partial derivatives $\frac{\partial f}{\partial a_i}$.
2. Start with the adjoint of the final result $= 1$ and zero all the other adjoints. This is called *seeding* the back-propagation algorithm.
3. For every operation in the evaluation, of the form:

   $$b = f(a_1, ..., a_n)$$

   conduct the n *adjoint operations*:

   $$\overline{a_i} += \frac{\partial f}{\partial a_i}\overline{b}$$

   (where the C style "+=" means "add the right-hand side to the value of the left-hand side") in the *reverse* order from the evaluation, last operation first, first operation last. This is the back-propagation phase, where all adjoints are computed from one another. Note that the entire adjoint differentiation algorithm is often called "back-propagation," especially in machine learning.

That this algorithm correctly computes all the adjoints, hence, all the derivatives sensitivities, of any calculation, is a direct consequence of the chain rule. The adjoints are zeroed first and *accumulated* during the back-propagation phase. For this reason, this algorithm is sometimes also called "adjoint accumulation."

The back-propagation algorithm has a number of important properties.

Analytic differentiation First, it is apparent that back-propagation produces *analytic* differentials, as long as the partial derivatives in the adjoint equations:

$$\overline{a_i} += \frac{\partial f}{\partial a_i}\overline{b}$$

are themselves analytic.

An important comment is that the partial derivatives of the operations involved in a calculation are *always* known analytically, provided that the calculation is broken down into sufficiently small pieces. In the limit, any calculation may be broken down to additions, subtractions, multiplications, divisions, powers, and a few mathematical functions like *log*, *exp*, *sqrt*, or the cumulative Normal distribution N, for which analytic partial derivatives are obviously known. This is exactly what the automatic algorithm in the next chapter does with the overloading of arithmetic operators and mathematical functions. We don't have to go that far when we apply AD manually, but we must break the calculation down to pieces small enough so their partial derivatives are known, explicit and easily derived.

Linear memory consumption We made the observation that, in order to compute the partial derivatives involved in the adjoint operations during the back-propagation phase, we must store the results of all the operations involved in the calculation as we evaluate it.

Therefore, manual AD requires (re-)writing the calculation code such that all intermediate results are stored and not overwritten. We coded our Black and Scholes function this way to start with, but in real life, to manually differentiate existing calculation code may take a lot of rewriting.

Automatic AD (AAD), as described from the next chapter onwards, does not require any rewriting. It is, however, subject, as any form of AD, to a vast memory consumption *proportional to the number of operations*. Memory consumption is the principal challenge of AD. If not addressed correctly, memory requirements make AD impractical and inefficient.[4]

Memory management is addressed in much detail in Chapters 10 and up. For now, we note that a linear (in the number of operations) memory consumption is a defining characteristic of back-propagation.

Constant time Finally, the most remarkable property of back-propagation, the unique property that differentiates it from all other differentiation methods, and the whole point of implementing it, is that it computes all the differentials of a calculation code in constant time *in the number of inputs*.

To see this, recall that for every operation in the evaluation of the form:

$$b = f(a_1, ..., a_n)$$

back-propagation conducts the *n adjoint operations*:

$$\overline{a_i} += \frac{\partial f}{\partial a_i} \overline{b}$$

Hence, the whole differentiation is conducted in less than $N(M+1)$ operations, where N is the number of operations involved in the calculation and M is the maximum number of arguments to an operation. The complexity of the adjoint accumulation is therefore proportional to the complexity of the calculation (with coefficient $M+1$), but independent from the number of inputs or required differentials.

Provided that the calculation was broken down to elementary pieces like addition, multiplication, logarithm, square root, or Gaussian density, $M = 2$ since all the elementary operations are either unary or binary. In this case, the adjoint algorithm computes the result together with all sensitivities in less than three times the complexity of the evaluation alone.

In our Black and Scholes code, we did *not* break the calculation down to elementary pieces: for instance, the ninth operation "v = df * (F * nd1 - K * nd2)" conducts four computations out of five arguments. This operation can be further broken down into elementary pieces as follows:

```
//  ...
//  const double v = df * (F * nd1 - K * nd2);

const double Fnd1 = F * nd1;
const double Knd2 = K * nd2;
```

[4]Because its working memory does not fit in the CPU cache – memory consumption is also the reason why AD struggles on GPUs.

```
    const double fwdPrice = Fnd1 - Knd2;
    const double v = df * fwdPrice;

    return v;
}
```

When we construct calculation graphs automatically in the next chapter, we always break calculations down to elementary operations. With manual adjoint differentiation, this is unnecessary and to group operations may help clarify the code.

A graphical illustration of back-propagation

The following chart shows the application of back-propagation to Black and Scholes: it displays all the direct operations conducted top to bottom, with the corresponding adjoint operations conducted bottom to top.

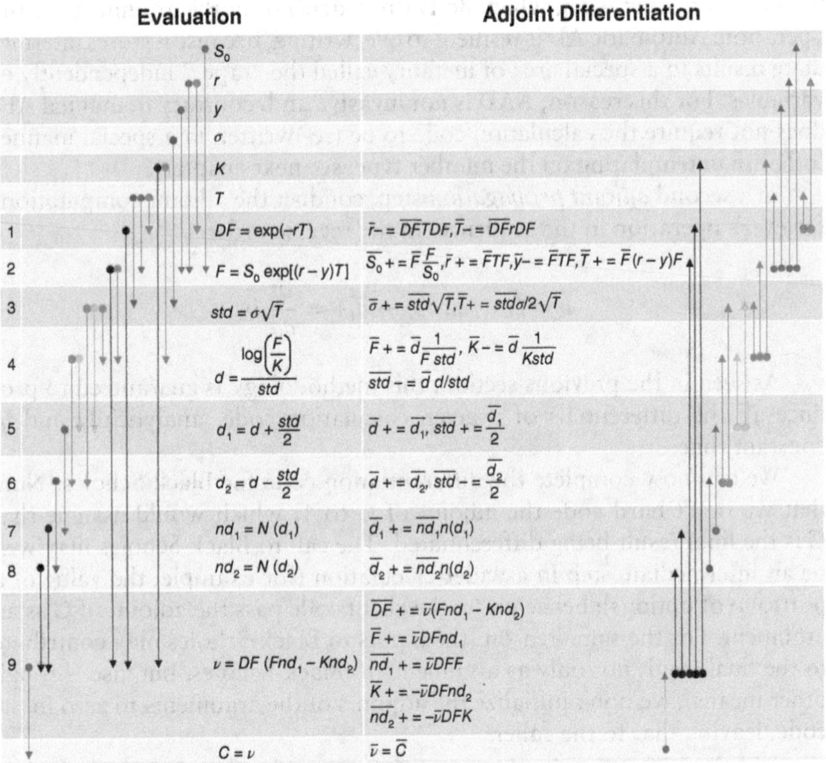
Evaluation
Adjoint Differentiation
S_0
r
y
σ
K
T
1 $DF = \exp(-rT)$
$\bar{r} -= \overline{DF} T DF, \bar{T} -= \overline{DF} r DF$
2 $F = S_0 \exp[(r-y)T]$
$\bar{S_0} += \bar{F}\frac{F}{S_0}, \bar{r} += \bar{F}TF, \bar{y} -= \bar{F}TF, \bar{T} += \bar{F}(r-y)F$
3 $std = \sigma\sqrt{T}$
$\bar{\sigma} += \overline{std}\sqrt{T}, \bar{T} += \overline{std}\sigma/2\sqrt{T}$
4 $d = \frac{\log\left(\frac{F}{K}\right)}{std}$
$\bar{F} += \bar{d}\frac{1}{F\, std}, \bar{K} -= \bar{d}\frac{1}{K std}$
$\overline{std} -= \bar{d}\, d/std$
5 $d_1 = d + \frac{std}{2}$
$\bar{d} += \bar{d_1}, \overline{std} += \frac{\bar{d_1}}{2}$
6 $d_2 = d - \frac{std}{2}$
$\bar{d} += \bar{d_2}, \overline{std} -= \frac{\bar{d_2}}{2}$
7 $nd_1 = N(d_1)$
$\bar{d_1} += \overline{nd_1} n(d_1)$
8 $nd_2 = N(d_2)$
$\bar{d_2} += \overline{nd_2} n(d_2)$
9 $v = DF(Fnd_1 - Knd_2)$
$\overline{DF} += \bar{v}(Fnd_1 - Knd_2)$
$\bar{F} += \bar{v} DF nd_1$
$\overline{nd_1} += \bar{v} DF F$
$\bar{K} += -\bar{v} DF nd_2$
$\overline{nd_2} += -\bar{v} DF K$
$C = v$
$\bar{v} = \bar{C}$

In the next chapter, we will learn to apply back-propagation automatically to any calculation code. For now, we apply it manually to the Black and Scholes code to earn a more intimate experience with the algorithm, and identify some programming rules for the differentiation of C++ calculation code.

8.2 ADJOINT DIFFERENTIATION BY HAND

We identified a general adjoint differentiation methodology, which for now we apply by hand, to compute all the derivatives of any calculation code in constant time.

First, conduct the *evaluation* step, that is, the usual sequence of operations to produce the result of the calculation, keeping track of all intermediate results. This means that the calculation must be (re-)coded in *functional* terms, where variables are immutable and memory is not overwritten. Otherwise, intermediate results are no longer accessible next, when we carry the adjoint operations, in the reverse order, where we must compute the partial derivatives of every operation, which derivatives depend on the arguments of the operation. Automatic AD is resilient to overwriting, because it stores intermediate results in a special area of memory called the "tape," independently of variables. For this reason, AAD is noninvasive and, contrary to manual AD, does not require the calculation code to be (re-)written in a special manner (other than templating on the number type, see next chapter).

In a second *adjoint propagation* step, conduct the adjoint computations for every operation in the sequence, in the reverse order:

$$b = f(a_1, ..., a_n) \rightarrow \overline{a_i} += \frac{\partial f}{\partial a_i}\overline{b}$$

As seen in the previous section, this methodology is guaranteed to produce all the differentials of a given calculation code, analytically and in constant time.

We can now complete the differentiation code for Black-Scholes. Note that we don't hard code the adjoint of C to 1, which would assume that C is the final result being differentiated. The call to Black-Scholes may well be an intermediate step in a wider calculation (for example, the value of a portfolio of options); hence we let the client code pass the adjoint of C as an argument. For the same reason (the inputs to Black-Scholes may contribute to the final result not only as arguments to Black-Scholes, but also through other means), we don't initialize the adjoints of the arguments to zero in the code, leaving that to the caller.

```
double C(
    // inputs
    const double S0,
    const double r,
    const double y,
```

```
    const double sig,
    const double K,
    const double T,
    //  adjoints
    const bool  calcAdjoints = false,
    double*     S0_ = nullptr,
    double*     r_ = nullptr,
    double*     y_ = nullptr,
    double*     sig_ = nullptr,
    double*     K_ = nullptr,
    double*     T_ = nullptr,
    //  adjoint of result in global calculation
    const double C_ = 1.0)
{
    //  Evaluation

    //  So we don't evaluate the sqrt multiple times
    const double sqrtT = sqrt(T);

    //  1
    const double df = exp(-r*T);
    //  2
    const double F = S0 * exp((r-y)*T);
    //  3
    const double std = sig * sqrtT;
    //  4
    const double d = log(F/K) / std;
    //  5, 6
    const double d1 = d + 0.5 * std, d2 = d - 0.5 * std;
    //  7, 8
    const double nd1 = normalCdf(d1), nd2 = normalCdf(d2);
    //  9
    const double v = df * (F * nd1 - K * nd2);

    if (!calcAdjoints) return v;

    //  Adjoint calculation

    //  Initialize
    double v_ = C_;

    //  9
    double df_ = v_ * (F * nd1 - K * nd2);
    double F_ = v_ * df * nd1;
    double nd1_ = v_ * df * F;
    if (K_) *K_ = - v_ * df * nd2;
    double nd2_ = - v_ * df * K;

    //  8, 7
    //  normalDens() = normal density from gaussians.h
    double d2_ = nd2_ * normalDens(d2);
    double d1_ = nd1_ * normalDens(d1);

    //  6, 5
    double d_ = d2_;
    double std_ = - 0.5 * d2_;
    d_ += d1_;
```

```
    std_ += 0.5 * d1_;

    // 4
    F_ += d_ / (F * std);
    if (K_) *K_ -= d_ / (K * std);
    std_ -= d_ * d / std;

    // 3
    if (sig_) *sig_ += std_ * sqrtT;
    if (T_) *T_ += 0.5 * std_ * sig / sqrtT;

    // 2
    if (S0_) *S0_ += F_ * F / S0;
    if (r_) *r_ += F_ * T * F;
    if (y_) *y_ -= F_ * T * F;
    if (T_) *T_ += F_ * (r - y) * F;

    // 1
    if (r_) *r_ += - df_ * df * T;
    if (T_) *T_ += - df_ * df * r;

    return v;
}
```

What we did not yet cover are some practical technicalities: how to deal with nested function calls and control flow. The automatic, noninvasive algorithm of the next chapter correctly deals with nested function calls and control flow, behind the scenes. For now, we discuss solutions in the context of manual AD.

Nested function calls

Suppose that the call to our function C() is part of a wider computation, for instance, the value and risk of a portfolio of options. We call $f()$ the top-level function that computes the value of the portfolio, and suppose that $f()$ involves a number of nested calls to C(), as in:

```
double f(
    // inputs
    // ...)
{
    // 1
    // ...

    // i
    const double c = C(S0, r, y, sig, K, T);

    // i + 1
    // do something with c, for instance:
    const double g = gamma * c;
    // ...

    // N
```

```
    //  const double v = ...;

    return v;
}
```

To differentiate $f()$ with respect to its inputs, we apply adjoint differentiation rules and perform the adjoint calculations in the reverse order:

```
double f(
    //  inputs
    //  ...,
    //  adjoints
    const bool  calcAdjoints = false,
    //  ...,
    //  adjoint of result in global calculation
    const double f_ = 1.0)
{
    //  Evaluation

    //  1
    // ...

    //  i
    const double c = c(S0, r, y, sig, K, T);

    //  i + 1
    //  do something with c, for instance:
    const double g = gamma * c;
    //  ...

    //  N
    //  const double v = ...;

    if (!calcAdjoints) return v;

    //  Adjoint calculation

    //  Initialize
    const double v_ = f_;

    //  N
    //  ...


    //  i + 1
    c_ += g_ * gamma;
    gamma_ += g_ * c;

    //  i
    C(S0, r, y, sig, K, T, true, S0_, r_, y_, sig_, K_, T_, c_);
    //  ...

    return v;
}
```

In the evaluation step, we call C() in evaluation mode to compute the value of the call option from the inputs. In the adjoint step, we call C() in adjoint mode, passing the adjoint $c\_ = \frac{\partial f}{\partial c}$ of the result of C(). We programmed C() so that this call correctly back-propagates c_ to S0_, r_, y_, sig_, K_, and T_, and the outer back-propagation process for $f()$ can continue from there.

When we write the adjoint code for a function, we must also write adjoint code for all the inner functions called from the function. This is, however, impossible when the function calls third-party libraries, for which the source may not be available or easily modifiable. In this case, we can wrap the third-party function in a function that satisfies the API of adjoint code, but conducts finite differences internally. Of course, we don't get the speed of adjoint differentiation for this part of the code, but adjoints do get propagated so the rest of the code can benefit from the acceleration. Concretely, if we have a function *g*() from a third-party library, we wrap it as follows:

```
#define EPS 1.0e-12

//  From the 3rd party lib header
double g(const double x, const double y);

//  Finite Difference wrapper for adjoint code

double ag(
    //  inputs
    const double x,
    const double y,
    //  adjoints
    const bool  calcAdjoints = false,
    double *x_  = nullptr,
    double *y_  = nullptr,
    //  adjoint of result in global calculation
    const double g_  = 1.0)
{
    const double g0 = g(x, y);
    if (!calcAdjoints) return g0;

    if (x_) *x_  += g_  *  (g(x+EPS, y)  - g0) / EPS;
    if (y_) *y_  += g_  *  (g(x, y+EPS)  - g0) / EPS;

    return g0;
}
```

Then, the wrapper function can be used in adjoint code in the same way we used our Black-Scholes adjoint code to differentiate a higher level calculation.

Conditional code and smoothing

Conditional code ("if this then that") is a more delicate subject. It must be understood that conditional code is discontinuous by nature, and not differentiable, with AD or otherwise. When AD deals with conditional code, it computes the differentials of the calculation *over the same branch* as the evaluation, but it does not differentiate *through* the condition itself. Concretely, the adjoint for the code:

```
// ...

if (something)
{
    y = f(x);
}
else
{
    y = g(x);
    z = h(x);
}

// ...
```

is:

```
// ...

if (something)
{
    x_ += y_ * dfdx;
}
else
{
    x_ += y_ * dgdx;
    x_ += z_ * dhdx;
}

// ...
```

The adjoint code simply repeats the conditions, and the adjoints are back-propagated along the same branch where the calculation effectively took place. The condition itself, and the alternative calculations, are ignored in the differentiation. Note that this is the desired behavior in some cases. Where this is not, and it is desired to differentiate *through* the condition, the code must be modified and the condition smoothed so as to remove the discontinuity. Note that this is not particular to AD. All types of differentiation algorithms, including finite differences, are at best unstable when differentiating through control flow. Therefore, the algorithms that make control flow differentiable, collectively known as smoothing algorithms, are not related to AD, although a code can be easily differentiated with AD after it is smoothed. It also follows that smoothing is out of the scope of this text; we refer interested readers to Bergomi's [78] and our dedicated talk [77] and publication [11].

Loops and linear algebra

Finally, operations made in a loop are no different than others. For instance, a loop like:

```
// ...

double x = 0.0;
for (size_t i=0; i<n; ++i) x += a[i] * y[i];

// ...
```

conducts $2n$ operations. In accordance with the rules of AD, we should separately store the result of each operation so we can back-propagate adjoints through the n additions and the n multiplications, last to first. Our automatic algorithm in the next chapter deals with loops in this manner exactly. When differentiating manually, however, we may indulge in shortcuts to improve both the clarity and the performance of the code. The loop above evidently implements:

$$x = \sum_{i=0}^{n-1} a_i y_i$$

so the adjoint operations are:

$$\overline{a_i} += \overline{x} y_i$$
$$\overline{y_i} += \overline{x} a_i$$

in a loop on i, or, using vector notations, we write the operation under the compact form:

$$x = a \cdot y$$

and using a basic matrix differentiation result, we can write the adjoint operations in a compact vector form, too:

$$\overline{a} += \overline{x} y$$
$$\overline{y} += \overline{x} a$$

which is evidently the same result as before, just with more compact notations.

More generally, with calculations involving matrix algebra, we have two options for the adjoint code: we can break all the matrix operations down to scalar operations and deal with the sequence of those as usual; or, we can use matrix differentiation results to derive and code the adjoint equations in matrix form, which is cleaner and sometimes more efficient, matrix operations often being optimized compared to hand-crafted loops.

For matrix differentiation results, we refer, for example, to [91] or [92]. For adjoint matrix calculus and applications to otherwise non-differentiable routines like singular value decomposition (SVD), we refer to [93]. For applications to calibration and FDM, we refer to [94].

8.3 APPLICATIONS IN MACHINE LEARNING AND FINANCE

Adjoint differentiation is known in the field of machine learning under the name "backpropagation," or simply "backprop." Backprop is what allows large, deep neural networks to learn their parameters in a reasonable time. In quantitative finance, in addition to risk sensitivities covered in detail in the following chapters, adjoint differentiation substantially accelerates calibration. Deep learning and calibration are both high-dimensional fitting problems that benefit from the quick computation of the sensitivities of error parameters.

We briefly introduce these applications and conclude before moving on to the automation of AD.

Deep learning and backpropagation

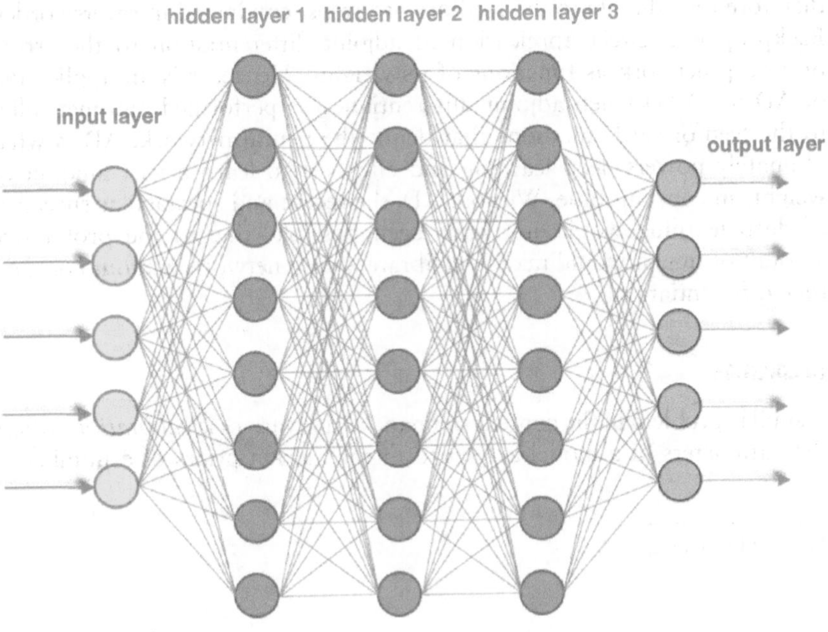

In the context of deep learning, multilayer artificial neural networks learn their weights (represented by the many edges in the chart above) by minimization of errors (difference between the targets and their predictors obtained by forward-feeding the corresponding inputs through the network) over a learning set, similarly to a regression problem, of which deep learning may be seen as an extension. Contrary to linear regression problems, however, deep learning has no closed-form solution, and an iterative procedure must be invoked to minimize the loss function (generally, the sum of squared errors).

The most widely used optimization algorithm in the field of machine learning[5] is the stochastic gradient descent (SGD), described in vast amounts of literature like [65] and online tutorials. SGD starts with a guess for the weights and updates them in the direction of the gradient of the error, iterating over the learning set. Hence, SGD requires the computation, at every step of the gradient, of the derivatives of the error function to the many weights in the network.

The derivatives of the error in a deep learning problem are generally analytic, and yet, the evaluation cost of all those closed-form derivatives at each step of the SGD would be prohibitive. Deep learning software computes all these derivatives in constant time with what they call *backprop*. Backprop equations are listed in Bishop [65], chapter 5.3, together with the feedforward equations that define the network. These equations are not reproduced here. What we see there is that the backprop equations are exactly the adjoints of the feedforward equations. Backprop steps are therefore the adjoints to the feedforward steps, conducted in reverse order. Backprop is a direct application of adjoint differentiation to the errors of a deep network as functions of its weights. Hence, it is an application of AD (or AAD when adjoint differentiation is performed automatically) to the feedforward equations that define the neural network. AD is what ultimately powers deep learning and allows it to learn a vast amount of weights in constant time. Without AD, the recent and spectacular successes of deep learning would not have been possible, due to the prohibitive amount of time it would take to calibrate a deep network without constant time differentiation.

Calibration

A similar problem in the context of financial derivatives is *calibration*, where the parameters of a model are set to fit the market prices of a number of

[5]To our knowledge.

liquid instruments. Dynamic models are typically dependent on curves and surfaces of parameters, like Dupire's local volatility or Hull and White's short rate volatility and mean-reversion. The parameters are generally time-dependent and their number is typically large. Perhaps against intuition, it is *not* customary to set those parameters through statistical estimates. For reasons explained for instance in [46], it is best practice to calibrate them to the market prices of traded instruments. The parameters of dynamic models are typically set to fit the market prices of European options, before the model is applied to produce the value and risk of more complex, less liquid exotics. Calibration is the process where a model *learns* its parameters on a sample consisting in a number of simple, standardized, actively traded instruments.

With the very notable exception of Dupire's model, calibration doesn't generally have an analytic solution in the form of a closed-form expression for the parameters as a function of option prices. A numerical optimization must be conducted to find the model parameters that minimize the errors. Derivatives practitioners generally favor the very effective Levenberg-Marquardt algorithm explained in *Numerical Recipes* [20], chapter 15.5. Like SGD, this is an iterative algorithm that starts with a guess for the parameters and updates them at each step from the gradient of the objective function (the sum of squared errors) and an internal approximation of its Hessian matrix.

Levenberg-Marquardt requires the computation, at each step, of the model prices of all the target instruments, along with their sensitivities to the model parameters. Model prices of European options are generally a closed form, either exact (Hull and White) or approximate (Libor Market Models or Multi-Factor Cheyette). The derivatives to parameters are therefore also analytic. However, the evaluation of a large number of sensitivities on each iteration is too expensive for a global calibration to converge fast enough. For this reason, it is common practice to localize calibrations (sequentially conduct "small" calibrations, where a small number of parameters is set to match a small number of market prices) or apply slower, less efficient alternatives to Levenberg-Marquardt that don't require derivative sensitivities, like the downhill simplex algorithm described in [20], chapter 10.4.

The performance of a global calibration can therefore considerably improve with the computation of the derivatives of the model prices to all the parameters in constant time with the adjoint method, with the added benefit that the recourse to less precise, less stable algorithms is no longer necessary. A possible implementation is by the manual adjoint differentiation of the pricing function for European options in the model, as we illustrated in the particularly simple case of Black and Scholes: code the adjoints of the sequence of all the operations in the evaluation in the reverse order. The process must be repeated for the multitude of pricing functions for various models across a changing library where new models are added and the implementation of existing models is typically improved over time. For this reason, hand-coded adjoint differentiation may not be the best option for calibration. Like market risks, the production of sensitivities for calibration is best conducted automatically.

Calibration is not limited to dynamic models. The construction of continuous interest rate curves and surfaces from a discrete number of market-traded swap and basis swap rates, deposits, and futures is also a calibration.[6] The construction of rate surfaces is a delicate, specialized field of quantitative finance and a prerequisite for trading any kind of interest rate derivative, including vanilla swaps, caps and swaptions, interest rate exotics and hybrids, or the calculation of xVA.

Swap markets are moving fast in near-continuous time; hence, swap traders need a reliable, near-instantaneous rate surface calibration, despite the complexities involved (see [39] for an introduction). Rate surfaces are subject to a vast number of parameters; hence, AD accelerates their calibration by orders of magnitude, as demonstrated in Scavenius' pioneering work [95].

Manual and automatic adjoint differentiation

It is apparent from our examples that manual adjoint differentiation is tedious, time consuming, and error prone. In addition, the maintenance of libraries with handwritten adjoint code may turn into a nightmare: every modification in the calculation code must be correctly reflected in the adjoint code in a manual process.

This is why our text focuses on *automatic* adjoint differentiation, whereby the adjoint calculations are conducted automatically and correctly, behind the scenes. Developers only write the calculation code. The adjoint "code" is generated at run time.

This being said, maximum performance is obtained with a manual differentiation. A naive implementation of AAD, like the play tool in the next chapter, carries a prohibitive overhead. The techniques explained in Chapters 10 and 13 mitigate the overhead and substantially accelerate AAD, but they still fall short of the performance of hand-coded AD. Only the advanced implementation of AAD in Chapter 15 approaches the performance of manual AD thanks to expression templates. It is still worth hand coding the derivatives of some short, unary, or binary low-level functions

[6] And so is the construction of implied volatility surfaces; see for instance [40] and [41].

that are frequently called in the calculation code. We will demonstrate how to do this in the context of our AAD library.

In this chapter, we covered manual adjoint differentiation mainly for a pedagogical purpose, in order to gently introduce readers to adjoint calculations before we delve into algorithms that automate them. Readers are encouraged to try hand-coded adjoint differentiation for various examples involving nested function calls, conditions, and loops to earn some experience and an intimate knowledge of the concepts covered here. They will also find a more complete coverage of manual adjoint differentiation in [88].

CHAPTER 9

Algorithmic Adjoint Differentiation

In this second introductory chapter, we discuss in detail how to *automatically* generate calculation graphs and conduct adjoint calculations through them to produce all the differentials of any calculation code in constant time, automatically and in a noninvasive manner.

We consider a mathematical calculation that takes a number of scalar inputs $(x_i)_{1 \leq i \leq I}$ and produces a scalar output y. We assume that the calculation is written in C++ (including perhaps nested function calls, overwriting of variables, control flow, and object manipulation). Our goal is to differentiate this calculation code in constant time, like we did in the previous chapter, but automatically. For this purpose, we develop *AAD code*, that is, code that runs adjoint differentiation over any calculation code. In order to do that, the AAD code must understand the sequence of operations involved in the calculation code, a notion known to programmers as "meta-programming" or "introspection." It is achieved through *operator overloading*, whereby all mathematical operators and functions are replaced by custom code when applied to specific types, as opposed to native C++ types like *double*. So the calculation code must be *templated* on the number type, the type used for the representation of numbers. Safe for templating, the calculation code remains untouched. All this will be clarified throughout the chapter.

The differentiation code in this chapter is only meant for pedagogical purposes. The implementation is incomplete, not optimal, and most of the code and concepts explored here are unnecessary for the implementation of AAD. Its purpose is to understand what exactly is going on inside AAD. We will also see that calculation graphs turn out to be unnecessary for adjoint differentiation, and that, instead, an optimal implementation uses "tapes," data structures that "record" all the mathematical operations involved in the calculation as they are executed. However, it is only by exploring calculation graphs in detail that we understand exactly how AAD works its magic and why tapes naturally appear as the better solution.

A complete, optimal AAD implementation is delivered in the next chapter, and further improved in the rest of the book. The resulting AAD library is available from our online repository.

9.1 CALCULATION GRAPHS

Definition and examples

The first notion we must understand to make sense of AAD is that of a *calculation graph*. Every calculation defines a graph. The *nodes* are all the elementary mathematical calculations involved: +, −, ∗, /, *pow*, *exp*, *log*, *sqrt*, and so forth.[1] The *edges* that join two nodes represent operation to argument relations: *child* nodes are the arguments to the calculation in the *parent* node. The *leaves* (childless nodes) are scalar numbers and inputs to the calculation.[2]

The structure of the graph reflects precedence among operators: for instance the graph for $xy + z$ is:

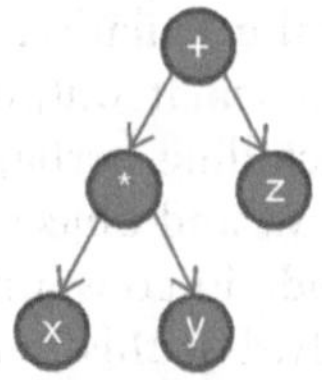

because multiplication has precedence, unless explicitly overwritten with parentheses. The graph for $x(y + z)$ is:

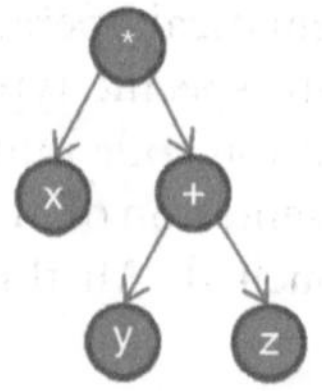

[1]We define elementary operations as the building blocks of any calculation: scalar numbers, unary functions like *exp* or *sqrt*, and binary arithmetic operators and functions like *pow*. In addition to the operators and mathematical functions available as standard in C++, we consider as building blocks all the fundamental unary or binary functions, repeatedly called in calculation code, and which derivatives are known analytically, for example, the Gaussian density *n*() and cumulative distribution *N*() from the previous chapter.

[2]Not all leaf nodes are inputs. Leaf nodes may also be constants. But that does not matter for now. We will compute the partial derivatives of the final result to all the leaves here, and optimize the constants away in the next chapter.

The graph also reflects the order of calculation selected by the compiler among equivalent ones. For instance, the graph for $x + y$ could be:

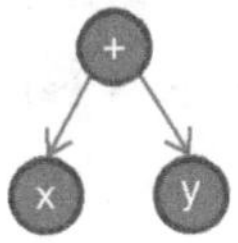

or:

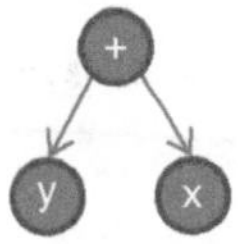

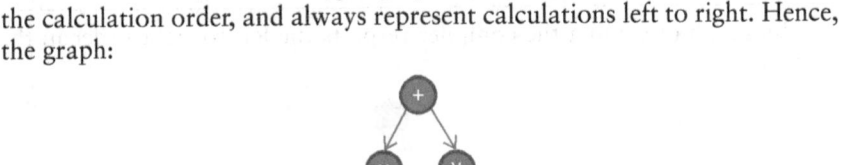

The compiler selects one of the possible calculation orders. Hence, the graph for the calculation *code* is not unique, but the graph for a given execution of the code is, the compiler having made a choice among the equivalent orders.[3] We work with the *unique* graphs that reflect the compiler's choice of the calculation order, and always represent calculations left to right. Hence, the graph:

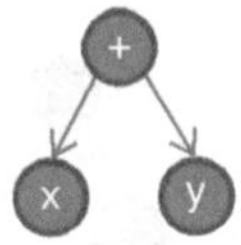

represents $x + y$ and:

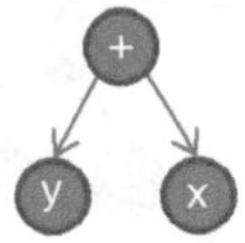

represents $y + x$.

The following example computes a scalar y out of 5 scalars $(x_i)_{0 \le i \le 4}$ (where x_4 is *purposely* left out of the calculation) with the formula:

$$y = (y_1 + x_3 y_2)(y_1 + y_2), y_1 = x_2(5x_0 + x_1), y_2 = \log(y_1)$$

We also compute and export the partial derivatives by finite differences for future reference. We will be coming back to this code throughout the chapter.[4]

---

[3]In all that follows, we neglect rounding errors that could, for example, make $x(y + z)$ different from $xy + xz$ and only consider *mathematical* equivalence.

[4]We use a particularly simple and practically irrelevant example in this chapter, in an attempt to demonstrate the main notions behind AAD with maximum simplicity and minimum code. In particular, this calculation only uses the operators +, * and *log*, so we can simplify the demonstration code with only four node types. The input $x[4]$ does not participate in the calculation; hence the sensitivity of $f()$ to $x[4]$ is 0, something our algorithm must reproduce correctly.

```
double f(double x[5])
{
    double y1 = x[2] * (5.0 * x[0] + x[1]);
    double y2 = log(y1);
    double y = (y1 + x[3] * y2) * (y1 + y2);
    return y;
}

int main()
{
    double x[5] = { 1.0, 2.0, 3.0, 4.0, 5.0 };
    double y = f(x);
    cout << y << endl;  //  797.751
    for (size_t i = 0; i < 5; ++i)
    {
        x[i] += 1.e-08;
        cout << 1.0e+08 * (f(x) - y) << endl;
        x[i] -= 1.e-08;
    }  //  950.736, 190.147, 443.677, 73.2041, 0
}
```

Its graph (assuming the compiler respects the left-to-right order in the code) is:

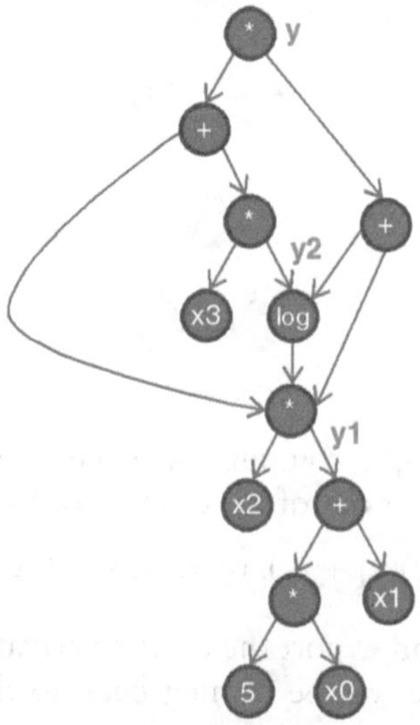

It is important to understand that complicated, compound calculations, even those that use OO (object-oriented) architecture and nested function calls, like the ones that compute the value of an exotic option with MC simulations or FDM, or even the value of the xVA on a large netting set, or any kind of valuation or other mathematical or numerical calculation, all define a similar graph, perhaps with billions of nodes, but a calculation graph all the same, where nodes are elementary operations and edges refer to their arguments.

Calculation graphs and computer programs

It is also important to understand that calculation graphs refer to *mathematical operations*, irrespective of how computer programs store them in *variables*. For instance, the calculation graph for:

```
y = y + x;
```

is most definitely *not*:

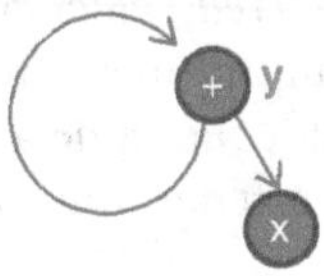

but:

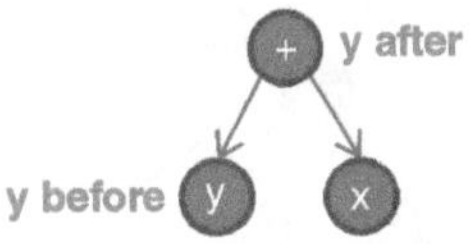

Although the variable y is reused and overwritten in the calculation of the sum, the sum defines a new calculation of its own right in the graph. Graph nodes are calculations, not variables.

Similarly, recursive function calls define new calculations in the associated graph: the following recursive implementation for the factorial:

```
unsigned fact(const unsigned n)
{
    return n == 1 ? 1 : n * fact(n - 1);
}
```

defines the following graph when called with 3 as argument:

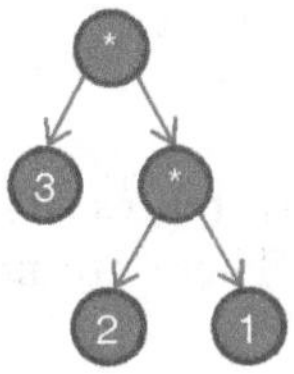

The fact that the function is defined recursively is irrelevant; the graph is only concerned with mathematical operations and their order of execution.

Directed Acyclic Graphs (DAGs)

Computer programs do not define just any type of graph.

First, notice that a calculation graph is not necessarily a *tree*, a particularly simple type of graph most people are used to working with, where every child has exactly one parent.[5] In the example on page 323, the node y_2 has two parents and the node y_1 has three. Calculation graphs are not trees, so reasoning and algorithms that apply to trees don't necessarily apply to them.

Second, the edges are *directed*, calculation to arguments. An edge from y to x means that x is an argument to the calculation of y, like in $y = \log(x)$, which is of course not the same as $x = \log(y)$. Graphs where all edges are directed are called directed graphs. Calculation graphs are always directed.

Finally, in a calculation graph, a node cannot refer to itself, either directly or indirectly. The following graph, for instance:

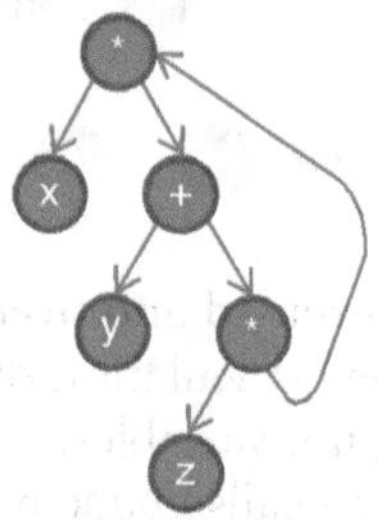

is *not* a calculation graph, and no code could ever generate a graph like this. The reason is that the arguments to a calculation must be evaluated prior to the calculation.

Graphs that satisfy such property are called *acyclic* graphs. Therefore, calculation graphs are *directed acyclic graphs* or DAGs. We will be applying some basic results from the graph theory shortly, so we must identify exactly the class of graphs that qualify as calculation graphs.

DAGs and control flow

The DAG of a calculation only represents the mathematical operations involved, not the control flow. There are no nodes for *if*, *for*, *while*, and

[5] We are not talking about recombining trees here.

friends. A DAG represents the chain of calculations under one particular control branch. For instance, the code

```
y = x > 0 ? x * a : x + a;
```

has the DAG:

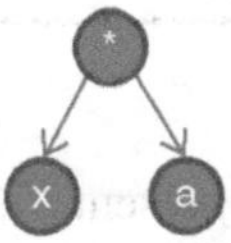

when executed with a positive input x and the different DAG:

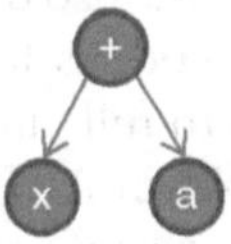

when executed with a negative input. It doesn't have a composite dag of the type:

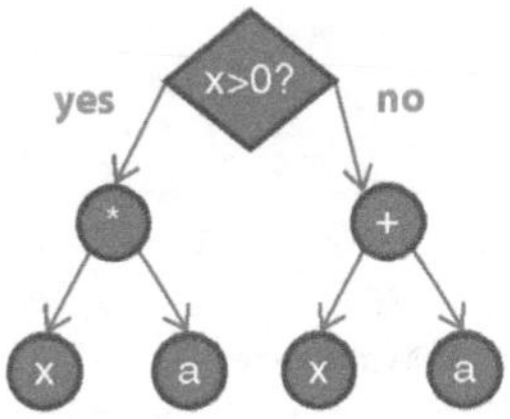

This is *not* a calculation DAG.

We work with *calculation* DAGs, not *code* DAGs. Calculation DAGs refer to one particular execution of the code, with a specific set of inputs, and a unique control branch, resulting in a specific output. They don't refer to the *code* that produced the calculation. AAD differentiates calculations, not code. It follows that AAD does not differentiate through the control flow itself: since there are no "if" nodes in the calculation DAG, the AAD "derivative" of an "if" statement is 0.

This may look at first sight like a flaw with AAD, but how is any type of differentiation supposed to differentiate through control flow, something that is by nature discontinuous and not differentiable?

We will discuss control flow and discontinuities further in Chapter 11. For now, we focus on building DAGs and applying them for the computation of differentials in constant time.

9.2 BUILDING AND APPLYING DAGs

Building a DAG

A DAG is not just a theoretical representation of some calculation. We apply *operator overloading* to explicitly build a calculation DAG in memory at run time, as follows. This code is incomplete, we only instantiate the +, $*$ and *log* nodes, and *extremely* inefficient because every operation allocates memory for its node on the DAG. We also don't implement const correctness and focus on functionality. This code is for demonstration purposes only. We suggest readers take the time to fully understand it. This inefficent code nonetheless demonstrates the DNA of AAD.

First, we create classes for the various types of nodes, +, $*$, *log*, and leaf, in an object-oriented hierarchy:

```
#include <memory>
#include <string>
#include <vector>

using namespace std;

class Node
{
protected:

    vector<shared_ptr<Node>> myArguments;

public:

    virtual ~Node() {}
};

class PlusNode : public Node
{
public:

    PlusNode(shared_ptr<Node> lhs, shared_ptr<Node> rhs)
    {
        myArguments.resize(2);
        myArguments[0] = lhs;
        myArguments[1] = rhs;
    }
};

class TimesNode : public Node
{
```

```
public:

    TimesNode(shared_ptr<Node> lhs, shared_ptr<Node> rhs)
    {
        myArguments.resize(2);
        myArguments[0] = lhs;
        myArguments[1] = rhs;
    }
};

class LogNode : public Node
{
public:

    LogNode(shared_ptr<Node> arg)
    {
        myArguments.resize(1);
        myArguments[0] = arg;
    }
};

class Leaf: public Node
{

    double myValue;

public:

    Leaf(double val)
        : myValue(val) {}

    double getVal()
    {
        return myValue;
    }

    void setVal(double val)
    {
        myValue = val;
    }
};
```

We apply a classical composite pattern (see GOF, [25]) for the nodes in the graph. Concrete nodes are derived for each operation type and hold their arguments by base class pointers. The systematic use of smart pointers guarantees an automatic (if inefficient) memory management. It has to be *shared* pointers because multiple parents may share a child. We can't have a parent release a child while other nodes are still referring to it. When all its parents are gone, the node is released automatically.

Next, we design a custom number type. This type holds a reference to the node on the graph corresponding to the operation last assigned to it. When such a number is initialized, it creates a leaf on the graph. We use this custom type in place of doubles in our calculation code.

```
class Number
{
    shared_ptr<Node> myNode;

public:

    Number(double val)
        : myNode(new Leaf(val)) {}

    Number(shared_ptr<Node> node)
        : myNode(node) {}

    shared_ptr<Node> node()
    {
        return myNode;
    }

    void setVal(double val)
    {
        //  Cast to leaf, only leaves can be changed
        dynamic_pointer_cast<Leaf>(myNode)->setVal(val);
    }

    double getVal()
    {
        //  Same comment here, only leaves can be read
        return dynamic_pointer_cast<Leaf>(myNode)->getVal();
    }
};
```

dynamic_pointer_cast() is a C++11 standard function associated with smart pointers that returns a copy of the smart pointer of the proper type with its stored pointer casted dynamically.[6] This is necessary because only leave nodes store a value.

Next, we *overload* the mathematical operators and functions for the custom number type so that when these operators and functions are executed with this type, the program does not evaluate the operator as normal, but instead constructs the corresponding node on the graph.

```
shared_ptr<Node> operator+(Number lhs, Number rhs)
{
   return shared_ptr<Node>(new PlusNode(lhs.node(), rhs.node()));
}

shared_ptr<Node> operator*(Number lhs, Number rhs)
{
   return shared_ptr<Node>(new TimesNode(lhs.node(), rhs.node()));
}

shared_ptr<Node> log(Number arg)
{
   return shared_ptr<Node>(new LogNode(arg.node()));
}
```

[6]See http://www.cplusplus.com/reference/memory/dynamic_pointer_cast/.

This completes our "AAD" code; we can now build graphs *automatically*. This is where the first A in AAD comes from. We don't build the DAG by hand. The code does it for us when we instantiate it with our number type. We could hard replace "double" by "Number" in the calculation code, or make two copies, one with doubles, one with Numbers. The best practice is *template* the calculation code on its number type, like this:

```
1 template <class T>
2 T f(T x[5])
3 {
4     T y1 = x[2] * (5.0 * x[0] + x[1]);
5     T y2 = log(y1);
6     T y = (y1 + x[3] * y2) * (y1 + y2);
7     return y;
8 }
```

and call it with our custom type like that:

```
1 int main()
2 {
3     Number x[5] = { 1.0, 2.0, 3.0, 4.0, 5.0 };
4 
5     Number y = f(x);
6 }
```

Because we called $f()$ with Numbers in place of doubles, line 5 does *not* calculate anything. All mathematical operations within $f()$ execute our overloaded operators and functions. So, in the end, what this line of code does is construct in memory the DAG for the calculation of $f()$.

We templated the code of $f()$ for its number representation type. To template calculation code in view of running of it with AAD is called *instrumentation*. In Chapter 12, we instrument our simulation code.

Traversing the DAG

Lazy evaluation What can we do with the DAG we just built?

First, we can calculate it. We evaluate the DAG itself, without reference to the function $f()$, the execution of which code built the DAG in the first place.

How do we proceed exactly? First, note that arguments to an operation must be computed before the operation. This means that the processing of a node starts with the processing of its children.

Second, we know that nodes may have more than one parent in a DAG. Those nodes will be processed multiple times if we are not careful. For this reason, we start the processing of a node with a check that this node was not processed previously. If this is verified, then we process its children, and then, and only then, we evaluate the operation on the node itself, mark it as processed, and cache the result of the operation on the node in case it is needed again from a different parent. If the node was previously processed, we simply pick its cached result.

In this sequence, the only step that is specific to evaluation is the execution of the operation on the node. All the rest is graph traversal logic and applies to evaluation as well as other forms of *visits* we explore next.

If we follow these rules, starting with the top node, we are guaranteed to visit (evaluate) each node exactly once and in the correct order for the evaluation of the DAG.

Hence, we have the following graph *traversal* and *visit* algorithm: starting with the top node,

1. Check if the node was already processed. If so, exit.
2. Process the child nodes.
3. Visit the node: conduct the calculation, cache the result.
4. Mark the node as processed.

The third line of the algorithm is the only one specific to evaluation. The rest only relates to the order of the visits. More precisely, the traversal strategy implemented here is well known to the graph theory. It is called "depth-first postorder" or simply *postorder*. Depth-first because it follows every branch down to a leaf before moving on to the next branch. Postorder because the children (arguments) of a node are always visited (evaluated) before the node (operation). It follows that visits start on the leaf nodes and end on the top node. Postorder implements a bottom-up traversal strategy and guarantees that every node in a DAG is processed exactly once, where children are always visited before their parents. It is the natural order for an evaluation. But we will see that other forms of visits may require a different traversal strategy.

We implement our evaluation algorithm, separating traversal and visit logic into different pieces of code. The traversal logic goes to the base class, which also stores a flag that tells if a node was processed and caches the result of its evaluation. For our information only, the order of calculation is also stored on the (base) nodes. Finally, we code a simple recursive method to reset the processed flags on all nodes.

```
class Node
{
protected:

    vector<shared_ptr<Node>> myArguments;

    bool        myProcessed = false;
    unsigned    myOrder = 0;
    double      myResult;

```

```
public:

    virtual ~Node() {}

    //  visitFunc:
    //  a function of Node& that conducts a particular form of visit
    //  templated so we can pass a lambda
    template <class V>
    void postorder(V& visitFunc)
    {
        //  Already processed -> do nothing
        if (myProcessed == false)
        {
            //  Process children first
            for (auto argument : myArguments)
                argument->postorder(visitFunc);

            //  Visit the node
            visitFunc(*this);

            //  Mark as processed
            myProcessed = true;
        }
    }

    //  Access result
    double result()
    {
        return myResult;
    }

    //  Reset processed flags
    void resetProcessed()
    {
        for (auto argument : myArguments) argument->resetProcessed();
        myProcessed = false;
    }
};
```

The *resetProcessed*() algorithm isn't optimal, as most of the code in this chapter. Next, we code the specific evaluation visitor as a virtual method: evaluation is different for the different concrete nodes (sum in a + node, product in a ∗ node, and so on):

```
//  On the base Node
    virtual void evaluate() = 0;

//  On the + Node
    void evaluate() override
    {
        myResult = myArguments[0]->result() + myArguments[1]->result();
    }

//  On the * Node
    void evaluate() override
```

```
    {
        myResult = myArguments[0]->result() * myArguments[1]->result();
    }

//  On the log Node
    void evaluate() override
    {
        myResult = log(myArguments[0]->result());
    }

//  On the Leaf
    void evaluate() override
    {
        myResult = myValue;
    }
```

We start the evaluation of the DAG from the Number that holds the result and, therefore, refers to the top node of the DAG. Recall that Numbers contain a shared pointer to the last assignment.

```
//  On the Number class
    double evaluate()
    {
        myNode->resetProcessed();
        myNode->postorder([](Node& n) {n.evaluate(); });
        return myNode->result();
    }
```

Now we can evaluate the DAG:

```
1  int main()
2  {
3      Number x[5] = { 1.0, 2.0, 3.0, 4.0, 5.0 };
4
5      Number y = f(x);
6
7      cout << y.evaluate() << endl;    // 797.751
8  }
```

The evaluation of the mathematical operations in the calculation did not happen on line 5. That just constructed the DAG. Nothing was calculated. The calculation happened on line 7 in the call to *evaluate*(), directly from the DAG after it was built. That it returns the exact same result as a direct function call (with double as the number type) indicates that our code is correct.

To see this more clearly, we can change the inputs on the DAG and evaluate it again, without any further call to the function that built the DAG. The new result correctly reflects the change of inputs, as readers may easily check:

```
int main()
{
    Number x[5] = { 1.0, 2.0, 3.0, 4.0, 5.0 };

```

```
    Number y = f(x);

    cout << y.evaluate() << endl;    // 797.751

    //  Change x0 on the DAG
    x[0].setVal(2.5);

    //  Evaluate the dag again
    cout << y.evaluate() << endl;    // 2769.76
}
```

This pattern to store a calculation in the form a DAG, or another form that lends itself to evaluation, instead of conducting the evaluation straightaway (or *eagerly*), and conducting the evaluation at a later time, directly from the DAG, possibly repeatedly and possibly with different inputs set directly on the DAG, is called *lazy evaluation*. Lazy evaluation is what makes AAD automatic; it is also the principle behind, for example, expression templates (see Chapter 15) and scripting (see our publication [11]).

Calculation order There is more we can do with a DAG than evaluate it lazily. For instance, we can store the calculation order number on the node, which will be useful for what follows. We identify the nodes by their calculation order and store it on the node. We reuse our postorder code; we just need a new visit function. In this case, the visits don't depend on the concrete type of the node; we don't need a virtual visit function, and we can code it directly on the base node.

```
//  On the base Node

    void setOrder(unsigned order)
    {
        myOrder = order;
    }

    unsigned order()
    {
        return myOrder;
    }

//  On the Number class

    void setOrder()
    {
        myNode->resetProcessed();
        myNode->postorder(
            [order = 0](Node& n) mutable {n.setOrder(++order); });
    }
```

The lambda defines a data member *order* that starts at 0. Since C++14, lambdas not only capture environment variables, but can also define other internal variables in the capture clause. The evaluation of the lambda modifies its *order* member, so the lambda must be marked mutable.

```
int main()
{
    //  Inputs
    Number x[5] = { 1.0, 2.0, 3.0, 4.0, 5.0 };

    //  Build DAG
    Number y = f(x);

    //  Number the nodes
    y.setOrder();
}
```

Our DAG is numbered after the execution of the code above, and each node is identified by its postorder. We can log the intermediate results in the evaluation as follows:

```
//  On the Number class

    void logResults()
    {
        myNode->resetProcessed();
        myNode->postorder([] (Node& n)
        {
            cout << "Processed node "
                 << n.order() << " result = "
                 << n.result() << endl;
        });
    }
```

after evaluation of the DAG. When we evaluate the DAG repeatedly with different inputs, we get different logs accordingly:

```
int main()
{
    //  Set inputs
    Number x[5] = { 1.0, 2.0, 3.0, 4.0, 5.0 };

    //  Build the dag
    Number y = f(x);

    //  Set order on the dag
    y.setOrder();

    //  Evaluate on the dag
    cout << y.evaluate() << endl;    // 797.751

    //  Log all results
    y.logResults();

    //  Change x0 on the dag
    x[0].setVal(2.5);

    //  Evaluate the dag again
    cout << y.evaluate() << endl;    // 2769.76

```

```
    // Log results again
    y.logResults();

}
```

For the first evaluation we get the log:

Processed node 1 result = 3
Processed node 2 result = 5
Processed node 3 result = 1
Processed node 4 result = 5
Processed node 5 result = 2
Processed node 6 result = 7
Processed node 7 result = 21
Processed node 8 result = 4
Processed node 9 result = 3.04452
Processed node 10 result = 12.1781
Processed node 11 result = 33.1781
Processed node 12 result = 24.0445
Processed node 13 result = 797.751

For the second evaluation, we get different logs reflecting $x_0 = 2.5$.

We can display the DAG, this time identifying nodes by their evaluation order:

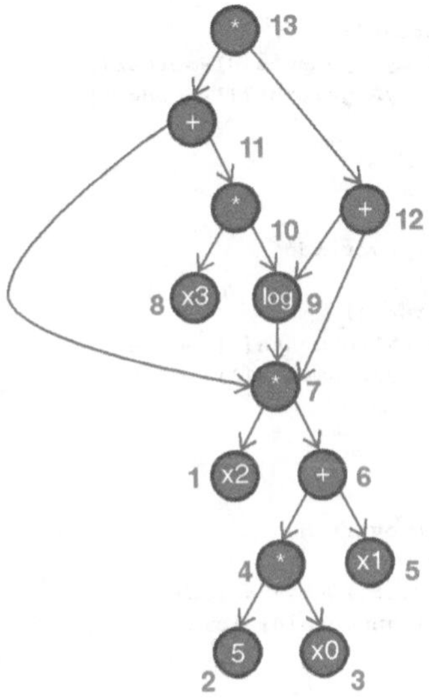

and intermediate results (for the first evaluation):

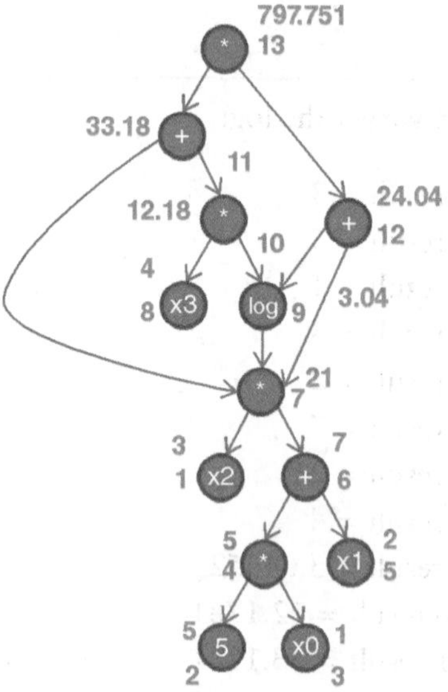

Reverse engineering There is even more we can do with the DAG. For instance, we can reverse engineer an equivalent calculation program:

```
//  On the base Node
    virtual void logInstruction() = 0;

//  On the + Node
    void logInstruction() override
    {
        cout << "y" << order()
            << " = y" << myArguments[0]->order()
            << " + y" << myArguments[1]->order()
            << endl;
    }

//  On the * Node
    void logInstruction() override
    {
        cout << "y" << order()
            << " = y" << myArguments[0]->order()
            << " * y" << myArguments[1]->order()
            << endl;
    }

//  On the log Node
    void logInstruction() override
    {
        cout << "y" << order() << " = log("
            << "y" << myArguments[0]->order() << ")" << endl;
    }

```

```
// On the Leaf
    void logInstruction() override
    {
        cout << "y" << order() << " = " << myValue << endl;
    }

//  On the number class
    void logProgram()
    {
        myNode->resetProcessed();
        myNode->postorder([] (Node& n) {n.logInstruction(); });
    }

int main()
{
    //  Inputs
    Number x[5] = { 1.0, 2.0, 3.0, 4.0, 5.0 };

    //  Build the DAG
    Number y = f(x);

    //  Set order
    y.setOrder();

    //  Log program
    y.logProgram();
}
```

We get the log:

y1 = 3
y2 = 5
y3 = 1
y4 = y2 * y3
y5 = 2
y6 = y4 + y5
y7 = y1 * y6
y8 = 4
y9 = log(y7)
y10 = y8 * y9
y11 = y7 + y10
y12 = y7 + y9
y13 = y11 * y12

We reconstructed an equivalent computer program from the DAG. Of course, we can change the inputs and generate a new computer program that reflects the new values of the leaves and produces different results.

Note that we managed to make visits (evaluation, instruction logging) depend on both the visitor and the concrete type of the visited node. Different things happen according to the visitor (evaluator, instruction logger) and the node (+, – and so forth). We effectively implemented GOF's *visitor* pattern; see [25]. This pattern is explained in more detail and systematically applied to scripts in our publication [11] (where we study trees, not DAGs, but the logic is the same).

We learned to automatically generate DAGs from templated calculation code with operator overloading. We learned to traverse DAGs in a specific order called postorder and visit its nodes in different ways: lazily evaluate the DAG or reverse engineer an equivalent sequence of instructions.

This covered the "Automatic" in "Automatic Adjoint Differentiation."[7] We now move on to "Adjoint Differentiation" and use the DAG to calculate *differentials*, all of them, in a single traversal; hence the key constant time property.

Before we do this, we need a detour through basic formalism and adjoint mathematics.

9.3 ADJOINT MATHEMATICS

We must now further formalize and generalize the adjoint mathematics introduced in the previous chapter.

In what follows, we identify the nodes n_i in the DAG by their postorder traversal number. We call y_i the result of the operation on n_i.

We denote $y = y_N$ the final result on the top node n_N.

We denote C_i the set of children (arguments) nodes of n_i.

Mathematical operations are either unary (they take one argument like *log* or *sqrt* or the unary – that changes a number into its opposite) or binary (two arguments, like *pow* or the operators $+, -, *, /$). Hence C_i is a set of either 0, 1, or 2 nodes. Besides, because of postorder, the numbers of the nodes in C_i are always less than i. The only nodes n_i for which C_i is empty are by definition the leaves.

The leaf nodes are processed first in postorder traversal, so their node number i is typically low. Leaves represent the inputs to the calculation.

We define the adjoint of the result y_i on the node n_i and denote a_i the partial derivative of the final result $y = y_N$ to y_i:

$$a_i \equiv \frac{\partial y}{\partial y_i} = \frac{\partial y_N}{\partial y_i}$$

[7]The first *A* in *AAD* sometimes stands for Automatic and sometimes for Algorithmic. Algorithmic sounds better but Automatic better describes the method.

Then, obviously:

$$a_N = \frac{\partial y_N}{\partial y_N} = 1$$

and, directly from the derivative chain rule, we have the fundamental adjoint equation:

$$a_j = \sum_{i/n_j \in C_i} \frac{\partial y_i}{\partial y_j} a_i$$

The adjoint of a node is the sum of the adjoints of its parents (calculations for which that node is an argument) weighted by the partial derivatives of parent to child, operation to argument. The operations on the nodes are elementary operations and their derivatives are known analytically.

Note the reverse order to evaluation: in an evaluation, the result of an operation depends on the values of the arguments. In the adjoint equation, it is the adjoint of the argument that depends on the adjoints of its operations.

Hence, just like evaluation computes results bottom-up, children first, adjoint computation propagates top down, parents first.

More precisely, the adjoint equation turns into the following algorithm that computes the adjoints for all the nodes in the DAG (all the differentials of the final result) and is a direct translation, in graph terms, of the back-propagation algorithm of the previous chapter.

Traversal starts on the top node, which adjoint is known to be 1, and makes its way toward the leaf nodes, which adjoints are the derivatives of the final result to the inputs.

The adjoints of every node are seeded to 0 (except the top node, which is seeded to 1) and the *visit* to node n_i consists in the propagation of its adjoint a_i:

$$a_j += \frac{\partial y_i}{\partial y_j} a_i$$

to the adjoints a_j of all its arguments (child nodes) $n_j \in C_i$. In the terms of the previous chapter, the visit of a node implements the adjoint of the operation represented on the node. After every node has been processed, exactly once and in the correct order, the adjoints of all nodes have correctly accumulated in accordance with the adjoint equation.

As we observed in the previous chapter, $\frac{\partial y_i}{\partial y_j}$ is known analytically, but depends on all the $(y_j)_{n_j \in C_i}$. This means that we must know all the intermediate results before we conduct adjoint propagation. Adjoint propagation is a *post-evaluation* step. It is only after we built the DAG and evaluated it that we can propagate adjoints through it. At this point, all the intermediate results are cached on the corresponding nodes, so the necessary partial derivatives can be computed.[8]

We may now formalize and code adjoint propagation: we set $a_N = 1$ and start with all other adjoints zeroed: for $i < N$, $a_i = 0$. We process all the nodes in the DAG, top to leaves. When visiting a node, we must know its adjoint a_i, so we can implement:

$$a_j + = \frac{\partial y_i}{\partial y_j} a_i$$

for all its kids $j \in C_i$. This effectively computes the adjoints of all the nodes in the DAG, hence all the partial derivatives of the final result, with a single DAG traversal hence, in constant time.

This point is worth repeating, since the purpose of AAD is constant time differentiation. All the adjoints of the calculation are accumulated from the top-down traversal of its DAG. In this traversal, every node is visited once and its adjoints are propagated to its child nodes, weighted by local derivatives. Hence, the adjoint accumulation, like the evaluation, consists in one visit to every node (operation) on the calculation DAG. Adjoint accumulation, which effectively computes all derivative sensitivities, is of the same order of complexity as the evaluation and constant in the number of inputs.

It is easy enough to code the visit routines. We store adjoints on the nodes and write methods to set and access them, like we did for evaluation results. We also need a method to reset all adjoints to 0, like we did for the processed flag. Finally, we log the visits so we can see what is going on.

```
//  On the base Node
    double      myAdjoint = 0.0;

    //  ...

public:

    //  ...

    //  Note: return by reference
    //  used to get and set
    double& adjoint()
    {
        return myAdjoint;
    }

```

[8]An alternative solution is to compute the partial derivatives *eagerly* during evaluation and store them on the parent node. Once the parent node caches all its partial derivatives, the intermediate results are no longer needed for back-propagation. This solution is actually the one implemented in Chapters 10 and 15.

```
    void resetAdjoints()
    {
        for (auto argument : myArguments) argument->resetAdjoints();
        myAdjoint = 0.0;
    }

    virtual void propagateAdjoint() = 0;

//  On the + Node
    void propagateAdjoint() override
    {
        cout << "Propagating node " << myOrder
            << " adjoint = " << myAdjoint << endl;

        myArguments[0]->adjoint() += myAdjoint;
        myArguments[1]->adjoint() += myAdjoint;
    }

//  On the * Node
    void propagateAdjoint() override
    {
        cout << "Propagating node " << myOrder
            << " adjoint = " << myAdjoint << endl;

        myArguments[0]->adjoint() += myAdjoint * myArguments[1]->result();
        myArguments[1]->adjoint() += myAdjoint * myArguments[0]->result();
    }

//  On the log Node
    void propagateAdjoint() override
    {
        cout << "Propagating node " << myOrder
            << " adjoint = " << myAdjoint << endl;

        myArguments[0]->adjoint() += myAdjoint / myArguments[0]->result();
    }

//  On the Leaf
    void propagateAdjoint() override
    {
        cout << "Accumulating leaf " << myOrder
            << " adjoint = " << myAdjoint << endl;
    }

//  On the number class

    //  Accessor/setter, from the inputs
    double& adjoint()
    {
        return myNode->adjoint();
    }

    //  Propagator, from the result
    void propagateAdjoints()
    {
        myNode->resetAdjoints();
        myNode->adjoint() = 1.0;
```

```
        //  At this point, we must propagate sensitivities,
        //      but how exactly do we do that?
    }
```

Notice how the propagation code uses the intermediate results stored on the child nodes to compute partial derivatives, as mentioned earlier. The propagation of adjoints on the unary and binary operator nodes implements the adjoint equation. Leaf nodes don't propagate anything; by the time propagation hits them, their adjoints are entirely accumulated, so their override of *propagateAdjoints*() simply logs the results.

We could easily code the visits but not start the propagation. It is clear that doing the same as for evaluation, something like:

```
myNode->postorder([](Node& n) {n.propagateAdjoint(); });
```

is not going to fly. Postorder traverses the DAG arguments first, but for adjoint propagation it is the other way around. Adjoints are propagated parent to child. We must find the correct traversal order.

9.4 ADJOINT ACCUMULATION AND DAG TRAVERSAL

Can we find a traversal strategy for adjoint propagation such that each node is visited exactly once and all adjoints accumulate correctly?

Preorder traversal

An intuitive candidate is *preorder*. Like postorder, preorder is a depth-first traversal pattern, but one that visits each node *before* its children. We reprint the postorder code and write the preorder code below (on the base Node class). We also add a method propagateAdjoints() on the Number class to start adjoint propagation visits in preorder from the result's node:

```
//  On the base Node
    template <class V>
    void postorder(V& visitFunc)
    {
        //  Already processed -> do nothing
        if (myProcessed == false)
        {
            //  Process children first
            for (auto argument : myArguments)
                argument->postorder(visitFunc);

            //  Visit the node
            visitFunc(*this);

            //  Mark as processed
            myProcessed = true;
```

```
        }
    }

    template <class V>
    void preorder(V& visitFunc)
    {
        //  Visit the node first
        visitFunc(*this);

        //  Process children
        for (auto argument : myArguments)
            argument->preorder(visitFunc);
    }

//  On the Number class
    //  Propagate from the result
    void propagateAdjoints()
    {
        //  Reset all to 0
        myNode->resetAdjoints();
        //  See result to 1
        myNode->adjoint() = 1.0;
        //  Propagate from top
        myNode->preorder([](Node& n) {n.propagateAdjoint(); });
    }

int main()
{
    //  Set inputs
    Number x[5] = { 1.0, 2.0, 3.0, 4.0, 5.0 };

    //  Build the DAG
    Number y = f(x);

    //  Set order on the dag
    y.setOrder();

    //  Evaluate
    y.evaluate();

    //  Propagate adjoints through the dag, from the result
    y.propagateAdjoints();

    //  Get derivatives
    for (size_t i = 0; i < 5; ++i)
    {
        cout << "a" << i << " = " << x[i].adjoint() << endl;
    }
}
```

Here is the result:

Propagating node 13 adjoint = 1
Propagating node 11 adjoint = 24.0445
Propagating node 7 adjoint = 24.0445

```
Accumulating leaf 1 adjoint = 168.312
Propagating node 6 adjoint = 72.1336
Propagating node 4 adjoint = 72.1336
Accumulating leaf 2 adjoint = 72.1336
Accumulating leaf 3 adjoint = 360.668
Accumulating leaf 5 adjoint = 72.1336
Propagating node 10 adjoint = 24.0445
Accumulating leaf 8 adjoint = 73.2041
Propagating node 9 adjoint = 96.1781
Propagating node 7 adjoint = 28.6244
Accumulating leaf 1 adjoint = 368.683
Propagating node 6 adjoint = 158.007
Propagating node 4 adjoint = 230.14
Accumulating leaf 2 adjoint = 302.274
Accumulating leaf 3 adjoint = 1511.37
Accumulating leaf 5 adjoint = 230.14
Propagating node 12 adjoint = 33.1781
Propagating node 7 adjoint = 61.8025
Accumulating leaf 1 adjoint = 801.3
Propagating node 6 adjoint = 343.414
Propagating node 4 adjoint = 573.555
Accumulating leaf 2 adjoint = 875.829
Accumulating leaf 3 adjoint = 4379.14
Accumulating leaf 5 adjoint = 573.555
Propagating node 9 adjoint = 129.356
Propagating node 7 adjoint = 67.9623
Accumulating leaf 1 adjoint = 1277.04
Propagating node 6 adjoint = 547.301
Propagating node 4 adjoint = 1120.86
Accumulating leaf 2 adjoint = 1996.69
Accumulating leaf 3 adjoint = 9983.43
Accumulating leaf 5 adjoint = 1120.86
a0 = 9983.43
a1 = 1120.86
```

a2 = 1277.04
a3 = 73.2041
a4 = 0

We conducted 35 visits through a DAG with 13 nodes. All nodes were visited, but many were visited multiple times. For example, node 7 propagated 4 times.

However, the derivatives are wrong. We recall that we computed them with finite differences earlier, and the correct values are: 950.736, 190.147, 443.677, 73.2041, and 0.

Actually, the values are incorrect *because* nodes are visited multiple times. Every time, their adjoint accumulated so far is propagated so their arguments accumulate adjoints multiple times. This is easily remedied by setting the adjoint to 0 after its propagation to child nodes. This way, only the part of the adjoint *accumulated since the previous propagation* is propagated the next time around:

```
//  On the + Node
    void propagateAdjoint() override
    {
        cout << "Propagating node " << myOrder
            << " adjoint = " << myAdjoint << endl;

        myArguments[0]->adjoint() += myAdjoint;
        myArguments[1]->adjoint() += myAdjoint;

        myAdjoint = 0.0;
    }

//  On the * Node
    void propagateAdjoint() override
    {
        cout << "Propagating node " << myOrder
            << " adjoint = " << myAdjoint << endl;

        myArguments[0]->adjoint() += myAdjoint * myArguments[1]->result();
        myArguments[1]->adjoint() += myAdjoint * myArguments[0]->result();

        myAdjoint = 0.0;
    }

//  On the log Node
    void propagateAdjoint() override
    {
        cout << "Propagating node " << myOrder
            << " adjoint = " << myAdjoint << endl;

        myArguments[0]->adjoint() += myAdjoint / myArguments[0]->result();

```

```
            myAdjoint = 0.0;
        }

    //  On the Leaf
        void propagateAdjoint() override
        {
            cout << "Accumulating leaf " << myOrder
                << " adjoint = " << myAdjoint << endl;

            //  No reset here, no propagation is happening, just accumulation
        }
```

Now we get the correct values, *exactly* the same derivatives we had with finite differences, but we still visited 35 nodes out of 13, and the remedy we implemented is really a hack: if every node was visited once, as it should, such remedy would not be necessary.

Note that AAD and finite differences agree on differentials, at least to the fourth significant figure. This illustrates that analytic derivatives produced by AAD are not different, in general, from the numerical derivatives produced by bumping. They are just computed a lot faster.

At least AAD is faster when implemented correctly. A poor implementation may well perform slower than finite differences. Our implementation so far certainly does. Efficient memory management and friends are addressed in the next chapter, but DAG traversal is addressed now. Our implementation will not fly unless we visit each node in the DAG once; 35 visits to 13 nodes is not acceptable. Preorder is not the correct traversal strategy.

Breadth-first traversal

Preorder is still a depth-first strategy in the sense that it sequentially follows paths on the DAG down to a leaf.

We may try a breadth-first strategy instead, where visits are scheduled by level in the hierarchy: top node first, then its children, left to right, then all the children of the children, and so forth.

Breadth-first traversal cannot be implemented recursively.[9] It is implemented as follows: start on the top node with an empty queue of (pointers on) nodes. Process a node in the following manner:

1. Send the children to the back of the queue.
2. Visit the node.
3. Process all nodes in the queue from the front until empty.

[9] To be honest, *nothing* should be implemented recursively in C++. It is best for performance to implement everything without recursion. But this is tedious; performance is not the concern of this chapter, and recursive code is terse and elegant.

Easy enough. The implementation takes a few lines:

```
#include <queue>

// On the base Node

    template <class V>
    void breadthFirst(
        V& visitFunc,
        queue<shared_ptr<Node>>& q = queue<shared_ptr<Node>>())
    {
        // Send kids to queue
        for (auto argument : myArguments) q.push(argument);

        // Visit the node
        visitFunc(*this);

        // Process nodes in the queue until empty
        while (!q.empty())
        {
            // Access the front node
            auto n = q.front();

            // Remove fom queue
            q.pop();

            // Process it
            n->breadthFirst(visitFunc, q);
        } // Finished processing the queue: exit
    }

// On the Number class

    // Propagator, from the result
    void propagateAdjoints()
    {
        myNode->resetAdjoints();
        myNode->adjoint() = 1.0;
        myNode->breadthFirst([](Node& n) {n.propagateAdjoint(); });
    }
```

The *main*() caller is unchanged.

Unfortunately, breadth-first is no better than preorder: it computes the correct results (providing we still zero adjoints after propagation) but still conducts 35 visits, and node 7, for instance, still propagates four times. The order is not the same, but the performance is identical.

Breadth-first is not the answer.

9.5 WORKING WITH TAPES

We believe it was useful to experiment with different unsuccessful traversal strategies to earn a good understanding of DAGs and how they work, and we took the time to introduce some essential pieces of graph theory and put them in practice in modern, modular (although inefficient) C++ code.

Now is the time to provide the correct answer.

What we need is a traversal strategy that guarantees that the entire, correct adjoints accumulate after a single visit to every node. The only way this is going to happen is that every node must have completed accumulating its adjoint *before* it propagates. In other terms, *all* the parents of a node must be visited *before* the node.

This particular order is well known to graph theory. It is called the *topological order*. In general, it takes specialized algorithms to sort a graph topologically. But for DAGs we have a simple result:

The topological order for a DAG is the reverse of its postorder.

This is simple enough, and the demonstration is intuitive: postorder corresponds to the evaluation order, where arguments are visited before calculations. Hence, in its reverse order, all the parents (operations) of a node (argument) are visited before the node. In addition, postorder visits each node once, hence, so does the reverse. Therefore, the reverse postorder *is* the topological order.[10]

The correct order of traversal in our example is simply: 13, 12, 11, . . . , 1.

What this means is that we don't need DAGs after all. We need *tapes*. We don't need a graph structure for the nodes; we need a *sequence* of nodes in the order they evaluate (postorder). After we completed evaluation and built the tape, we traversed it in the reverse order to propagate adjoints. One single (reverse) traversal of the tape correctly computes the adjoints for all the nodes. We computed all the derivative sensitivities in constant time.

We propagate adjoints once through each node just as we evaluate each node once. The complexity of the propagation in terms of visits is therefore the same as for the evaluation. Both conduct the same number of visits, corresponding to the number of operations. Propagation visits are slightly more complex, though; for example, to propagate through a multiplication takes two multiplications. So the total complexity of AAD is around three evaluations (depending on the code): one evaluation + (one propagation = two evaluations). In addition, we have a significant *administrative* overhead: with our toy code, allocations alone could make AAD slower than bumping. We will reduce admin costs to an absolute minimum in the next chapter, and further in Chapter 15.

For now, we refactor our code to use a tape.

Conceptually, a tape is a DAG sorted in the correct order.

---

[10]Actually, this proves that reverse postorder is *a* topological order but let us put uniqueness considerations aside.

Programmatically, this simplifies data structures. The tape is a sequence of nodes, so it is natural to store the nodes in a vector of (unique pointers). That way, a node's lifespan is the same as the tape's. We also save the overhead of shared pointers. Nodes still need references to their arguments to propagate adjoints, but they don't need to own their memory: we can use dumb pointers for that.

We also simplify the code and remove lazy evaluation and numbering: these were for demonstration and we don't need them anymore. We need the tape only for adjoint propagation. Therefore, we modify the code so it evaluates calculations (eagerly) as it builds the tape, and, when the calculation is complete, propagates adjoints through the tape in the reverse order.

The complete refactored code is listed below:

```
#include <memory>
#include <string>
#include <vector>
#include <queue>

using namespace std;

class Node
{
protected:

    vector<Node*> myArguments;

    double      myResult;
    double      myAdjoint = 0.0;

public:

    //  Access result
    double result()
    {
        return myResult;
    }

    //  Access adjoint
    double& adjoint()
    {
        return myAdjoint;
    }

    void resetAdjoints()
    {
        for (auto argument : myArguments) argument->resetAdjoints();
        myAdjoint = 0.0;
    }

    virtual void propagateAdjoint() = 0;
};

```

```
class PlusNode : public Node
{
public:

    PlusNode(Node* lhs, Node* rhs)
    {
        myArguments.resize(2);
        myArguments[0] = lhs;
        myArguments[1] = rhs;

        //  Eager evaluation
        myResult = lhs->result() + rhs->result();
    }

    void propagateAdjoint() override
    {
        myArguments[0]->adjoint() += myAdjoint;
        myArguments[1]->adjoint() += myAdjoint;
    }
};

class TimesNode : public Node
{
public:

    TimesNode(Node* lhs, Node* rhs)
    {
        myArguments.resize(2);
        myArguments[0] = lhs;
        myArguments[1] = rhs;

        //  Eager evaluation
        myResult = lhs->result() * rhs->result();
    }

    void propagateAdjoint() override
    {
        myArguments[0]->adjoint() += myAdjoint * myArguments[1]->result();
        myArguments[1]->adjoint() += myAdjoint * myArguments[0]->result();
    }
};

class LogNode : public Node
{
public:

    LogNode(Node* arg)
    {
        myArguments.resize(1);
        myArguments[0] = arg;

        //  Eager evaluation
        myResult = log(arg->result());
    }
```

```

    void propagateAdjoint() override
    {
        myArguments[0]->adjoint() += myAdjoint / myArguments[0]->result();
    }
};

class Leaf: public Node
{
public:

    Leaf(double val)
    {
        myResult = val;
    }

    double getVal()
    {
        return myResult;
    }

    void setVal(double val)
    {
        myResult = val;
    }

    void propagateAdjoint() override {}
};

class Number
{
    Node* myNode;

public:

    //  The tape, as a public static member
    static vector<unique_ptr<Node>> tape;

    //  Create node and put it on tape
    Number(double val)
        : myNode(new Leaf(val))
    {
        tape.push_back(unique_ptr<Node>(myNode));
    }

    Number(Node* node)
        : myNode(node) {}

    Node* node()
    {
        return myNode;
    }

```

```
    void setVal(double val)
    {
        //  Cast to leaf, only leaves can be changed
        dynamic_cast<Leaf*>(myNode)->setVal(val);
    }

    double getVal()
    {
        //  Same comment here, only leaves can be read
        return dynamic_cast<Leaf*>(myNode)->getVal();
    }

    //  Accessor for addjoints
    double& adjoint()
    {
        return myNode->adjoint();
    }

    //  Adjoint propagation
    void propagateAdjoints()
    {
        myNode->resetAdjoints();
        myNode->adjoint() = 1.0;

        //  Find my node on the tape, searching from last
        auto it = tape.rbegin();    //  last node on tape
        while (it->get() != myNode)
            ++it;   //  reverse iter: ++ means go back

        //  Now it is on my node
        //  Conduct propagation in reverse order
        while (it != tape.rend())
        {
            (*it)->propagateAdjoint();
            ++it;   //  Really means --
        }
    }
};

vector<unique_ptr<Node>> Number::tape;

Number operator+(Number lhs, Number rhs)
{
    //  Create node: note eagerly computes result
    Node* n = new PlusNode(lhs.node(), rhs.node());
    //  Put on tape
    Number::tape.push_back(unique_ptr<Node>(n));
    //  Return result
    return n;
}

Number operator*(Number lhs, Number rhs)
{
    //  Create node: note eagerly computes result
    Node* n = new TimesNode(lhs.node(), rhs.node());
    //  Put on tape
    Number::tape.push_back(unique_ptr<Node>(n));
```

```
    //  Return result
    return n;
}

Number log(Number arg)
{
    //  Create node: note eagerly computes result
    Node* n = new LogNode(arg.node());
    //  Put on tape
    Number::tape.push_back(unique_ptr<Node>(n));
    //  Return result
    return n;
}

template <class T>
T f(T x[5])
{
    auto y1 = x[2] * (5.0 * x[0] + x[1]);
    auto y2 = log(y1);
    auto y = (y1 + x[3] * y2) * (y1 + y2);
    return y;
}

int main()
{
    //  Set inputs
    Number x[5] = { 1.0, 2.0, 3.0, 4.0, 5.0 };

    //  Evaluate and build the tape
    Number y = f(x);

    //  Propagate adjoints through the tape
    //  in reverse order
    y.propagateAdjoints();

    //  Get derivatives
    for (size_t i = 0; i < 5; ++i)
    {
        cout << "a" << i << " = " << x[i].adjoint() << endl;
    }

    //  950.736, 190.147, 443.677, 73.2041, 0
}
```

This code returns the correct derivatives, having traversed the 13 nodes on the tape exactly once.

We just completed our first implementation of AAD.

This code is likely slower than bumping, the Number type is incomplete: it only supports sum, product, and logarithm. The code is not optimal in many ways, and uses shortcuts such as a static tape and unsafe casting. In the next chapter, we produce professional code, work on memory management, and implement various optimizations and conveniences. We make AAD orders of magnitude faster than bumping.

But the *essence* of AAD lies in this code and the explanations of this chapter.

CHAPTER 10

Effective AAD and Memory Management

In this chapter, we turn the ideas and toy code from the introductory chapters into a professional, efficient AAD library.

We have seen that, to differentiate calculation code with AAD, we instantiate the calculation code with a custom type for the representation of numbers, where all arithmetic operators and mathematical functions are overloaded, so that, in addition to being evaluated, all operations are recorded on a tape. After the calculation is complete and the entire sequence of its operations is recorded, adjoints are back-propagated over the tape to produce all its differentials in constant time.

The AAD library provides the custom number type, the tape data structure, and the object that represents an operation on tape, called *node*, together with back-propagation utilities to propagate adjoints over the nodes on tape in the reverse order from evaluation. Calculation code must be templated on its number type. Differentiation code executes the calculation code, instantiated with the AAD number type, which records the calculation on tape, calls the back-propagation utilities to accumulate adjoints, and reads differentials on the inputs' adjoints. This is called *instrumentation*. The AAD library is independent from any calculation code; it is the instrumentation code that applies it to differentiate a particular calculation.

With AAD, more so than other algorithms, efficient implementation is crucial, and algorithmic correctness insufficient. Our toy code from the previous chapter implements the correct algorithm and computes differentials in constant time. However, if we tried to differentiate a complex calculation, implemented with sophisticated code, like the simulations of Part II, it would take a very long constant time to compute the differentials. In all likelihood, a naive AAD differentiation of such code would be in practice slower than finite differences. In addition, the recording of a large number of operations could saturate RAM and result in a system-wide slowdown or a crash.

For these reasons, it is sometimes heard that the benefits of AAD are purely theoretical and cannot be achieved in a practical context. Some results seem to indicate that AAD would only be applicable in restricted contexts, and often produce sensitivities slower than cleverly implemented bumps. All of this could not be more wrong. AAD is universally applicable; it produces derivatives with unthinkable speed, but only when both the AAD library and the instrumented calculation code are written efficiently. We discuss the AAD library here, and instrumentation in Chapter 12.

We work in sequence on the Node (object that stores one operation on tape), the Tape (data structure that holds the sequence of nodes), and the Number (custom number representation type that overloads operators and functions so they are recorded on tape), including operator overloading and adjoint propagation utilities. These are the same classes and routines we manipulated in the previous chapter, refactored for practicality and performance. The result is a self-contained AAD library, independent of calculation code and that may be used to instrument any kind of computation. The code is found in our repository. The AAD library consists in the AAD*.* files, with dependency on gaussians.h[1] and blocklist.h.[2]

An efficient AAD library is mainly about memory management. Everything that touches the tape is executed on every evaluation of a mathematical operation. On every sum, multiplication, square root, and so forth, an additional set of instructions are executed to put that operation on tape. AAD only performs when every aspect of that code is optimized to the extreme. For instance, we cannot burden every mathematical operation with the significant overhead of a memory allocation, so the process that puts an operation on tape cannot involve allocation. Memory on tape must be preallocated so operations are recorded with minimal overhead.

What we develop here is a "traditional" AAD library, together with memory management techniques that allow it to perform with remarkable speed in contexts of practical relevance. We will revisit parts of the library and improve its performance in Chapter 15, resulting in an additional acceleration by a factor around two. We discuss some important characteristics and limitations in the next chapter, then, in Chapter 12, we instrument the parallel Monte Carlo library of part II, like we instrumented our toy function $f()$ in the previous chapter, but on the larger scale of a financial library and in a parallel context.

[1]For the instrumentation of the Gaussian density and cumulative distribution, which we consider as standard mathematical functions, even though they are not part of standard C++.

[2]For memory management.

10.1 THE NODE CLASS

In the toy implementation of the previous chapter, the partial derivatives of every operation were computed at back-propagation time. As a result, nodes were implemented as a polymorphic class, one concrete class for every mathematical operation, overriding the propagation method. In addition, the results of every operation were stored on the corresponding node, so that the partial derivatives could be computed at propagation time.

In the real code, we compute the local derivatives *eagerly* on evaluation and store them on the parent node. It follows that we no longer need polymorphic nodes, and we don't have to store the results of the operations. We have one single Node class to represent all operations, something like this:

```
//  Not acceptable code, for demonstration only
struct Node
{
    //  0: leaf, 1: unary operation, 2: binary operation
    size_t              n;

    //  This node's adjoint, initialized to 0
    double              adjoint = 0.0;
    //  Local derivatives to arguments
    vector<double>      derivatives;
    //  Pointers to the adjoints of the arguments
    vector<double*>     argAdjoints;

    void propagate()
    {
        for (size_t i = 0; i < n; ++i)
        {
            *argAdjoints[i] += derivatives[i] * adjoint;
        }
    }
};
```

where the method *propagate*() correctly implements the adjoint equation of the previous chapter. Note that we store pointers on the child adjoints in *argAdjoints*, not pointers on the child Nodes. Child adjoints are all we need for propagation.

There are a number of advantages to this approach:

Simplicity The Node class is simpler and the code is considerably shorter. Adjoint propagation from node n_i adds $a_i \frac{\partial y_i}{\partial y_j}$ to the adjoints of its arguments $n_j \in C_i$. The knowledge of the specific operation and its arguments was only necessary to compute the partial derivatives $\frac{\partial y_i}{\partial y_j}$ at propagation time. With the partial derivatives known and stored on the parent node n_i, polymorphism and the storage of intermediate results are no longer necessary and we have a single Node type to represent all operations on tape.

Performance Polymorphic classes are not free. They store hidden data members, called *vtable pointers*, necessary for the run-time resolution of virtual methods. Calls to virtual functions involve the additional overhead of run-time resolution, and the compiler cannot optimize and inline them as easily as nonvirtual methods. The single-Node design saves run-time polymorphism overhead at the cost of additional storage space on the Node for the partial derivatives.

In addition, the partial derivatives stored on the Nodes can be computed once and reused for multiple back-propagations, for example, for the accumulation of the adjoints of multiple results.

It is also visible that this Node design is not limited to unary and binary operations. The data member n, which specifies the number of arguments to the operation on the node, may be larger than two. The adjoint propagation implemented in *propagate*() remains correct with arbitrary n. This will alllow us to implement a substantial optimization in Chapter 15, where we represent entire *expressions* on a single multinomial node, resulting in around twice the speed in cases of practical relevance.

Convenience It is no longer necessary to keep results on Nodes, so the value of a Number is stored on the Number itself. This makes it considerably simpler to reason about, design, and debug instrumented code.

There is also a drawback to this design. The storage of the derivatives on the Node increases its memory footprint. The benefits outweigh the drawbacks, but memory considerations are not to be underestimated with AAD. Every single mathematical operation is recorded on a node, so tapes store a large amount of nodes and consume vast amounts of RAM.

The code above correctly illustrates our desired *design* for the Node, but not the implementation. The data stored on the Node – the partial derivatives and the pointers to the arguments' adjoints – are stored in vectors. Vectors are dynamically sized data structures. They allocate, free, and manage dynamic memory. We can't have that on the Node. Dynamic memory management is expensive, and a Node is created every time a mathematical operation is executed.

A simplistic solution could be to store data in static arrays of size two instead, since we know that the mathematical operations recorded on tape have zero, one, or two arguments. Not only would that result in an unacceptable waste of memory in leaf and unary nodes, but it would forbid multinomial nodes, those with more than two arguments, necessary for the optimizations of Chapter 15.

Another solution could be to derive the Node type depending on the number of its arguments, but it would reintroduce dynamic polymorphism and associated overheads. The solution we implemented[3] is to store dynamically sized data *outside* of the Node, in a separate space in memory. Where and how exactly we store this data will be discussed in the next section, together with the Tape data structure and memory management constructs. The Node doesn't store its data, but holds pointers on its place in memory.

```
//  Final code
class Node
{
    //  Number of children (arguments)
    const size_t n;

    //  The adjoint
        double          mAdjoint = 0;

    //  Data lives in separate memory

    //  the n derivatives to arguments
    double*             pDerivatives;

    //  the n pointers to the adjoints of the arguments
    double**            pAdjPtrs;

public:

    Node(const size_t N = 0) : n(N) {}

    //  Access adjoint for read and write
    double& adjoint()
    {
        return mAdjoint;
    }

    //  Back-propagate adjoints to arguments adjoints
    void propagateOne()
    {
        //  Nothing to propagate
        if (!n || !mAdjoint) return;

        for (size_t i = 0; i < n; ++i)
        {
            *(pAdjPtrs[i]) += pDerivatives[i] * mAdjoint;
        }
    }
};
```

[3]After testing many possible solutions and carefully measuring their performance.

We named the back-propagation function *propagateOne*(), not simply *propagate*(), for a reason that will be made apparent in Chapter 14, where we upgrade the library and make minor additions to the Node class to efficiently differentiate not one, but multiple results in a calculation.

Note the check on line 32. We don't propagate anything unless there is a non-zero adjoint to propagate on the parent node. This is a simple, yet strong optimization. In cases of practical relevance, zero adjoints are frequent, due to control flow in the instrumented code. Empirically, this single optimization accelerates differentiation by an average 15% in practical situations.

The code is found in AADNode.h.

10.2 MEMORY MANAGEMENT AND THE TAPE CLASS

The toy implementation of the previous chapter stored Nodes in a vector. It had the merit of simplicity. Nodes were stored in the sequence where operations are evaluated, and the STL vector made it particularly simple and convenient to iterate over the sequence of the Nodes, either forward, or backwards for the purpose of adjoint propagation.

But this is both inefficient and impractical. It is inefficient because we don't know in advance the number of operations in a calculation. The size of the vector holding the Nodes grows during the calculation, causing it not only to reallocate larger chunks of memory, but also *to copy all its contents* from its previous memory. It is also impractical because the reallocation of a vector invalidates pointers and iterators to the elements in the vector. We have seen that Numbers hold a pointer to their Node on tape. All these pointers would be invalidated every time the vector holding the Nodes reallocates. The toy code walked around this particular difficulty by storing in vectors, not the Nodes themselves, but (unique) pointers on the Nodes. This way, the Nodes always remained in the same memory space, it is the (unique) pointers that were copied every time the vector reallocated, so the pointers on the Nodes held in the Numbers remained valid, at the cost of yet another overhead.

For this reason, we are not going to hold the Nodes in a vector. We are going to develop a custom container, purposely designed for maximum efficiency and practicality in the context of AAD. We will be following a pattern known as "memory pool" or "small object allocator," whereby we will preallocate large blocks of memory and use them to store the sequence of Nodes side by side as operations in the instrumented code are recorded on tape. We baptized our data structure "blocklist" and implemented it in the file blocklist.h. Our implementation mainly follows the standard directives of STL container requirements, although we don't implement *all* of them.

The blocklist

The blocklist essentially follows the small object allocator pattern. It preallocates (large) blocks of memory and stores objects in the preallocated space so as to prevent costly allocations every time a new object is created. The design of such memory pools revolves around two considerations:

1. Manage what happens when a preallocated memory block fills out.
2. Handle the release of memory when individual objects are destroyed.

The second problem is typically hard to resolve in an efficient manner, but, luckily, it is not an issue for us. *We never need to release Nodes individually*. We don't free specific Nodes. We store the Nodes on tape in the sequence of the evaluation of their operations, and, when we no longer need the tape, we dispose of it and release all its Nodes simultaneously. In addition, Nodes don't manage external memory: they hold pointers to dynamic memory, but they don't own that memory, and it is not their responsibility to release it. It follows that Nodes don't need to be individually destroyed, either. They are just wiped out of memory altogether without any destruction overhead.

We store the Nodes in a sequence side by side in preallocated blocks of memory. When a block fills out, we preallocate another block for the storage of the subsequent Nodes. When this happens, the existing Nodes remain where they are. We don't copy anything, and, more importantly, all pointers and iterators to the existing Nodes remain valid.

The natural data structure for a memory block is a static array. C++11 introduced the standard array data structure, defined in the <array> header, as a thin, zero-cost wrapper around a static array, with the benefit of STL-compliant iterators. The ideal data structure to hold the multiple blocks recording a sequence of Nodes is a list: the standard guarantees that to insert a new element in a list does not affect the existing elements in any way, in particular, any pointer or iterator on existing elements remains valid. It follows that the container for the storage of the Nodes is a list of arrays, hence, "blocklist." Our data structure wraps a list of arrays and provides functionality to completely encapsulate the underlying data structure and let client code navigate, read, and write its data *as if it was stored in a single container*.

```
#include <array>
#include <list>
#include <iterator>
using namespace std;

template <class T, size_t block_size>
class blocklist
{
    //  Container = list of blocks
    list<array<T, block_size>>  data;
```

```
using list_iter = decltype(data.begin());
using block_iter = decltype(data.back().begin());

//  Current block
list_iter           cur_block;

//  Last block
list_iter           last_block;

//  Next free space in current block
block_iter          next_space;

//  Last free space (+1) in current block
block_iter          last_space;

//  ...
```

The iterator cur_block points to the current block, and the iterator next_space points to the first free space in that block. Together, these two iterators indicate the memory space where the next object is stored. The iterator last_space guards the end of the current block. When that space is reached, we jump to the next block. If the current block is the last block, as indicated by the iterator last_block, there is no next block, so we create another one at the back of the list.

For example, the following figure illustrates the state of a blocklist <Node, 10> after the storage of 15 Nodes:[4]

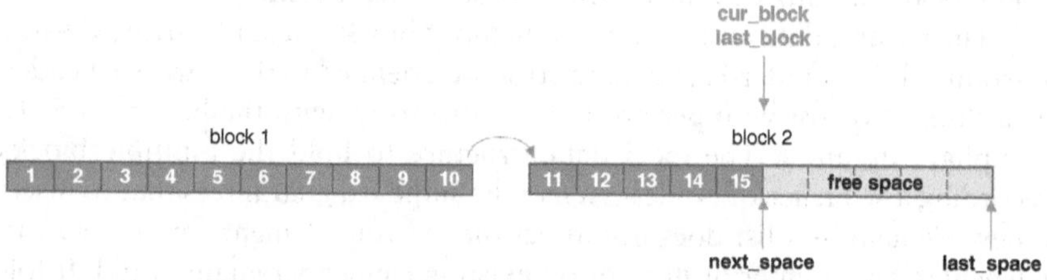

When a blocklist is reused repeatedly, we don't have to dispose of all the blocks, only to recreate them for the storage of subsequent objects. Instead, we *rewind* the blocklist with a reassignment of its iterators. The blocks remain in memory and their contents are overwritten with the storage of new objects.

The following figure illustrates the state of the previous blocklist after a rewind. The next object will overwrite the first space in the first block, moving next_space to slot two. The storage of 10 objects will move the

[4]Of course, we have much larger blocks in practice, default being 13,384 Nodes in a block.

next_space to the first slot of the second block, without creation of a block, since we already have two. It is only after the storage of 20 objects, when the two blocks are full, that a third block will be created on the list.

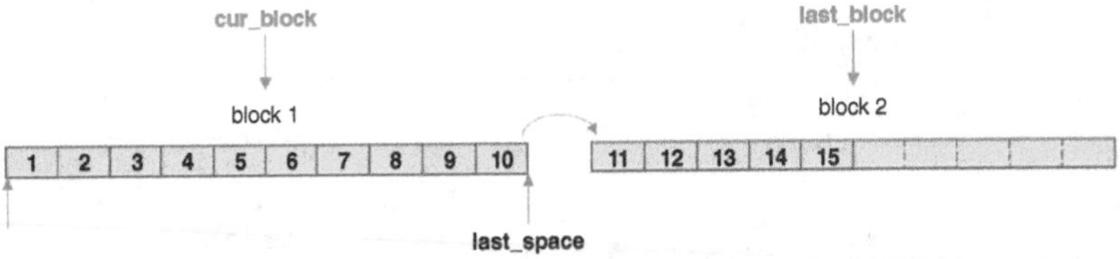

A very useful feature is to rewind a blocklist, not to the beginning, but to a mark left at a previous stage. We add a couple of iterators to the data structure so we can mark its state and subsequently revert to that marked state:

```
//  ...
//  Mark
list_iter          marked_block;
block_iter         marked_space;
//  ...
```

The following figure illustrates the state of the blocklist when the mark is set after the storage of five elements:

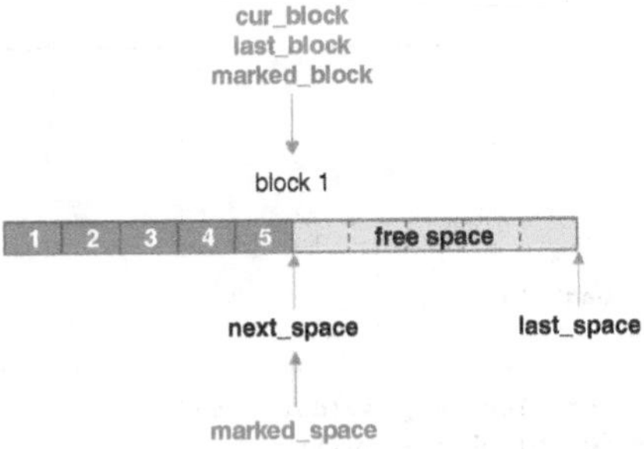

After the storage of another 10 elements, the state is illustrated below:

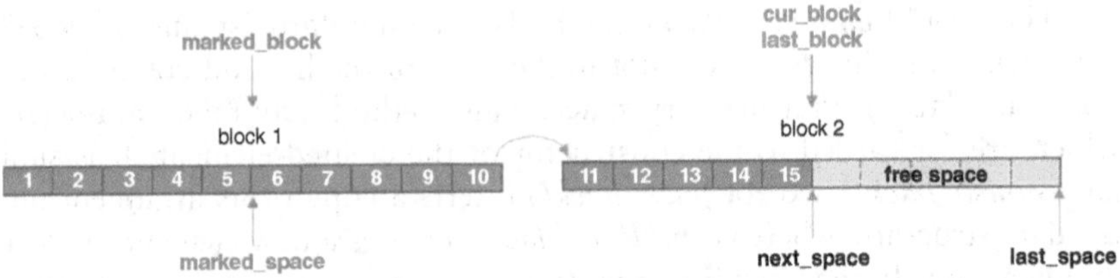

At this point, we rewind to the mark. Again, we just reassign iterators. Nothing is deleted; no memory is allocated or deallocated. The next stored object will overwrite the sixth slot of the first block:

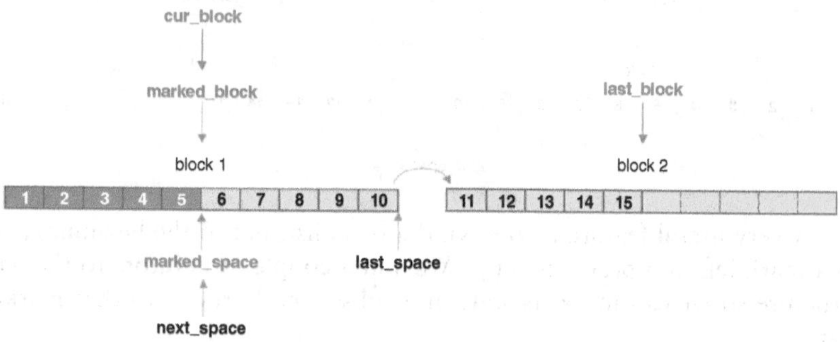

These iterator gymnastics permit us to navigate the blocklist for reading and writing. The methods implemented in the blocklist essentially manipulate these iterators to store objects side by side in the preallocated memory blocks, and rewind the blocklist in various ways by reassignment of the iterators and without any modification to the contents or memory. First, we have a private method to create a new block and position the iterators for subsequent storage:

```
    //  ...
private:
    //  Create new array
    void newblock()
    {
        //  Create array
        data.emplace_back();

        //  Re-assign iterators to the created array
        cur_block = last_block = prev(data.end());
        next_space = cur_block->begin();
        last_space = cur_block->end();
    }
    //  ...
```

The C++11 *emplace_black*() method on the standard list (and other STL data structures) inserts a new slot in the back of the list and creates a new object in place in that memory space. This method may take parameters, which are forwarded to the constructor of the created element. It is similar to *push_back*(), except *push_back*() inserts a copy of its argument into the data structure, whereas *emplace_black*() creates a new element in place, saving a copy. In the blocklist code, *emplace_black*() creates a new array on the list *data*. The blocklist's iterators are assigned to the newly created array. We have another method to move to the next block, and create a new one if necessary:

```
    //  ...
    //  Move on to next array
    void nextblock()
    {
        //  This is the last array: create new
        if (cur_block == last_block)
        {
            newblock();
        }

        //  This is not the last array: move to next
        else
        {
            ++cur_block;
            next_space = cur_block->begin();
            last_space = cur_block->end();
        }
    }
    //  ...
```

If there is no next block, we create one and reassign iterators to the newly created array. Otherwise, we reassign the iterators to the next array on the list.

Next, we have the public interface, starting with the constructor (which creates the first block and sets iterators) and a method *clear*() to factory reset the blocklist, as is customary for STL containers:

```
    //  ...
public:
    //  Create first block on construction
    blocklist()
    {
        newblock();
    }

    //  Factory reset
    void clear()
    {
        data.clear();
        newblock();
    }
    //  ...
```

We have methods to set a mark, rewind to the mark, or rewind to the beginning, by reassignment of the iterators, as explained earlier:

```
    //  ...
    //  Rewind but keep all blocks
    void rewind()
    {
        cur_block = data.begin();
```

```
        next_space = cur_block->begin();
        last_space = cur_block->end();
    }

    //  Set mark on current position
    void setmark()
    {
        marked_block = cur_block;
        marked_space = next_space;
    }

    //  Rewind to mark
    void rewind_to_mark()
    {
        cur_block = marked_block;
        next_space = marked_space;
        last_space = cur_block->end();
    }
    //  ...
```

The main purpose of the blocklist is to store objects. We implement a method *emplace_back*(), with STL consistent semantics to create a new object of type T in the next available memory space, as indicated by the iterators, and return a pointer on the object:

```
    //  ...
    //  Overload for default constructed
    T* emplace_back()
    {
        //  No more space in current array
        //      move to next (if none, create)
        if (next_space == last_space)
        {
            nextblock();
        }

        //  Current space
        auto old_next = next_space;

        //  Advance
        ++next_space;

        //  Return pointer
        //      on PREVIOUS space where the new object is stored
        return &*old_next;
    }
    //  ...
```

This overload is for default constructed objects. We don't even need to explicitly create the object, since arrays are populated with default objects on construction. We simply advance the iterator *next_space* and skip over the current slot. Note that in case new objects overwrite existing ones, they are not reset to default state. This is contrary to STL standard, but it prevents wasting CPU time resetting state, something we don't need.

Also note the check on the first line: if we stand at the end of a block, we move on to the next one, which also creates a new block if necessary. The method returns a pointer on the memory space where the new object lives, as specified by the *previous* state of the iterator *next_space*.

The whole purpose of the blocklist is to store objects in memory with minimum overhead. The method *emplace_back*() visibly achieves this goal. The method *nextblock*() is only called in the rare situations where the current block fills out (one operation out of 13,384 by default). Besides, *nextblock*() only involves an expensive allocation when a new block is created. When the same blocklist is repeatedly reused and rewound, *nextblock*() typically involves a cheap reassignment of pointers. In the vast majority of situations where there is space available on the block, the storage of an object involves a check (next_space == last_space), an iterator assignment (old_next = next_space), and a skip (++next_space), three insignificantly cheap low-level instructions. The method *emplace_back*(), and its overloads below, are a crucial factor in AAD performance, because they record Nodes in memory virtually for free.

We have two overloads to store more than one object, following the exact same logic. The first overload may be used when the number of objects is known at compile time and passed as a template parameter. When this is not the case, the second overload must be used. Both overloads implement the same code, virtually identical to the first overload above, and also with virtually zero run-time overhead, returning a pointer on the first created element.

```
//  ...
//  Stores n default constructed elements
//      and returns a pointer on the first

//      Version 1: n known at compile time
template <size_t n>
T* emplace_back_multi()
{
    //  No more space in current array
    if (distance(next_space, last_space) < n)
    {
        nextblock();
    }

    //  Current space
    auto old_next = next_space;

    //  Advance next
    next_space += n;

    //  Return
    return &*old_next;
}
```

```
//  Version 2: n unknown at compile time
//      same code
T* emplace_back_multi(const size_t n)
{
    //  No more space in current array
    if (distance(next_space, last_space) < n)
    {
        nextblock();
    }

    //  Current space
    auto old_next = next_space;

    //  Advance next
    next_space += n;

    //  Return
    return &*old_next;
}
//  ...
```

We mentioned that default constructed objects are not necessarily reset to factory state with our implementation of *emplace_back*(). We find it more efficient to provide instead a method *memset*() that resets all the bytes in all the arrays to some value, usually zero, simultaneously.

```
//  ...
//  Memset
void memset(unsigned char value = 0)
{
    for (auto& arr : data)
    {
        std::memset(&arr[0], value, block_size * sizeof(T));
    }
}
//  ...
```

All it does is call the standard *memset*() repeatedly on all the blocks. Finally, we have an overload for the creation of objects that are not default constructible. This overload takes arguments and forwards them to the constructor of the new element:

```
//  ...
//  Construct object of type T in place
//      in the next free space and return a pointer on it
//  Implements perfect forwarding of constructor arguments
template<typename ...Args>
T* emplace_back(Args&& ...args)
{
    //  No more space in current array
    if (next_space == last_space)
    {
        nextblock();
    }
    //  Placement new, construct in memory pointed by next
```

```
        //  memory pointed by next as T*
        T* emplaced = new (&*next_space)
            //  perfect forwarding of ctor arguments
            T(forward<Args>(args)...);

        //  Advance next
        ++next_space;

        //  Return
        return emplaced;
    }
    //  ...
```

This overload follows the same logic, except it doesn't only skip over a slot in the array; instead, it does construct a new object in place, forwarding its arguments to the object's constructor, with the *placement new* syntax. Contrary to what the operator *new* may suggest, no allocation takes place. Placement new is a traditional C++ construct, albeit a rarely used one, with a perhaps peculiar syntax: *new*(*mem*)*T*(*params*...) that creates an object of type T *in place* in the memory location *mem* with a call to T's constructor with arguments *params*.... The arguments to *emplace_back*() are *forwarded* to the object's constructor.

In addition, the number of arguments to *emplace_back*() is not fixed and depends on the number of arguments of the object's constructor. C++11 deals with variable number of arguments with a construct called *variadic templates*, which offers a specific syntax to deal with arguments not only of unknown type, like all templates, but also of unknown number. We don't discuss variadic templates in detail here, referring to any up-to-date C++ textbook instead. The arguments to *emplace_back*() are forwarded to the object's constructor with the standard C++11 function *forward*(), which, as the name indicates, passes the arguments through, in the same number, order, type, and conserving properties like constness or rvalue and lvalue reference, something known in C++11 as *perfect forwarding*.

Blocklist iterators

Blocklists encapsulate block creation and navigation logic, so that client code may act *as if* its elements were stored in one contiguous array, and seamlessly navigate the data, the blocklist correctly and automatically jumping from block to block, creating new blocks when necessary behind the scenes. For this purpose, we develop an *iterator* on the blocklist as a public nested class within the blocklist class. Like all iterators, the blocklist::iterator acts as a pointer on an element in the blocklist. It can be dereferenced to access the element with the usual prefix (*) or post-fix (->). Like any STL *bidirectional* iterator, a blocklist iterator can be skipped to the next element with the prefix (++) or reverted to the previous element with the prefix (–), its internal mechanics taking care of the blocklist logic and seamlessly jumping from block to block when necessary.

A blocklist iterator has an internal state that specifies one slot in the blocklist:

```
//  ...
//  Iterator
class iterator
{
    //  List and block
    list_iter   cur_block;   // current block
    block_iter  cur_space;   // current space
    block_iter  first_space; // first space in block
    block_iter  last_space;  // last (+1) space in block
//  ...
```

The iterator's state consists of a current block and a current space in that block, as explained earlier in detail. It also holds iterators on the first and last space in its block so it can jump blocks when necessary. Although we don't implement all the features of a fully STL-compliant iterator, we write a few definitions, called *iterator traits*, that identify the iterator as bidirectional, so it may be used with basic STL algorithms:

```
//  ...
public:

    //  iterator traits
    using difference_type = ptrdiff_t;
    using reference = T&;
    using pointer = T*;
    using value_type = T;
    using iterator_category = bidirectional_iterator_tag;
//  ...
```

We implement a default constructor for the iterator, and another constructor that sets its state to a specific slot in the blocklist:

```
//  ...
    //  Default constructor
    iterator() {}

    //  Constructor
    iterator(list_iter cb, block_iter cs, block_iter fs, block_iter ls) :
        cur_block(cb), cur_space(cs), first_space(fs), last_space(ls) {}
//  ...
```

Next, we have the prefix (++) operator, skipping the iterator to the next slot, jumping to the next block when necessary:

```
//  ...
    //  Pre-increment (we do not provide post)
    iterator& operator++()
```

```
    {
        //  Next slot
        ++cur_space;

        //  Skip over block?
        if (cur_space == last_space)
        {
            //  Next block
            ++cur_block;
            //  Reset block iterators
            first_space = cur_block->begin();
            last_space = cur_block->end();
            cur_space = first_space;
        }

        return *this;
    }
//  ...
```

and the prefix (–) that implements the same logic to point to the previous element:

```
//  ...
    //  Pre-decrement
    iterator& operator--()
    {
        //  Jump to previous block?
        if (cur_space == first_space)
        {
            //  Previous block
            --cur_block;
            //  Reset block iterators
            first_space = cur_block->begin();
            last_space = cur_block->end();
            cur_space = last_space;
        }

        --cur_space;

        return *this;
    }
//  ...
```

Finally, we have the dereference operators expected from any pointer or iterator to access its element:

```
//  ...
    //  Access elements
    T& operator*()
    {
        return *cur_space;
    }
    const T& operator*() const
    {
        return *cur_space;
    }
```

```
        T* operator->()
        {
            return &*cur_space;
        }
        const T* operator->() const
        {
            return &*cur_space;
        }
    };
```

and some comparison operators:

```
    //  ...
        //  Check equality
        bool operator ==(const iterator& rhs) const
        {
            return (cur_block == rhs.cur_block && cur_space == rhs.cur_space);
        }
        bool operator !=(const iterator& rhs) const
        {
            return (cur_block != rhs.cur_block || cur_space != rhs.cur_space);
        }
    };

    //  End of iterator code, blocklist code continues
    //  ...
```

so we can use these iterators in a loop like:

```
blocklist<T> b;
//...

for(auto it = b.begin(); it != b.end(); ++it)
{
    do_something_with_element(*it);
}
```

where we note that we implemented (a sufficient part of) the bidirectional iterator requirements so we can use STL algorithms like *for_each*(), and rewrite the loop above in a more proper manner:

```
for_each(b.begin(), b.end(), do_something_with_element);
```

or even use the C++11 range syntax:

```
for (T& elem : b) do_something_with_element(elem);
```

Our blocklist data structure is *almost* complete. We must provide methods for client code to access iterators on the first and (one after) last element in the blocklist, like all STL containers:

```
    //  blocklist code continues
    //  ...
    //  Access to iterators
```

```
    iterator begin()
    {
        return iterator(data.begin(), data.begin()->begin(),
            data.begin()->begin(), data.begin()->end());
    }

    iterator end()
    {
        auto last_block = prev(data.end());
        return iterator(cur_block, next_space,
            cur_block->begin(), cur_block->end());
    }
    \\ ...
```

begin() returns an iterator on the first slot of the first block; *end*() returns an iterator on the *next available slot for the storage of an element.*[5] We also provide an iterator on the mark:

```
    //  ...
    //  Iterator on mark
    iterator mark()
    {
        return iterator(marked_block, marked_space,
            marked_block->begin(), marked_block->end());
    }
    \\  ...
```

and a method *find*() that locates an element in the blocklist *by address* and returns an iterator on this element if found, *end*() otherwise, as is customary with STL containers:

```
    //  ...
    //  Find element, by pointer, searching sequentially from the end
    iterator find(const T* const element)
    {
        //  Search from the end
        iterator it = end();
        iterator b = begin();

        while (it != b)
        {
            --it;
            if (&*it == element) return it;
        }

        if (&*it == element) return it;

        return end();
    }
};
```

[5]When the blocklist is being overwritten, *end*() points to the *current* slot, not necessarily the last one in memory.

Note that the semantics of *find*() differ from usual STL semantics, where elements are found by value. It so happens that it best suits the needs of AAD to find elements by address. Also note that the search starts *from the end*, which is also AAD specific.

The code is found in blocklist.h in our repository.

The Tape class

We have encapsulated all the memory management difficulties in a custom data structure, the blocklist, which stores data in an extremely efficient manner, never invalidates its iterators (unless, of course, the blocklist is erased or rewound), and exposes a simple, STL-style interface to navigate data. The blocklist data structure was purposely designed to provide maximum efficiency and convenience, specifically in the context of AAD, so that the development of the tape presents no additional difficulty. Its code is in AADTape.h.

The tape mainly wraps a blocklist of Nodes. Remember from Section 10.1 that Nodes don't store their dynamically sized data (local derivatives and pointers on arguments adjoints). They only store the size of the data and pointers on its memory location. The actual Node data lives *in its own dedicated blocklists on the tape*:

```
#include "blocklist.h"
#include "AADNode.h"

constexpr size_t BLOCKSIZE  = 16384;    //  Number of Nodes in a block
constexpr size_t DATASIZE   = 65536;    //  Size of data in a block

class Tape
{
    //  Storage for the Nodes
    blocklist<Node, BLOCKSIZE>          myNodes;

    //  Storage for Node data:
    //        derivatives and child adjoint pointers
    blocklist<double, DATASIZE>          myDers;
    blocklist<double*, DATASIZE>         myArgPtrs;

    //  Padding so tapes in a vector don't interfere
    char                                 myPad[64];

public:

    //  Build Node in place and return a pointer
    //  N : number of children (arguments)
    template <size_t N>
    Node* recordNode()
    {
        //  Construct the Node in place on the blocklist
        Node* node = myNodes.emplace_back(N);

```

```
        //  Store Node data unless leaf
        //       note constexpr if
        if constexpr(N)
        {
            //  derivatives
            node->pDerivatives = myDers.emplace_back_multi<N>();
            //  child adjoints
            node->pAdjPtrs = myArgPtrs.emplace_back_multi<N>();
        }

        //  Return pointer on the new node
        return node;
    }

    void resetAdjoints()
    {
        for (Node& node : myNodes)
        {
            node.mAdjoint = 0;
        }
    }

    //  Clear, all blocklists
    void clear()
    {
        myDers.clear();
        myArgPtrs.clear();
        myNodes.clear();
    }

    //  Rewind, all blocklists together
    void rewind()
    {
        myDers.rewind();
        myArgPtrs.rewind();
        myNodes.rewind();
    }

    //  Set mark, all blocklists in sync
    void mark()
    {
        myDers.setmark();
        myArgPtrs.setmark();
        myNodes.setmark();
    }

    //  Rewind to mark, all blocklists
    void rewindToMark()
    {
        myDers.rewind_to_mark();
        myArgPtrs.rewind_to_mark();
        myNodes.rewind_to_mark();
    }

    //  Iterators

    using iterator = blocklist<Node, BLOCKSIZE>::iterator;
```

```
87
88      auto begin()
89      {
90          return myNodes.begin();
91      }
92
93      auto end()
94      {
95          return myNodes.end();
96      }
97
98      auto markIt()
99      {
100         return myNodes.mark();
101     }
102
103     auto find(Node* node)
104     {
105         return myNodes.find(node);
106     }
107 };
```

The code should be mostly self-explanatory, since it mainly wraps blocklists. The padding on line 17 avoids false sharing when multiple tapes stored in a vector are manipulated concurrently from different threads, as explained in Section 3.5.

The node recorder on line 22 takes the number of child nodes as a template parameter, not as an argument. Leaf nodes, by definition, don't have children. They represent the inputs to a calculation. Unary functions like *log*() or *sqrt*() have one child. Binary operators like + or $*$ have two. All this is always known at compile time. We will see in Chapter 15 that the number of inputs to an entire *expression* can also be determined at compile time with expression templates. It follows that we *can* pass it as a template parameter instead of an argument, which may reduce run-time overhead.

To record a node means store it in the node blocklist. The local derivatives and child adjoint pointers are stored in their own blocklists, where their node can find them by pointer, set on lines 32 and down. There is nothing to set when the node is a leaf, and this is something known at compile time. C++17 implements a new construct for these situations, called *constexpr if*, which we use on line 32. A constexpr if, as its name indicates, evaluates at compile time. A call to *recordNode*() instantiated with $N = 0$ does not *compile* the block of code starting line 32, in a similar manner as if it was surrounded by pragmas like "#if 0 ... #endif." This code is executed every time an operation happens, so we must optimize it to the extreme, evaluating all we can at compile time to eliminate as much run-time overhead as possible.

We have a method to reset all adjoints to zero (which we will not use), one to clear the tape (by clearing all its blocklists), followed by methods to rewind the tape, set a mark, and rewind to the mark. We explained these notions in detail earlier, and their implementation is entirely delegated to the underlying blocklists. The only comment is that all the blocklists must be rewound and marked synchronously.

Finally, the tape provides iterators to the usual first and (one after) last records, as well as the marked record and a record found by address. This is all entirely delegated to the blocklist iterator, explained in detail earlier.

10.3 THE NUMBER CLASS

The last component of the AAD library is the custom Number class, including operator overloading and adjoint propagation utilities. Remember its purpose: when calculation code instantiated with this number type is executed, its mathematical operations are not only evaluated, but also recorded on tape, so that adjoints may be back-propagated through the tape at a later stage to compute differentials in constant time. The code is found in AADNumber.h.

Data members

The code mainly follows the logic of the toy code from the previous chapter. One difference, explained in Section 10.1, is that the toy code stored values on the Node for simplicity, whereas it is both more convenient and more efficient to store it on the Number. It follows that a Number stores its value in addition to a pointer on its Node. A Number doesn't have any other non-static data members.

A Number also provides accessors on its value and adjoint.

```
class Number
{
    double myValue;
    Node* myNode;

    // ...

public:

    double& value()
    {
        return myValue;
    }
    double value() const
    {
        return myValue;
    }
```

```
    double& adjoint()
    {
        return myNode->adjoint;
    }
    double adjoint() const
    {
        return myNode->adjoint;
    }

    // ...
```

For the convenience of (friend) operator and mathematical function overloads, we also provide some (private) shortcut accessors to the data stored on a Number's Node:

```
    // ...
private:

    // Convenient access for friends

    // Access to local derivatives

    // For unary functions
    double& derivative() { return myNode->pDerivatives[0]; }
    // For binary operators and functions
    double& lDer() { return myNode->pDerivatives[0]; }
    double& rDer() { return myNode->pDerivatives[1]; }

    // Access to child adjoints

    // For unary functions
    double*& adjPtr() { return myNode->pAdjPtrs[0]; }
    // For binary operators and functions
    double*& leftAdj() { return myNode->pAdjPtrs[0]; }
    double*& rightAdj() { return myNode->pAdjPtrs[1]; }

    // ...
```

Thread local tape

Like in the toy code, the tape is a static member of the Number class, but with two twists. First, the tape itself is a global instance in the application. What the Number class (statically) holds is a *pointer* on it.

Second, the static Tape pointer is *thread local*, see Section 3.4 in Chapter 3. This means that every thread has *its own copy* of it. Where exactly Number::tape points depends on the thread who's asking. We need this feature to differentiate parallel code in Chapter 12. More generally, this is what makes AAD applicable to parallel algorithms. All the operations conducted on a given thread are recorded *on that thread's tape*. In particular, concurrent threads don't interfere when recording operations.

```
class Number
{
    //  ...

public:

    //  Static access to tape
    static thread_local Tape* tape;

//  ...
```

The main thread's tape is global to the application and declared in AAD.cpp. The static pointer on the Number type is initialized *on the main thread* to refer to that global tape. The following code is in AAD.cpp:

```
#include "AADnumber.h"

//  Declaration of the main thread's tape, global to the application
Tape globalTape;
//  Set the Number's static thread-local pointer to the global tape
//  from the main thread
thread_local Tape* Number::tape = &globalTape;
```

It is important to understand that the result of the instruction:

```
#include "AADnumber.h"

Tape& useThatTape = *Number::tape;
```

depends on the thread who executes it. If it is executed on the main thread, it returns a reference to globalTape, because the main thread's pointer is initialized to point to globalTape in AAD.cpp when the application opens. On another thread, Number::tape refers to a different tape, whichever was assigned *on that thread* by the execution of:

```
#include "AADnumber.h"

//  Called on some thread
Number::tape = &tapeForThatThread;
```

This point will be further clarified in Chapter 12, when we instrument our parallel simulation code. In the meantime, we continue with the Number class.

Node creation

We put a private method *createNode*() on the Number class to create a new Node on a tape for a Number:

```
    //  ...

private:

    //  Create Node on tape,
    //       with the right number of children (arguments)
    template <size_t N>
    void createNode()
    {
        myNode = tape->recordNode<N>();
    }

//  ...
```

and use it to create Nodes instead of calling *recordNode*() on the tape directly. This clarifies code and highlights the situations where a Node is created.

Leaf nodes Leaf nodes are the inputs to a calculation. They belong to Numbers that are not themselves the result of an operation involving other Numbers. These Numbers are the initial values of a calculation out of which the calculation is conducted and with respect to which the sensitivities of the final result are computed. It follows that leaf nodes are created in three situations, listed below:

```
    //  ...

public:

    Number() {}

    //  Construction out of a numeric value
    //       produces a leaf record on tape
    explicit Number(const double val) :
        myValue(val)
    {
        createNode<0>();
    }

    //  Assignment of a numeric value
    //  (i.e. overwriting)
    //       also produces a leaf record on tape
    Number& operator=(const double val)
    {
        myValue = val;
        createNode<0>();

        return *this;
    }

    //  putOnTape() explicitly called
    //       to produce a leaf record on tape
    void putOnTape()
```

```
    {
        createNode<0>();
    }

//  ...
```

where the *explicit* qualifier in the constructor will be explained shortly. Leaf nodes are created in three circumstances: when a Number is constructed with an initial value, when a Number is reinitialized with an initial value, and when the client code explicitly invokes *putOnTape*() on a Number instance. These situations are illustrated in the client example below:

```
#include "AAD.h"

int main()
{
    //  Creates a leaf on tape for x
    //      from Number's constructor
    Number x(1.0);

    //  Use x in a calculation
    //      OK: x is on tape
    Number y = sqrt(x);

    //  Now y is also on tape as a unary node
    //      whose child is x's leaf node

    //...

    //  Starting another calculation

    //  Wipe or rewind tape first

    Number::tape->rewind();

    //  Reuse the variable x
    x = 2.0;
    //  This call also creates a leaf on tape for x
    //      from Number's assignment operator

    //  Use x in a calculation
    //      this works because x is on tape
    //      from the previous line
    Number z = 2 * x;

    //  ...

    //  Starting a third calculation
    //  Wipe or rewind tape first
    Number::tape->rewind();

    //  This calculation reuses x without changing its value
    //  Number t = exp(x);
    //  This can't work: x is not on tape,
    //      t's Node would have a dangling pointer
```

```
    //  In case we reuse an existing variable with its existing value,
    //      we must put it on tape explicitly

    x.putOnTape();

    //  Now x is on tape (as a leaf), we can use it in calculations

    Number t = exp(x);

    //  ...
}
```

This code illustrates the situations where a Number goes on tape as a leaf. Operator overloading records all mathematical operations on tape as unary or binary nodes with pointers on the arguments of the operation on tape. It follows that the arguments to any operation must be on tape when the operation is evaluated, or we have dangling pointers. When the arguments are themselves the results of previous operations, they would be on tape automatically, as unary or binary nodes. But it is the client code's responsibility to put initial values on tape, explicitly, as leaf nodes.

This is performed in the constructor when a Number instance is created and given a numeric value. When an existing Number instance is reinitialized with another numeric value, this is performed in the assignment operator.

But it also happens frequently that a calculation would reuse existing Numbers as inputs, as opposed to creating or reinitializing Numbers purposely. For instance, in the simulation code of Chapter 6, an existing model's parameters are the initial values to a pricing. Those parameters were created and assigned a value on the construction of the model; they are not recreated or reinitialized for the purpose of the simulation. Before the calculation starts, all its inputs must be on tape. Numbers newly created or overwritten for the occasion are recorded as leaves from the constructor or assignment operator. But existing Numbers input to the calculation, like the parameters of a model prior to simulation, must be recorded manually, explicitly, with a call to *putOnTape*().

We also develop a free function to put a collection of Numbers on tape (in AAD.h):

```
template <class IT>[numbers = none]
inline void putOnTape(IT begin, IT end)
{
    for_each(begin, end, [](Number& n) {n.putOnTape(); });
}
```

Operations on Numbers with arguments not recorded on tape are by far the primary cause of bugs in AAD instrumented code. Tapes are wiped or rewound in between calculations and existing inputs are frequently reused. When an operation is evaluated with Number arguments not on tape, the resulting Node gets dangling pointers on the arguments' adjoints.

As a result, the back-propagation step from that node writes into random memory instead of correctly accumulating adjoints. The result is random: the application can crash, corrupt memory, or return incorrect results. All inputs to a calculation must be recorded as leaf nodes on tape before the first operation in the calculation is evaluated. When the tape is wiped or rewound, all the inputs must be put back on tape. When the inputs are not created for the occasion, it is the responsibility of the client code to put them on tape with a call to *putOnTape*(), either individually or as a collection.

This is fundamental and a major design concern for instrumented code, as we will see when we instrument simulations in Chapter 12. Furthermore, with multi-threaded code, we have one tape *per thread*. The inputs must be registered *on all the tapes*.

Note that we need not implement custom copy, assignment, or move semantics on the Number class. The default ones do the right thing. When a variable of type Number is assigned to another:

```
Number y;
//  ...
Number x = y;
```

the value of y is copied into x, but x does not get a *new* Node. The pointer y.myNode is copied into x.myNode, so that x refers to y's Node. This is the desired behavior: a Number refers to the last mathematical operation that produced its value. Assignment means that the same operation that created y's Node is responsible for the value of x.

This is worth repeating: *Assignment is not a mathematical operation, and it is not recorded on tape*. It simply redirects the pointer of the left-hand side to the Node of the right-hand side.

Operator nodes Operator nodes record operations on tape: unary functions like *log*() or *sqrt*(), binary operators like + or ∗, or binary functions like *pow*(). Operations are not limited to standard C++ mathematical functions and operators. For example, the Gaussian (exact) density and (approximate) cumulative distribution are building blocks in financial libraries. We consider them as standard mathematics and create a Node for them, even though they are not part of C++ standard and are defined in our header gaussians.h.[6] It is advisable to overload all frequently called, low-level mathematical functions in the same manner. It results in a more efficient, more accurate differentiation.

This is best seen in the case of the cumulative Gaussian distribution, which does not admit a closed-form and is typically computed with

[6]This is the reason why the AAD library, despite being self-contained, has a dependency to gaussians.h.

polynomial approximations.[7] Our code in gaussians.h implements Zelen and Severo's approximation of 1964:

```
//  Normal density
inline double normalDens(const double x)
{
    return x<-10.0 || 10.0<x ? 0.0 : exp(-0.5*x*x) / 2.506628274631;
}

//  Normal CDF (N in Black-Scholes)
//  Zelen and Severo's approximation (1964)
inline double normalCdf(const double x)
{
    if (x < -10.0) return 0.0;
    if (x > 10.0) return 1.0;
    if (x < 0.0) return 1.0 - normalCdf(-x);

    static constexpr double p = 0.2316419;
    static constexpr double b1 = 0.319381530;
    static constexpr double b2 = -0.356563782;
    static constexpr double b3 = 1.781477937;
    static constexpr double b4 = -1.821255978;
    static constexpr double b5 = 1.330274429;

    const auto t = 1.0 / (1.0 + p*x);

    const auto pol = t*(b1 + t*(b2 + t*(b3 + t*(b4 + t*b5))));

    const auto pdf = normalDens(x);

    return 1.0 - pdf * pol;
}
```

The normal density involves four operations, the cumulative distributions 18 (including the call to *normalDens*()). If we simply instrumented *normalCdf*() by templating its code, its evaluation with the Number type would create 18 Nodes on tapes. Its differentiation would propagate adjoints backward through these 18 Nodes. The result would not necessarily be accurate:

The differential of an accurate approximation is not necessarily an accurate approximation of the differential.

This being said, we *know* that the derivative of *normalCdf*(), by definition, is *normalDens*(). By considering it as a standard mathematical function, we create on overload, which, when evaluated, like all other standard functions and operators, creates a single Node on tape with the

[7]See https://en.wikipedia.org/wiki/Normal_distribution#Numerical_approximations_for_the_normal_CDF.

exact derivative. As a result, the differentiation is both measurably faster and significantly more accurate.

The code of the overloaded operators and functions will be discussed shortly. Functions and operators are overloaded so they compute not only a result, but also its derivatives to the arguments, and create an operator node on tape. All frequently called, low-level functions should be overloaded along with the standard C++ operators and functions. The standard functions are all unary or binary. The Gaussian functions are also unary, so all our overloads admit one or two arguments. It doesn't have to be the case. There is nothing in the code that prevents overloading functions with three or more arguments. To some extent, the expression templates of Chapter 15 overload an entire *expression* and store it on a single Node with an arbitrary number of arguments.

To help overloading, we write two private constructors on the Number class for the results of the overloads. These constructors are only called from the overloads and all overloads are friends of the Number class; hence, we declare them private to prevent accidental calls. The unary operator constructor is listed below:

```
//  ...
private:

    //  Unary
    Number(Node& arg, const double val) :
        myValue(val)
    {
        createNode<1>();
        myNode->pAdjPtrs[0] = &arg.mAdjoint;
    }
//  ...
```

It sets the created Number's value, constructs a unary node on tape for it, and sets the pointer on the Node to correctly refer to the adjoint of the argument. Note that the private constructor does *not* set or calculate the derivative. This is done directly in the different overloads. The binary constructor is very similar:

```
//  ...
    //  Binary
    Number(Node& lhs, Node& rhs, const double val) :
        myValue(val)
    {
        createNode<2>();
        myNode->pAdjPtrs[0] = &lhs.mAdjoint;
        myNode->pAdjPtrs[1] = &rhs.mAdjoint;
    }

//  ...
```

Similar constructors could be developed for n-ary operators if required.

Conversions

We now address the *explicit* qualifier in the Number constructor. It prevents *implicit* conversion, like in:

```
Number g(const Number x);
double y;
//  ...
Number z = g(y);
```

If the Number's constructor was not marked explicit, this code would compile and silently convert the double y into a Number x when g is invoked. The explicit qualifier causes a compilation error, attracting attention to the conversion and preventing accidental conversions. When a conversion is indeed desired, it compiles when made explicit:

```
Number z = g(Number(y));
```

Accidental conversions are the second-most-common cause of bugs and inefficiencies in instrumented code. It is best forcing explicit conversions on client code. For the same reason, conversions from Numbers to doubles is implemented with explicit conversion operators:

```
//  ...
//  Explicit coversion to double
explicit operator double& () { return myValue; }
explicit operator double() const { return myValue; }
//  ...
```

When doubles are used in place of Numbers, operations are not recorded on tape on evaluation, and adjoints are not propagated on back-propagation. The chain rule is broken, resulting in wrong sensitivities.

When Numbers are used in place of doubles, their evaluation is subject to AAD overhead: nodes are created on tape and local derivatives are computed and stored with the Node. In the situations where Numbers are not necessary, they cause substantial and unnecessary overhead.

Instrumented code must be very clearly written to use doubles or Numbers where relevant. We will see in Chapter 12 that this is even a key design consideration for an efficient instrumentation. Explicit constructors and conversion operators force the client code to express these choices explicitly. They prevent compilation of nonexplicit conversions, attracting developers' attention to the places where conversions occur, hence mitigating a frequent cause of bugs and inefficiencies.

We also provide (in AAD.h) a function that converts a *collection* of Numbers into a collection of doubles and vice versa, with STL iterator syntax:

```
template<class It1, class It2>
inline void convertCollection(It1 srcBegin, It1 srcEnd, It2 destBegin)
{
```

```
    using destType = remove_reference_t<decltype(*destBegin)>;
    transform(srcBegin, srcEnd, destBegin,
        [](const auto& source) { return destType(source); });
}
```

where *decltype*(*expression*) is a standard C++11 compile-time utility that gives the type of the expression, so *decltype*(∗ *destBegin*) is the type of the elements in the destination collection pointed by the iterator *destBegin*. The compile-time function *remove_reference_t*(), as the name indicates, gives the raw type, turning *T*& or *T*&& into *T*. We can apply this conversion utility, for example, to convert a vector of templated types into a vector of other templated types, and back:

```
template<class FROM, class TO>
vector<TO> convertVector(const vector<FROM>& src)
{
    //  Allocate
    vector<TO> dest(src.size());

    //  Convert
    //  If TO = FROM, we get a copy
    //  If not, we get the correct conversion
    convertCollection(src.begin(), src.end(), dest.begin());

    return dest;
}
```

Back-propagation

Numbers store a pointer to their Node on tape. Back-propagation consists in the propagation of adjoints through the tape in reverse order from the final result's Node to the first Node on tape. Prior to back-propagation, all adjoints are set to 0 (although adjoints are always initialized to zero, so it is generally not necessary to reset them), and the adjoint of the final result is set to one. This is called *seeding the tape*. We write a convenient method on the Number class to propagate adjoints from the caller Number's Node:

```
//  ...
    //  Reset all adjoints on the tape
    //      note we don't use this method
    void resetAdjoints()
    {
        tape->resetAdjoints();
    }

    //  Propagation

    //  Propagate adjoints
    //      from and to both INCLUSIVE
    static void propagateAdjoints(
        Tape::iterator propagateFrom,
```

```
        Tape::iterator propagateTo)
    {
        auto it = propagateFrom;
        while (it != propagateTo)
        {
            it->propagateOne();
            --it;
        }
        it->propagateOne();
    }

    //  Convenient overloads

    //  Set the adjoint on this Node to 1,
    //      then propagate from this Node to propagateTo
    void propagateAdjoints(
        //  We start on this number's Node
        Tape::iterator propagateTo)
    {
        //  Set this adjoint to 1
        adjoint() = 1.0;
        //  Find node on tape
        auto propagateFrom = tape->find(myNode);
        //  Propagate
        propagateAdjoints(propagateFrom, propagateTo);
    }

    //  Other overloads to propagate
    //      from this Node, after setting its adjoint to 1,
    //      to either start or mark
    void propagateToStart()
    {
        propagateAdjoints(tape->begin());
    }
    void propagateToMark()
    {
        propagateAdjoints(tape->markIt());
    }
//  ...
```

The first method *resetAdjoints*() resets all adjoints on tape to zero. We don't use this method; it is only provided for completeness. The second is a static method that propagates adjoints backwards through the tape, from a starting point to an endpoint, both represented by tape iterators given as arguments. The third *nonstatic* method propagates adjoints from the Number instance's Node, which adjoint it first initializes to one, to a specified endpoint. The last two convenient overloads propagate in the same manner, to, respectively, the beginning or the mark on the tape. The code uses tape iterators and the Node's propagation method and should be self-explanatory.

Const incorrectness

Before we discuss the code of the operator and function overloads, we illustrate how they work on a simple example and expose one of the main inherent flaws of AAD. Consider the following code:

```
#include "AAD.h"

//  A simple Multiplier class
template <class T>
struct Multiplier
{
    const T multiplyBy;

    Multiplier(const double m) : multiplyBy (m) {}

    T operator () (const T y) const
    {
        return multiplyBy * y;
    }
};

//  Instrumented code

//  Initialize tape first
Number::tape->rewind();

//  Initialize inputs, including data members
//  Note that puts inputs on tape, as leave nodes
Multiplier<Number> doubler(2.0);
Number y(5.0);

//  The calculation itself, in this case, a simple product
Number z = doubler(y);

//  Adjoint propagation
z.propagateToStart();

//  Display results
cout << z.value() << endl;                          //  10.0
cout << y.adjoint() << endl;                        //  2.0
cout << doubler.multiplyBy.adjoint() << endl;       //  5.0
```

This code illustrates, in a simple example, the general pattern for AAD instrumentation.

The Multiplier class is a simple function object that multiplies numbers by a constant. It is templated on its number type, but its constructor takes a double to initialize its templated data member *myltiplyBy*. Its operator (), marked const, returns the product of the argument by its data member *myltiplyBy*, without apparent modification of the Multiplier's internal state or the argument.

The rest of the code is a simple instrumentation that evaluates and differentiates a product. Line 20 initializes the tape: it is faster to rewind than wipe it. Line 24 initializes a Multiplier<Number>, baptized "doubler," with a multiplicative constant 2. This invokes the Multiplier's constructor, which in turn invokes the Number's constructor on line 9 to build *doubler*'s *multiplyBy* data member. *This puts multiplyBy on tape.* Line 25 initializes a Number y with value 5, which puts y on tape.

We see that the order of the instructions is important here: first, initialize the tape, then, initialize the inputs, including data members, so they are recorded on tape, and only then perform the calculation.

Line 28 conducts the calculation. The call to *doubler*'s operator () evaluates the product on line 13. The operator $*$, like all other arithmetic operators and mathematical functions, is overloaded for the Number type, as we will see shortly, so line 13 not only evaluates the product, it also compute its derivatives to the arguments y and *multiplyBy*, and builds a binary node on tape, correctly linking it to the arguments' Nodes. The result is held by the Number variable z, which node is the binary node created by the overloaded product.

Line 31 propagates adjoints, from the result z, which adjoint is set to one, to the start of the tape. After the propagation completes, the adjoint of every node i on tape correctly accumulated to $\partial z \, \partial y_i$. We can read the final result $z = multiplyBy * y = 5 \times 2 = 10$), as well as $\partial z / \partial y = 2$ in y's adjoint and $\partial z / \partial multiplyBy = 5$ in *multiplyBy*'s adjoint.

This works nicely and illustrates the typical instrumentation pattern: initialize the tape, put inputs on tape, calculate with the Number type, propagate, pick differentials.

But it should be apparent that the instrumented Multiplier's operator () on line 11 *is no longer correctly const*, and neither is its argument y. As a result of the product on line 13, the adjoints of *multiplyBy* and y's Nodes are held *by non-const pointers* on z's Node, and during the adjoint propagation on line 31 the adjoints of *multiplyBy* and y are *modified*. Although the Multiplier's operator () and its argument are marked const on line 11, its data member and argument's adjoints are modified as an indirect result of its evaluation.

AAD is inherently const incorrect.

This rather subtle characteristic is not without consequence for the design of instrumented code, especially in a parallel context. The Multiplier's operator () is const. It may be called safely from concurrent threads for evaluation when instantiated with the double type, but it cannot be called concurrently for differentiation when instantiated with the Number type. Contrary to a Multiplier<double>, a Multiplier<Number> is a *mutable object* in the sense of Section 3.11. For the exact same reason, a model is mutable in the context of an instrumented simulation. We have seen in Chapter 7 that it is safe to simulate paths concurrently with the same Model<double> for evaluation. It is no longer the case for differentiation with a Model<Number>. An instrumented model is mutable, so each thread must use its own copy of it, or data races will occur. Our instrumented simulation code of Chapter 12 is designed accordingly.

It is very annoying that we can no longer trust const qualifiers in the context of AAD. We must constantly remember that AAD modifies the adjoints of all Numbers involved in calculations, so all Numbers, even the ones that are only read, never explicitly written into, are always mutable when instrumented. An intrusive alternative would be to modify instrumented code and remove all const qualifiers of all Numbers involved in calculations and all objects that store such Numbers. This solution is impractical and we don't recommend it. One of the difficulties of programming with AAD is not being able to trust constness.

In practice, this means that we need another private accessor on the Number class to access a Number's Node, by non-const reference, even for a const Number. Overloads use it to set the child adjoint pointers on the operator nodes:

```
//  Number class
//  ...

private:

    //  Access Node (friends only)
    //  Note const incorectness
    Node& node() const
    {
        return const_cast<Node&>(*myNode);
    }

//  ...
```

Operator overloading

Finally, we implement operator overloading for instances of the Number class. Remember that every mathematical operation that involves Numbers should implement all of the following:

1. Evaluate the operation and store its result on the resulting Number.
2. Create an operator node on the tape, unary or binary depending on the operation. Set the child adjoint pointers on the Node to the adjoints of the operation's arguments.
3. Evaluate the derivatives of the operation to its arguments, store them on tape, and link the pointers on the operator node.

We implemented private unary and binary constructors on the Number class to facilitate overloading and implement the second item on the list. This considerably simplifies the development of the overloads. For instance, we list below the overloaded operator $*$:

```
1      //  Number class
2      //  ...
3
4  public:
5
6      inline friend Number operator*(const Number& lhs, const Number& rhs)
7      {
8          const double e = lhs.value() * rhs.value();
9          //  Eagerly evaluate and put on tape
10         Number result(lhs.node(), rhs.node(), e);
11         //  Eagerly compute derivatives and set on the Node
12         result.lDer() = rhs.value();
13         result.rDer() = lhs.value();
14
15         return result;
16     }
```

This code is executed every time two Numbers are multiplied with the operator $*$. Line 8 computes the result. Line 10 calls the private binary constructor on the Number class, constructing a Number *result* to hold the result of the multiplication together with its binary node on tape. The constructor also correctly sets the child adjoints' pointers on the operator node to the adjoints or *lhs* and *rhs*, which Node is accessed (and const-casted) by calls to the private method *node*(). Finally, lines 12 and 13 set the two local derivatives on the operator node:

$$\frac{\partial xy}{x} = y, \frac{\partial xy}{y} = x$$

and return the result Number. This Number stores the result of the multiplication and points to the newly created operator node on tape. When this Number, at a later stage, becomes an argument to a subsequent operation, the operator node of that operation will point to its adjoint on tape. As the calculation evaluates its sequence of operations, a network of nodes is recorded on tape, linked to one another with the argument adjoint pointers. These pointers are the edges of the DAG held on tape, the veins where adjoint propagation flows in reverse after the calculation completes.

This overload is called when two Numbers are multiplied. What happens when a Number is multiplied by a double, say a constant, as here?

```
Number y;
//  ...
Number x = y * 2.0;
```

Note than any double is a constant as far as AAD is concerned. It doesn't matter if the double is the result of a calculation. That calculation is out of the tape. Only the operations involving at least one Number, and resulting in a Number, are *active*. The multiplication above is not a binary operation anymore, not in the sense of AAD. This is the *unary* operation (*2), an operation that doubles its argument. We must implement that properly for the sake of performance, with a different overload for the operator *:

```
1   // ...
2   inline friend Number operator*(const Number& lhs, const double& rhs)
3   {
4       const double e = lhs.value() * rhs;
5       // Eagerly evaluate and put on tape
6       Number result(lhs.node(), e);
7       // Eagerly compute derivatives and set on the Node
8       result.derivative() = rhs;
9
10      return result;
11  }
12  // ...
```

Note that although the code is similar, it is the unary private Number constructor that is invoked now on line 6 so a unary node is recorded on tape, linked to *lhs* Number but not the *rhs* constant. We even need a third overload when the double is on the left:

```
// ...
inline friend Number operator*(const double& lhs, const Number& rhs)
{
    return rhs * lhs;
}
// ...
```

All the binary operators and functions must implement the three overloads. This is a strong optimization with measurable impact in cases of practical relevance. It is substantially faster to construct a unary node and propagate through one than with binary nodes. Although we don't list the code here in the interest of conciseness, AADNumber.h implements the binary operators and functions +, −, /, *pow*, *max*, and *min* in the exact same way.

We must also overload the operators a la *=, both with Number of double arguments. This is easy; the code for the / = overload, for example, is listed below. The others are found in AADNumber.h.

```
// ...
Number& operator/=(const Number& arg)
{
    *this = *this / arg;
    return *this;
}
```

```
Number& operator/=(const double& arg)
{
    *this = *this / arg;
    return *this;
}
//  ...
```

Unary operators and functions follow the same pattern, although (evidently) they don't need an overload for doubles. We list below the overloads for *exp*(), *log*(), *sqrt*(), and *fabs*(), as well as the trivial unary + and –:

```
//  ...
//  Unary +/-
Number operator-() const
{
    return 0.0 - *this;
}
Number operator+() const
{
    return *this;
}

//  Unary functions

inline friend Number exp(const Number& arg)
{
    const double e = exp(arg.value());
    //  Eagerly evaluate and put on tape
    Number result(arg.node(), e);
    //  Eagerly compute derivatives and set on the Node
    result.derivative() = e;

    return result;
}

inline friend Number log(const Number& arg)
{
    const double e = log(arg.value());
    //  Eagerly evaluate and put on tape
    Number result(arg.node(), e);
    //  Eagerly compute derivatives and set on the Node
    result.derivative() = 1.0 / arg.value();

    return result;
}

inline friend Number sqrt(const Number& arg)
{
    const double e = sqrt(arg.value());
    //  Eagerly evaluate and put on tape
    Number result(arg.node(), e);
    //  Eagerly compute derivatives and set on the Node
    result.derivative() = 0.5 / e;

    return result;
}
```

```

    inline friend Number fabs(const Number& arg)
    {
        const double e = fabs(arg.value());
        //  Eagerly evaluate and put on tape
        Number result(arg.node(), e);
        //  Eagerly compute derivatives and set on the Node
        result.derivative() = arg.value() > 0.0 ? 1.0 : -1.0;

        return result;
    }
    //  ...
```

Following the earlier discussion regarding the Gaussian density and cumulative distribution, we overload these in the same manner:

```
    //  ...
    inline friend Number normalDens(const Number& arg)
    {
        const double e = normalDens(arg.value());
        //  Eagerly evaluate and put on tape
        Number result(arg.node(), e);
        //  Eagerly compute derivatives and set on the Node
        result.derivative() = - arg.value() * e;

        return result;
    }

    inline friend Number normalCdf(const Number& arg)
    {
        const double e = normalCdf(arg.value());
        //  Eagerly evaluate and put on tape
        Number result(arg.node(), e);
        //  Eagerly compute derivatives and set on the Node
        result.derivative() = normalDens(arg.value());

        return result;
    }
    //  ...
```

Finally, we need comparison operators since calculation code often compares Numbers by value:

```
    //  ...
    inline friend bool operator==(const Number& lhs, const Number& rhs)
    {
        return lhs.value() == rhs.value();
    }
    //  ...
```

We have ==, !=, >, >=, <, and <= to overload. Each overload in fact consists of three overloads in case we compare Numbers to doubles, where the double may be on the left or right of the comparison. This is a total of 18 overloads in AADNumber.h.

This concludes the design of the Number type and our AAD library. The code is in our repository, files AADNode.h, AADTape.h, AADNumber.h., and, of course, blocklist.h. The static variables are defined in AAD.cpp. All the headers are listed in AAD.h, so client code may include that header alone. With the exception of a dependency on gaussians.h, these files form a self-contained, professional AAD library in C++. Its performance is excellent and even impressive compared to traditional differentiation, although somewhat short of handwritten adjoint code. Performance will be further improved in Chapter 15.

10.4 BASIC INSTRUMENTATION

The library may be used identically to the toy code in the previous chapter. Include AAD.h, template calculation code, initialize the tape, put the inputs on tape, evaluate with Numbers, propagate adjoints, and pick derivatives. We described instrumentation in deep detail in a simple example on page 391. We provide another example here, and, of course, we fully instrument our simulation code in Chapter 12.

```
#include "AAD.h"

//  Instrumented calculation code
template <class T>
T f(T x[5])
{
    T y1 = x[2] * (5.0 * x[0] + x[1]);
    T y2 = log(y1);
    T y = (y1 + x[3] * y2) * (y1 + y2);
    return y;
}

int main()
{
    //  No need to initialize the tape here,
    //      since the application just started
    //  It is however a good habit so we do it anyway
    Number::tape->rewind();

    //  Set inputs
    //  Note this puts all the inputs on tape
    //      see Number constructor
    Number x[5] = { 1.0, 2.0, 3.0, 4.0, 5.0 };

    //  Evaluate and build the tape
    Number y = f(x);

    //  Propagate adjoints through the tape
    //      in reverse order from y's Node
    //      y's adjoint is set to 1.0 in propagateAdjoints()
    //      prior to propagation
```

```
    y.propagateAdjoints();

    //  Display the derivatives
    for (size_t i = 0; i < 5; ++i)
    {
        cout << "a" << i << " = " << x[i].adjoint() << endl;
    }

    //  950.736, 190.147, 443.677, 73.2041, 0

}
```

The full definition of the Number class in AADNumber.h contains 650 lines of code. We don't list it here, referring to our repository for reference instead.

CHAPTER 11

Discussion and Limitations

Before we apply our AAD library to instrument our simulation code from Part II, we briefly discuss some important aspects and limitations of AAD that will guide our work.

11.1 INPUTS AND OUTPUTS

We minimized AAD overhead as much as possible thanks to efficient memory management, and we minimize it further in Chapter 15 with expression templates, but AAD overhead cannot be eliminated. Even if we did not apply operator overloading or a tape, and manually coded adjoint computations along with the calculation code, AAD differentiation would still take around three times the cost of one calculation, as seen in Chapter 8. We come close to this theoretical limit with an automatic approach in Chapter 15.

AAD is only relevant when the number of inputs is large (certainly larger than four), as is almost always the case for financial valuation. With a low number of inputs, bumping is generally faster.

Further, AAD computes the differentials of *one* scalar result in constant time in the number of differentials, but the differentials of *multiple* results take linear time in the number of results. This is the case even when the calculation itself takes constant time or close in the number of the results, when all results are computed together and share a vast fraction of the computations, like when pricing multiple products in a simulation over the same paths. Differentiation still takes linear time, because the back-propagation of adjoints must be conducted separately for each result.

In the context of financial simulations, we should therefore expect AAD to produce constant time risk reports *for one transaction*, or one aggregated value of multiple transactions, but, when we need an itemized risk report for every transaction in a book, we should expect linear time in the number of transactions.

This is somewhat frustrating, because we designed our simulation library in such a way that multiple transactions may be valued simultaneously in virtually constant time by sharing the simulated paths. This being said, AAD's linear time in the number of outputs is theoretical and *asymptotic*. While it is inescapable to spend time $\alpha + \beta m$ to differentiate m results with AAD, if we manage to make β orders of magnitude smaller than α, we will be done in "almost constant time" in situations of practical relevance. How to do this, and whether it is at all possible, depends on the differentiated code. In the context of financial simulations, we achieve such "almost constant time" in Chapter 14 with the development of special support in the AAD code. In the meantime, we only consider the differentiation of a single result.[1]

11.2 HIGHER-ORDER DERIVATIVES

AAD also computes first-order derivatives. To compute second-order derivatives would take "AAD over AAD," where the adjoint calculation of first-order derivatives becomes part of the calculation code that is itself instrumented with AAD to produce second-order derivatives. That is not viable in practice.

Despite active research in the production of higher-order derivatives with AAD, and the claim from some commercial AAD frameworks to produce second- and higher-order AAD risk efficiently, there exists, to date and to our knowledge, no solution of practical relevance. It is generally admitted among AAD academics and professionals that second derivatives are best produced by "bumping over AAD," where the first derivatives are produced with AAD but higher-order derivatives are produced by bumping.

This means that AAD provides acceleration by one order in the production of derivatives, but no more: constant instead of linear time for the production of first-order derivatives, linear instead of quadratic for the second-order derivatives, and so forth.

11.3 CONTROL FLOW

We have already pointed out that AAD differentiates mathematical calculations *along a fixed control path*. It does not differentiate control flow: "if this then that" statements and friends. Control flow introduces discontinuities, which cannot be differentiated with AAD or otherwise, so this is not necessarily a problem, but something that must be kept in mind when instrumenting code.

One example is how we defined the payoff of a barrier option in Section 6.3. The payoff of a barrier is discontinuous, so the MC estimate

[1]Even though that single result may be the aggregated value of a large number of transactions.

of its price is not differentiable. We smoothed its payoff with a well-known "smooth barrier approximation" to remove the discontinuity.

The differentiation of discontinuous functions, including control flow, is not resolved in the differentiation library with differentiation methods, but in the valuation library with smoothing methods. To smooth a function means to approximate it by a close continuous function. The work relates to the differentiated function, not its differentiation. Smoothing in general terms, and smoothing of financial cash-flows in particular, is introduced and discussed in [77] and [78].

11.4 MEMORY

Finally, and most importantly, it should be clear that AAD may consume an *insane* amount of RAM. *Every single mathematical operation*, every sum, difference, product, division, and so forth, is recorded on tape. The billions of such operations involved in a practical valuation take considerable space in memory. It has been estimated through both theoretical and empirical means (see [96] and [90]) that tapes consume around 5 GB of RAM per second per core in the non-instrumented calculation. That would limit AAD to calculations faster than around 12 seconds of CPU time on a 64 GB workstation, and 2 seconds.cpu on a 16 GB laptop (we must leave *some* RAM for the OS).

It gets worse: in order to traverse the tape *efficiently* for back-propagation, it should fit in the L3 cache, that is around 50 MB for the best mainstream CPUs available. That would limit AAD to calculations under 0.01 seconds.cpu and preclude it for pretty much all cases of practical relevance.

Path-wise AAD simulations

The solution consists in computing derivatives separately on parts of the calculation and aggregating them in the end. In the case of MC simulations, the obvious solution is to differentiate path-wise. The differential of the average payoff among N paths is the average of the differentials over each path. Path-wise differentials can be computed by conducting both the computation and the adjoint differentiation for each path separately. The final differentials are averages of path-wise differentials.

Mathematically: one path number i consists in the vector $(S^i_{T_j})_{1 \le j \le J}$; see 5.2. This vector is a function of the Gaussian random vector G_i for the path, and a set of model parameters a: initial asset prices, volatility, and so forth:

$$(S^i_{T_j})_{1 \le j \le J} = h(G_i, a)$$

A payoff is a function g of the path. The MC estimate of the value of the corresponding instrument is:

$$\frac{1}{N}\sum_{i=1}^{N} g(S^i_{T_j})$$

Hence:

$$\frac{\partial}{\partial a}\left(\frac{1}{N}\sum_{i=1}^{N} g(S^i_{T_j})_{1\le j\le J}\right)$$

$$= \frac{1}{N}\sum_{i=1}^{N}\frac{\partial}{\partial a}\left[g\left(S^i_{T_j}\right)_{1\le j\le J}\right]$$

$$= \frac{1}{N}\sum_{i=1}^{N}\frac{\partial}{\partial a}\{g[h(G_i,a)]\}$$

The generation and evaluation of one path takes a few microseconds.cpu in general; see our results from Chapter 6. Even in the case of an xVA on a monster netting set, it takes less than a millisecond, and, in extreme cases, perhaps up to 0.01 s. Hence, memory consumption for path-wise derivatives is limited to 50 MB in the worse case, and always fits in the L3 cache of high-end modern CPUs.

Practically, we evaluate path-wise derivatives

$$\frac{\partial}{\partial a}\{g[h(G_i,a)]\}$$

with AAD over one path, then we wipe the tape for the processing on the next path. The size of the tape is related to the processing time of one single path, hence, always below 50 MB and generally substantially less than that. The path-wise derivatives are stored and averaged in the end.

Introduction to check-pointing

As a side note, we separated the generation h from the evaluation g of the paths in our notations, and we can further split the expression of our path-wise derivative into:

$$\frac{\partial}{\partial a}\{g[h(G_i,a)]\} = \frac{\partial h(G_i,a)}{\partial a}\frac{\partial g[h(G_i,a)]}{\partial h}$$

This means that we *could* reduce RAM usage even further by computing the derivatives for the generation and the evaluation of the path separately, and then use this formula for the computation of the complete path-wise derivative. This is a multidimensional formula, $\frac{\partial h(G_i,a)}{\partial a}$ is the Jacobian of the components of the path to model parameters, and the whole operation is expensive. The check-pointing methodology discussed in Chapter 13 conducts this calculation *in constant time*, without the production of a Jacobian, or a matrix product, while limiting RAM usage to recording h and g separately.

In practice, there is no need to check-point this calculation or split path processing into generation and evaluation for the purpose of differentiation. Tapes for the full processing of a path are small enough. We only made this comment to show that we *could* further split the production of derivatives, and introduce the important check-pointing technique, which we develop in the dedicated Chapter 13. Check-pointing is also applied to its full extent in our publication [31] in order to efficiently differentiate through the Longstaff-Schwartz algorithm.

MC is a "lucky" case where the RAM consumption problem is easily overcome with path-wise differentiation. In addition, such path-wise computation of derivatives may be conducted in parallel over paths, as we demonstrate in the next chapter.

Evidently, outside of MC, it is not always possible to separate computations in such a trivial manner. For example, multidimensional FDM is too slow for a self-contained AAD differentiation, but not as easily separable as MC. FDM is step-wise, generally back to front, and the calculations for each time step depend on the results for the previously calculated time step.

Check-pointing provides solutions for all such cases. For the AAD differentiation of FDM with check-pointing, see [90]. In the next chapter, we differentiate our simulation code with path-wise AAD without the need for check-pointing. In Chapter 13, we implement check-pointing and apply it to the problem of differentiating through the *calibration* of the local volatilities of Dupire's model to market data and discuss the general case of the differentiation of any model through its calibration.

CHAPTER 12

Differentiation of the Simulation Library

12.1 ACTIVE CODE

In this chapter, we apply the AAD library from the previous chapter to differentiate our simulation library from Part II. More precisely, we apply AAD to compute the differentials of prices to *model parameters*. In the next chapter, we discuss differentials of prices to *market variables*, also called market risks or *hedge coefficients*.

What will clearly appear is that an efficient AAD library is certainly necessary, but absolutely not sufficient for the efficient differentiation of complex valuation code. It takes work, art, and skill to correctly instrument calculation code. An efficient AAD library provides all the necessary pieces, but to articulate them together in an efficient instrumentation requires an intimate knowledge of the instrumented algorithms (known as *domain-specific knowledge*) and a comprehensive understanding of the inner mechanics of the AAD library. The notion that one can template calculation code, link to an efficient AAD library, and obtain differentials with AAD speed is a myth. Instrumentation takes at least as much skill and effort as the production of the AAD library itself. A naive instrumentation does produce differentials in constant time, but it may be a very long constant time, often slower than finite differences. While the notion of an efficient instrumentation depends on the differentiated code, this chapter provides general advice and battle-tested methodologies while instrumenting the simulation code of Part II with remarkable results.

It is especially hard to instrument calculation code that was not written with AAD in mind to start with. We wrote our simulation library knowing that we would eventually instrument it for AAD. This will make its instrumentation substantially easier. In fact, we wrote templated code to start with throughout the simulation library. But not all of it. In particular, we did *not* template the code for the generation of uniform and Gaussian random numbers. All that part of the code uses doubles to represent numbers.

Why did we code it this way? We know that the final result, the average payoff across the simulated paths, depends on the random numbers. Those numbers have non-zero adjoints. However, our goal is to differentiate the final result with respect to the model parameters: spot, local volatilities, and so forth. We know that the random numbers do *not* depend on those parameters. Hence, their adjoints do not propagate all the way back to them.

Formally, with the notations of previous chapters, we differentiate:

$$V_0[(G_i)_{1 \leq i \leq N}, a] = \frac{1}{N} \sum_{i=1}^{N} g[h(G_i, a)]$$

with respect to the vector of model parameters a. G_i is the vector of Gaussian random numbers for path i, h is the path generation function, and g is the payoff function. It is clear from those notations that G_i is another argument to h and has no dependency on a. The derivatives we want to compute are:

$$\frac{\partial V_0[(G_i)_{1 \leq i \leq N}, a]}{\partial a} = \frac{1}{N} \sum_{i=1}^{N} \frac{\partial g[h(G_i, a)]}{\partial h} \frac{\partial h(G_i, a)}{\partial a}$$

The G_is do not depend depend on a. In particular,

$$\frac{\partial G_i}{\partial a} = 0$$

In AAD lingo, we call the G_is *inactive*.

We know that because we know exactly how MC simulations work: this is domain-specific knowledge in action. Because we know that, we don't need to propagate from the random numbers at all. Hence, we don't need to propagate *to* them, either. Their adjoints don't contribute to the final derivatives, so we don't need to compute these adjoints at all. To instantiate the code that produces random numbers with the Number type would result in unnecessary AAD overhead at evaluation time (to put operations and local derivatives on tape) and propagation time (conducting unnecessary propagations to and from the random numbers). Because we know that, we can easily leave the random numbers and their generation outside of the adjoint logic. All it takes is use *doubles* for those calculations in place of a templated number type.

Code that uses the templated number type is said to be *instrumented.* This code is recorded on tape and participates in the adjoint propagation at the cost of AAD overhead. Code that uses native types like doubles is said to be noninstrumented. It is not recorded or propagated and free of AAD overhead. Calculations that depend directly or indirectly on the inputs and, therefore, must participate in adjoint propagation, are called *active*. Code that is independent of inputs and, therefore, does not contribute to their adjoints, is called *inactive*. We have a first and probably most important rule for the efficient differentiation of calculation code:

Only instrument active code.

We call this major optimization *selective instrumentation*. On the other hand, if we forget to instrument active code, we produce wrong derivatives. Active adjoints are not propagated and their contribution is lost. If we instrument inactive code, like random number generators, we produce unnecessary AAD overhead. We must instrument all the active code, but only the active code. The identification of inactive code is a key step of an efficient differentiation.

Readers are encouraged to review our simulation code and verify that we correctly applied selective instrumentation.

12.2 SERIAL CODE

We start with the serial code. We have nothing to do about random numbers, and the models and products are already instrumented (although we will need minor extensions on models). Therefore, our work focuses on the template algorithm of Chapter 6, which we recall here:

```
//  MC simulator: free function that conducts simulations
//      and returns a matrix (as vector of vectors) of payoffs
//          (0..nPath-1 , 0..nPay-1)
inline vector<vector<double>> mcSimul(
    const Product<double>&      prd,
    const Model<double>&        mdl,
    const RNG&                  rng,
    const size_t                nPath)
{
    //  Work with copies of the model and RNG
    //      which are modified when we set up the simulation
    //  Copies are OK at high level
    auto cMdl = mdl.clone();
    auto cRng = rng.clone();

    //  Allocate results
    const size_t nPay = prd.payoffLabels().size();
    vector<vector<double>> results(nPath, vector<double>(nPay));
    //  Init the simulation timeline
    cMdl->allocate(prd.timeline(), prd.defline());
    cMdl->init(prd.timeline(), prd.defline());
    //  Init the RNG
    cRng->init(cMdl->simDim());
    //  Allocate Gaussian vector
```

```
    vector<double> gaussVec(cMdl->simDim());
    //  Allocate and initialize path
    Scenario<double> path;
    allocatePath(prd.defline(), path);
    initializePath(path);

    //  Iterate over paths
    for (size_t i = 0; i<nPath; i++)
    {
        //  Next Gaussian vector, dimension D
        cRng->nextG(gaussVec);
        //  Generate path, consume Gaussian vector
        cMdl->generatePath(gaussVec, path);
        //  Compute payoffs
        prd.payoffs(path, results[i]);
    }

    return results; //  C++11: move
}
```

We now differentiate this code, instantiating valuation code with the Number type in place of doubles and inserting additional logic for the back-propagation of adjoints and, more generally, the management of the tape. Following the discussion of the previous chapter, we back-propagate adjoints path-wise to avoid unreasonable tape growth.

```
#include "AAD.h"

//  AAD instrumentation of mcSimul()

//  returns the following results:
struct AADSimulResults
{
    AADSimulResults(
        const size_t nPath,
        const size_t nPay,
        const size_t nParam) :
            payoffs(nPath, vector<double>(nPay)),
            aggregated(nPath),
            risks(nParam)
    {}

    //  matrix(0..nPath - 1, 0..nPay - 1) of payoffs
    //      same as mcSimul()
    vector<vector<double>>  payoffs;

    //  vector(0..nPath) of aggregated payoffs
    vector<double>          aggregated;

    //  vector(0..nParam - 1) of risk sensitivities
    //      of aggregated payoff, averaged over paths
    vector<double>          risks;
};

//  Default aggregator = 1st payoff = payoff[0]
const auto defaultAggregator = [](const vector<Number>& v) {return v[0]; };
```

```
template<class F = decltype(defaultAggregator)>
inline AADSimulResults
mcSimulAAD(
    const Product<Number>&  prd,
    const Model<Number>&    mdl,
    const RNG& rng,
    const size_t            nPath,
    const F&                aggFun = defaultAggregator)
{
    //  Work with copies of the model and RNG
    //      which are modified when we set up the simulation
    //  Copies are OK at high level
    auto cMdl = mdl.clone();
    auto cRng = rng.clone();

    //  Allocate path and model
    //      do not initialize yet
    Scenario<Number> path;
    allocatePath(prd.defline(), path);
    cMdl->allocate(prd.timeline(), prd.defline());

    //  Dimensions
    const size_t nPay = prd.payoffLabels().size();
    const vector<Number*>& params = cMdl->parameters();
    const size_t nParam = params.size();

    //  ...
```

The function has the same inputs as the original, except they are now instantiated with Numbers instead of doubles and an additional "aggregator." Following the discussion of the previous chapter, we are differentiating one result for now. In case the product has multiple payoffs, we differentiate an aggregate provided by client code as a function of the payoffs, represented by a lambda or another callable type. For example, the product may be a portfolio of transactions, and we may want to differentiate the value of the entire book, the sum of the values of the transactions weighted by the notionals. In this case, we would apply an aggregator that computes the "payoff" of the portfolio as the sum of the payoffs of the products, weighted by the notionals:

```
vector<double> notionals;
//  fill notionals
//  ...
const auto portfolioAggregator = [&notionals](const vector<Number>& v)
    {
        return inner_product(
            notionals.begin(),
            notionals.end(),
            v.begin(),
            Number(0.0));
    }
//  call AADSimulResults() with portfolioAggregator() as aggregator
```

where *inner_product*() is a standard C++ library algorithm from header <numeric> that does exactly what its names says. Note that it is not the values, but the payoffs that are aggregated. The aggregator effectively turns multiple payoffs into one aggregated payoff and it is this aggregated payoff that is differentiated. The default aggregator simply picks the first payoff.

The return type is different: we return a matrix of path-wise payoffs as previously, but we also return a vector of path-wise aggregated payoffs and the vector of all the sensitivities of its value (path-wise average) to all the model parameters.

Like in valuation, we start by cloning the model and the RNG, modified by the subsequent code. We allocate the path and the model's working memory *but we don't initialize them yet* and pick the dimensions: number of payoffs and number of parameters. We take a reference on the vector of *pointers to* the model parameters so we easily address it thereafter under the name *params*.

What follows is new and deserves an explanation:

```
1   //  ...
2
3   //  AAD - 1
4
5   //  Access tape
6   Tape& tape = *Number::tape;
7
8   //  Clear and initialize tape
9   tape.clear();
10
11  //  Put parameters on tape
12  //      note this also initializes all adjoints
13  cMdl->putParametersOnTape();
14
15  //  Init the model
16  //  CAREFUL: initialization is recorded on tape
17  //  Model parameters must be on tape prior to initialization
18  cMdl->init(prd.timeline(), prd.defline());
19
20  //  Initialize path
21  initializePath(path);
22
23  //  Mark the tape straight after initialization
24  tape.mark();
25
26  //
27
28  //  ...
```

We take a reference to the tape and initialize it.

The next thing we do (line 13) is put all the model parameters on tape. They are the inputs to our calculation with respect to which we compute derivatives. They must go on tape before any calculation is executed. See the discussion on page 391 for details and a simple example.

This is the modification we need in our simulation code: we must develop a method on the base Model class to put all the parameters on tape. The modification is minor, because we can implement it directly on the base class without modification to concrete models. The (small) difficulty is that the Model class is templated in its number type, but only Numbers go on tape. We must implement some simple template specialization magic so that *Model < Number >: : putParametersOnTape*() puts its parameters on tape, whereas for any other type T, *Model < T >: : putParametersOnTape*() does nothing.[1] Readers unfamiliar with template specialization may learn all about it in [76].

The code below completes the base Model class in mcBase.h.

```
    //  Model base class
    //  ...

    //  Put parameters on tape, only valid for T = Number
    void putParametersOnTape()
    {
        putParametersOnTapeT<T>();
    }

private:

    //  Template specialization here

    //  If T is not Number : do nothing
    template<class U>
    void putParametersOnTapeT()
    {

    }

    //  If T is Number : put all parameters on tape
    template <>
    void putParametersOnTapeT<Number>()
    {
        for (Number* param : parameters()) param->putOnTape();
    }
```

Back to our simulation code, the next thing we do (lines 18 and 21) is initialize the model and the scenario. It is crucial to do this here and not before. Initialization must be conducted *after* the parameters are on tape. The initialization of a model typically conducts calculations with the model parameters. In Black and Scholes, we pre-calculate deterministic amounts for the subsequent generation of the paths. In Dupire, we pre-interpolate

[1]Or we could throw an exception, or even implement a static assert that would result in a compilation error when code attempts to call the method on a Model<T> other than Model <Number>, although we implement the simple solution here.

the local volatilities in time. So the parameters must be on tape, or these calculations result in nodes with dangling pointers to nonexisting parameter adjoints. We again refer to the detailed discussion on page 391.

At this point the parameters, and the initial calculations, those that are not repeated path-wise, are all on tape and correctly linked. We *mark* the tape there (line 24). Remember, we are going to compute path-wise derivatives. We will construct the tape for a path, propagate adjoints through it, and *rewind* it for the next path. We are going to rewind it *up to the mark* so that the inputs and the initialization remain on tape throughout the whole simulation, their adjoints accumulating contributions across simulations.

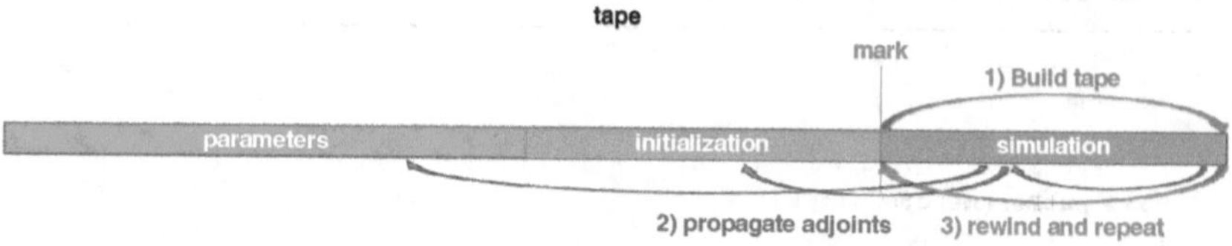

Next in our simulation code, we initialize the RNG, preallocate working memory for the simulations, and allocate results. This is admin code; it is not recorded and no different than valuation.

```
//  ...

//  Init the RNG
cRng->init(cMdl->simDim());

//  Allocate workspace
vector<Number> nPayoffs(nPay);
//  Gaussian vector
vector<double> gaussVec(cMdl->simDim());

//  Results
AADSimulResults results(nPath, nPay, nParam);

//  ...
```

The important initialization phase is complete; we now proceed with the simulations:

```
//  ...

//  Iterate through paths
for (size_t i = 0; i<nPath; i++)
{
    //  AAD - 2
    //  Rewind tape to mark
    //      parameters and initializers stay on tape
    //      but the simulation is overwritten
    tape.rewindToMark();
    //

```

```
13        //  Next Gaussian vector, dimension D
14        cRng->nextG(gaussVec);
15        //  Generate path, consume Gaussian vector
16        cMdl->generatePath(gaussVec, path);
17        //  Compute payofffs
18        prd.payoffs(path, nPayoffs);
19        //  Aggregate
20        Number result = aggFun(nPayoffs);
21
22        //  AAD - 3
23        //  Propagate adjoints
24        result.propagateToMark();
25        //  Store results for the path
26        results.aggregated[i] = double(result);
27        convertCollection(
28            nPayoffs.begin(),
29            nPayoffs.end(),
30            results.payoffs[i].begin());
31        //
32    }
33
34    //  ...
```

A path is processed in three stages:

1. Rewind the tape *to the mark* (line 10) so as to reuse the simulation piece of the tape across simulations and avoid growth while leaving the model parameters and initialization on tape throughout all simulations so that their adjoints correctly accumulate contributions from all the simulations.
2. Generate and evaluate the path (lines 13–20). This code is identical to the valuation code. It executes the three steps of the processing of a path: generate Gaussian numbers, simulate a path, evaluate payoffs over the path. The difference is that this code is now instantiated with Numbers. So it goes on tape when executed and builds the simulation piece of the tape, overwriting the previous simulation so that the size of the tape remains fixed to the size of one simulation; see figure on page 414.

   We have an additional line 20, where the payoffs for the current path are aggregated in one result, with an aggregator provided by the client code, as explained earlier.
3. Propagate adjoints from the result (which adjoint is seeded to 1) *to the mark*. We back-propagate the adjoints of the current simulation, but only those. As a result, the adjoints on the persistent piece of tape, where the parameters and initialization were recorded, accumulate their contribution to the current simulation.

After the simulation loop completes, the adjoints on the persistent piece of tape accumulated their contribution to all the simulations. What is left is back-propagate adjoints through the persistent piece so that the adjoints of the entire Monte-Carlo valuation accumulate into the model parameters.

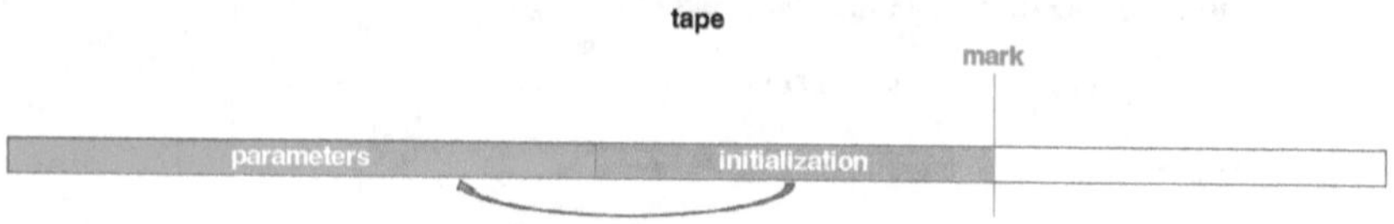

This takes *one* line of code:

```
//  ...


//  AAD - 4
//  Mark = limit between pre-calculations and path-wise operations
//  Operations above mark have been propagated and accumulated
//  We conduct one propagation mark to start
Number::propagateMarkToStart();
//


//  ...
```

It follows that we back-propagate simulation adjoints repeatedly, but initialization adjoints only once. This is another one of the strongest optimizations in our code. If we propagated adjoints all the way, repeatedly over each simulation, to differentiate a Monte-Carlo valuation would be *many times* slower. This optimization, to a large extent, is the dual of the strongest optimization we included in our simulation code of Chapter 6. We gathered as much work as possible into the initialization, conducted once, so as to minimize work during the repeated simulations. It follows that we have large a initialization tape and small simulation tapes. In the differentiation algorithm, we back-propagate repeatedly over the small tapes and once over the large tape, reaping the benefits of our optimization of Chapter 6 for differentiation, too.

All that remains is pick sensitivities to parameters on the parameters adjoints, pack,[2] clean, and return:

```
//  ...


//  Pick sensitivities, summed over paths, and normalize
transform(
    params.begin(),
    params.end(),
    results.risks.begin(),
    [nPath](const Number* p) {return p->adjoint() / nPath; });
```

[2]The parameters' adjoint accumulated the *sum* of contributions over all the paths; we must divide by N to obtain the sensitivities of values, which are averages of payoffs across paths.

```

    //  Clear the tape
    tape.clear();

    return results;
}
```

The code is found in mcBase.h.

12.3 USER INTERFACE

We must upgrade our user interface in main.h to run the differentiation algorithm. We develop two functions: *AADriskOne*(), which differentiates one selected payoff in a product, and *AADriskAggregate*(), which differentiates a portfolio of payoffs with given notionals. Both functions take a model and a product in the memory store, developed in Section 6.5, run the differentiation algorithm, and return:

1. The labels of the payoffs in a vector<string>
2. The corresponding values in a vector<double>
3. The value of the differentiated aggregate, as a double
4. The names of all the parameters in a vector<string>
5. The corresponding risk sensitivities of the aggregate to parameters in a vector<double>

```
    struct
    {
        vector<string>  payoffIds;
        vector<double>  payoffValues;
        double          riskPayoffValue;
        vector<string>  paramIds;
        vector<double>  risks;
    } results;
```

The differentiation algorithm works with instrumented models and products, Model <Number> and Product <Number>. We upgrade the memory store in store.h so we can store and retrieve both instrumented and non-instrumented models and products. In practice, we store a pair of models or products, one instrumented and one not:

```
#include "mcBase.h"
#include "mcMdl.h"
#include "mcPrd.h"
#include <unordered_map>
#include <memory>
using namespace std;

```

```
using ModelStore =
unordered_map<string,
    pair<unique_ptr<Model<double>>, unique_ptr<Model<Number>>>>;
using ProductStore =
unordered_map<string,
    pair<unique_ptr<Product<double>>, unique_ptr<Product<Number>>>>;

ModelStore modelStore;
ProductStore productStore;
```

We upgrade the functions to initialize models and products in memory accordingly:

```
void putBlackScholes(
    const double            spot,
    const double            vol,
    const bool              qSpot,
    const double            rate,
    const double            div,
    const string&           store)
{
    //  We create 2 models, one for valuation and one for risk
    unique_ptr<Model<double>> mdl = make_unique<BlackScholes<double>>(
        spot, vol, qSpot, rate, div);
    unique_ptr<Model<Number>> riskMdl = make_unique<BlackScholes<Number>>(
        spot, vol, qSpot, rate, div);

    //  And move them into the map
    modelStore[store] = make_pair(move(mdl), move(riskMdl));
}

//  Same pattern for putDupire

void putEuropean(
    const double            strike,
    const Time              exerciseDate,
    const Time              settlementDate,
    const string&           store)
{
    //  We create 2 products, one for valuation and one for risk
    unique_ptr<Product<double>> prd = make_unique<European<double>>(
        strike, exerciseDate, settlementDate);
    unique_ptr<Product<Number>> riskPrd = make_unique<European<Number>>(
        strike, exerciseDate, settlementDate);

    //  And move them into the map
    productStore[store] = make_pair(move(prd), move(riskPrd));
}

//  Same pattern for other products
```

We template the functions to retrieve models and products, so client codes calls $getModel < double > ()$ for pricing or $getModel < Number > ()$ for risk:

```
template<class T>
const Model<T>* getModel(const string& store);

template<>
const Model<double>* getModel(const string& store)
{
    auto it = modelStore.find(store);
    if (it == modelStore.end()) return nullptr;
    else return it->second.first.get();
}

template<>
const Model<Number>* getModel(const string& store)
{
    auto it = modelStore.find(store);
    if (it == modelStore.end()) return nullptr;
    else return it->second.second.get();
}

//  Same pattern for products
```

The entire code is found in store.h. With the store upgraded to deal with instrumented objects, we develop the user interface for the differentiation of simulations in main.h below. We write two functions, one for the differentiation of one named payoff, and one for the differentiation of a portfolio. The project in our repository exports these functions to Excel.

```
//  Generic risk
inline auto AADriskOne(
    const string&           modelId,
    const string&           productId,
    const NumericalParam&   num,
    const string&           riskPayoff = "")
{
    //  Get instrumented model and product
    const Model<Number>* model = getModel<Number>(modelId);
    const Product<Number>* product = getProduct<Number> (productId);

    if (!model || !product)
    {
        throw runtime_error(
            "AADrisk() : Could not retrieve model and product");
    }

    //  Random Number Generator
    unique_ptr<RNG> rng;
    if (num.useSobol) rng = make_unique<Sobol>();
    else rng = make_unique<mrg32k3a>(num.seed1, num.seed2);

    //  Find the payoff to differentiate
    size_t riskPayoffIdx = 0;
    if (!riskPayoff.empty())
    {
        const vector<string>& allPayoffs = product->payoffLabels();
```

```
        auto it = find(allPayoffs.begin(), allPayoffs.end(), riskPayoff);
        if (it == allPayoffs.end())
        {
            throw runtime_error("AADriskOne() : payoff not found");
        }
        riskPayoffIdx = distance(allPayoffs.begin(), it);
    }

    //  Simulate
    const auto simulResults = num.parallel
        ? mcParallelSimulAAD(*product, *model, *rng, num.numPath,
            [riskPayoffIdx](const vector<Number>& v)
            {return v[riskPayoffIdx]; })
        : mcSimulAAD(*product, *model, *rng, num.numPath,
            [riskPayoffIdx](const vector<Number>& v)
            {return v[riskPayoffIdx]; });

    //  We return: a number and 2 vectors :
    //  -   The payoff identifiers and their values
    //  -   The value of the aggreagte payoff
    //  -   The parameter idenitifiers
    //  -   The sensitivities of the aggregate to parameters
    struct
    {
        vector<string>  payoffIds;
        vector<double>  payoffValues;
        double          riskPayoffValue;
        vector<string>  paramIds;
        vector<double>  risks;
    } results;

    const size_t nPayoffs = product->payoffLabels().size();
    results.payoffIds = product->payoffLabels();
    results.payoffValues.resize(nPayoffs);
    for (size_t i = 0; i < nPayoffs; ++i)
    {
        results.payoffValues[i] = accumulate(
            simulResults.payoffs.begin(),
            simulResults.payoffs.end(),
            0.0,
            [i](const double acc, const vector<double>& v)
                { return acc + v[i]; }
        ) / num.numPath;
    }
    results.riskPayoffValue = accumulate(
        simulResults.aggregated.begin(),
        simulResults.aggregated.end(),
        0.0) / num.numPath;
    results.paramIds = model->parameterLabels();
    results.risks = move (simulResults.risks);

    return results;
}

inline auto AADriskAggregate(
    const string&           modelId,
    const string&           productId,
```

```
    const map<string, double>&   notionals,
    const NumericalParam&   num)
{
    //  Get instrumented model and product
    const Model<Number>* model = getModel<Number>(modelId);
    const Product<Number>* product = getProduct<Number> (productId);

    if (!model || !product)
    {
        throw runtime_error(
        "AADriskAggregate() : Could not retrieve model and product");
    }

    //  Random Number Generator
    unique_ptr<RNG> rng;
    if (num.useSobol) rng = make_unique<Sobol>();
    else rng = make_unique<mrg32k3a>(num.seed1, num.seed2);

    //  Vector of notionals
    const vector<string>& allPayoffs = product->payoffLabels();
    vector<double> vnots(allPayoffs.size(), 0.0);
    for (const auto& notional : notionals)
    {
        auto it = find(allPayoffs.begin(), allPayoffs.end(), notional.first);
        if (it == allPayoffs.end())
        {
            throw runtime_error("AADriskAggregate() : payoff not found");
        }
        vnots[distance(allPayoffs.begin(), it)] = notional.second;
    }

    //  Aggregator
    auto aggregator = [&vnots](const vector<Number>& payoffs)
    {
        return inner_product(
            payoffs.begin(),
            payoffs.end(),
            vnots.begin(),
            Number(0.0));
    };

    //  Simulate
    const auto simulResults = num.parallel
        ? mcParallelSimulAAD(*product, *model, *rng, num.numPath, aggregator)
        : mcSimulAAD(*product, *model, *rng, num.numPath, aggregator);

    //  We return: a number and 2 vectors :
    //  -   The payoff identifiers and their values
    //  -   The value of the aggreagte payoff
    //  -   The parameter idenitifiers
    //  -   The sensititivities of the aggregate to parameters
    struct
    {
        vector<string>  payoffIds;
        vector<double>  payoffValues;
        double          riskPayoffValue;
        vector<string>  paramIds;
```

```
        vector<double>  risks;
    } results;

    const size_t nPayoffs = product->payoffLabels().size();
    results.payoffIds = product->payoffLabels();
    results.payoffValues.resize(nPayoffs);
    for (size_t i = 0; i < nPayoffs; ++i)
    {
        results.payoffValues[i] = accumulate(
            simulResults.payoffs.begin(),
            simulResults.payoffs.end(),
            0.0,
            [i](const double acc, const vector<double>& v)
                { return acc + v[i]; }
        ) / num.numPath;
    }
    results.riskPayoffValue = accumulate(
        simulResults.aggregated.begin(),
        simulResults.aggregated.end(),
        0.0) / num.numPath;
    results.paramIds = model->parameterLabels();
    results.risks = move(simulResults.risks);

    return results;
}
```

where *mcParallelSimulAAD*(), the parallel version of the differentiation algorithm, is developed next.

For test purposes, and to assess the speed and the accuracy of our derivatives, we also implement bump risk. Bumps are implemented in a trivial manner to produce itemized risk reports[3] in a trivial manner. Bump risk is up to thousands of times slower than AAD, and the resulting risk sensitivities are expected to be virtually identical. The following self–explanatory function is found in main.h:

```
//  Returns a vector of values and a matrix of risks
//      with payoffs in columns and parameters in rows
//      along with ids of payoffs and parameters

struct RiskReports
{
    vector<string> payoffs;
    vector<string> params;
    vector<double> values;
    matrix<double> risks;
};

```

[3]One risk report for each payoff, as opposed to one risk report for an aggregate–we produce itemized risk reports with AAD in Chapter 14.

```
inline RiskReports bumpRisk(
    const string&           modelId,
    const string&           productId,
    const NumericalParam&   num)
{
    auto* orig = getModel<double>(modelId);
    const Product<double>* product = getProduct<double> (productId);

    if (!orig || !product)
    {
        throw runtime_error(
            "bumpRisk() : Could not retrieve model and product");
    }

    RiskReports results;

    //  Base values
    auto baseRes = value(*orig, *product, num);
    results.payoffs = baseRes.identifiers;
    results.values = baseRes.values;

    //  Make copy so we don't modify the model in memory
    auto model = orig->clone();

    results.params = model->parameterLabels();
    const vector<double*> parameters = model->parameters();
    const size_t n = parameters.size(), m = results.payoffs.size();
    results.risks.resize(n, m);

    //  Bump parameters one by one
    for (size_t i = 0; i < n; ++i)
    {
        //  Bump
        *parameters[i] += 1.e-08;
        //  Reval
        auto bumpRes = value(*model, *product, num);
        //  Unbump
        *parameters[i] -= 1.e-08;

        // Compute finite differences for all payoffs
        for (size_t j = 0; j < m; ++j)
        {
            results.risks[i][j] = 1.0e+08 *
                (bumpRes.values[j] - baseRes.values[j]);
        }
    }

    return results;
}
```

12.4 SERIAL RESULTS

We can now differentiate the results of our simulation library with respect to model parameters. For instance, we compute the sensitivities of a European or barrier option to all local volatilities. Sensitivities to local volatilities, what Dupire calls a *microbucket*, is valuable information for research purpose. For the purpose of trading, however, we really need the sensitivities to tradable market variables, not model parameters. In the context of Dupire's model, we want the sensitivities to the *market-implied* volatilities, what Dupire calls a *superbucket*, because those are risks that may be directly hedged by trading European options in the market. Local volatilities are calibrated to implied volatilities, so superbuckets can be computed from microbuckets, as discussed in detail in Chapter 13. In this chapter, we focus on the production of microbuckets.

We differentiate the example of Section 6.6: a 3y European call of strike 120, together with a 150 barrier, in Dupire's model with a local volatility of 15% over 60 times and 30 spots, simulating 500,000 paths over 156 weekly steps, where a pricing took around 3 seconds with Sobol. With 1,801 sensitivities to compute, we need 1,802 prices with finite differences, so the bump risk takes around 1.5 hours.[4]

Our AAD implementation produces the entire risk in around 8 seconds (8,250 ms), less than three times a single valuation. With the improvements of Chapter 15, when we further accelerate AAD by factor around 2, we produce this report in less than 5 seconds (4,500 ms). With the parallel algorithm, it takes *half a second* (575 ms).

The results are displayed on the chart below, first for the European, then for the barrier, with a smoothing of 5. The charts display the differentials to local volatilities on seven different times, as functions of the spot.

[4]To be fair, with a 3y maturity, we "only" have 1,080 active local volatilities, so we would be done in an hour. More generally, our bump risk implementation is particularly dumb, only for reference against the AAD code. Before AAD made its way to finance, derivatives systems typically implemented a "smart bump" risk, a conscientious, invasive work on models to recalculate as little as possible in the bumped scenarios. We don't do that here because it is not the subject, but we acknowledge that a smart bump risk could complete in this example in around 20 minutes. With multi-threading over the 8 cores of our iMac Pro, it could complete in less than 3 minutes. As a comparison, AAD risk computes in around 8 seconds with the current implementation, 5 seconds with the improvements of Chapter 15, and *half a second* with multi-threading. This is still at least 300 times faster than the smartest possible implementation with bumps.

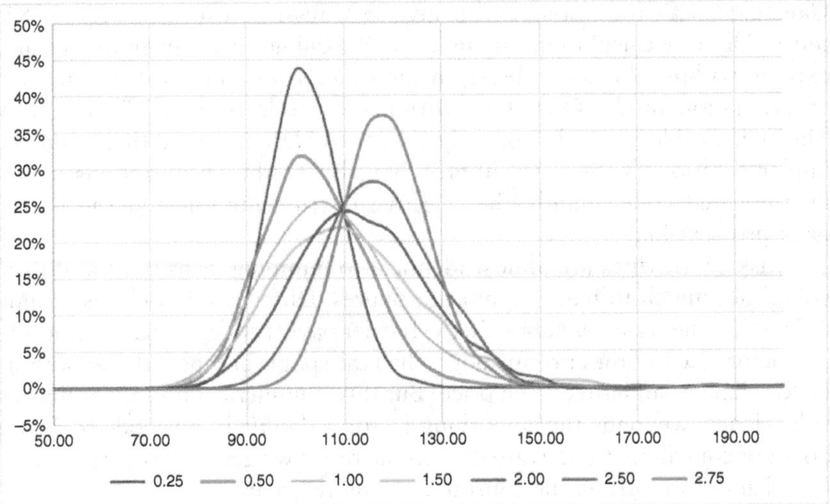

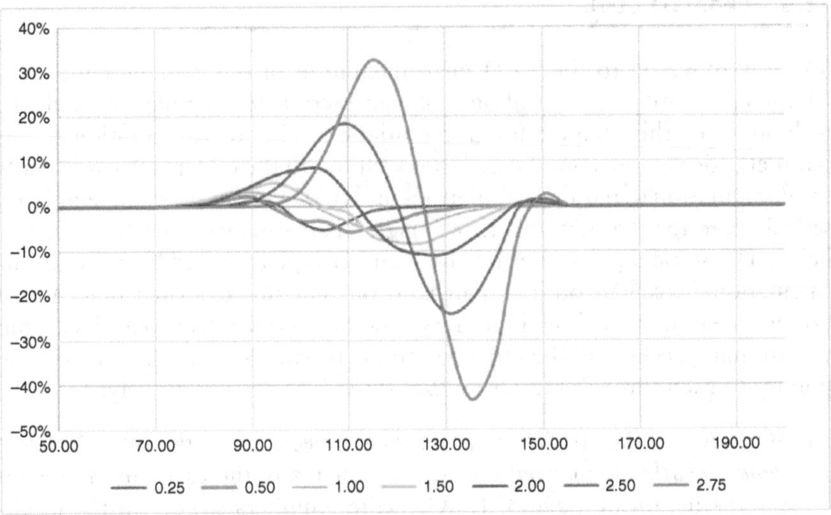

The speed is remarkable and may look suspect. Didn't we establish the theoretical maximum speed for AAD at three times a valuation? With the improvements of Chapter 15, our example computes in the time of 1.5 valuations, *half the theoretical minimum*.

This is mainly due to our selective code instrumentation. A lot of the valuation time is spent generating random numbers and turning them into

Gaussians with the not-so-cheap inverse Gaussian, executed $ND = 78M$ times. All those calculations are inactive. We did not instrument them. They execute without ADD overhead, in the exact same time in differentiation or evaluation mode. Only the instrumented code bears AAD overhead. The remarkable speed is not only due to AAD's constant time, and the implementation of the AAD library, but also to the optimizations in our instrumented code, mainly the selective instrumentation and the partial back-propagation.

Risk sensitivities are almost identical to bumping. Sensitivities to local volatilities match to many significant figures. Difference in delta is around $2.5e-05$. The reason is subtle. In the barrier payoff code, we used a smoothing factor of $x\%$ of the current spot. When the spot is bumped, the smoothing factor changes and affects the price. But this is numerical noise, not market risk. Hence, we made the smoothing factor a *double* in our code so that it does not contribute to derivatives, and the result we get is market risk alone. Our delta is not off; on the contrary, it is more accurate.

12.5 PARALLEL CODE

We now move on to the AAD instrumentation of our *parallel* code from Chapter 7, combining the idioms of that section for parallelism with the techniques of this chapter for differentiation. The instrumentation of the parallel code is more complicated, but well worth the effort, since it is really the combination of parallel simulations and AAD (and expression templates, see Chapter 15) that gives us this "magic" performance. The AAD library helps. The static tape pointer on the Number type is thread local, so all the calculations executed on a thread land on that thread's own tape. But it doesn't mean our work is done. First, the tape *pointer* is thread local, but we still must declare all the tapes for all the threads and set the thread local pointer to each thread's tape *from that thread*. More importantly:

1. With AAD, the model is a *mutable* object. Even though *Model* :: *generatePath*() is marked const, it is not really the case any more; see the discussion on page 391. When the Numbers stored on the model (parameters and initialized amounts) are *read*, their adjoints on tape are *written* into during back-propagation. So, we must make copies of the model for every thread like other mutable objects.[5]
2. In the single-thread version, we were careful to put the model parameters on tape and initialize the model so the initialized amounts are recorded, too. In the parallel version, we must put the parameters and initializers, not on one tape, but on every thread's tape. The only way to do this

[5]Products are unaffected because they are not supposed to store Numbers in their state, although this is not enforced, so we must be very careful there.

is call *Model* : : *putParametersOnTape*() and *Model* : : *init*() from *each thread* before the first simulation on that thread.

3. Adjoints are accumulated on thread local tapes, so after the simulation loop, the adjoints of the parameters and initializers are scattered across different tapes. We must back-propagate over initialization on multiple tapes and aggregate all the sensitivities to parameters.

The code is essentially a combination of the parallel code of Section 7 with the differentiation code of this chapter, with a special response to these specific challenges.

To address number 2, we separate the initialization in a dedicated function:

```
//  Put parameters on tape and init model
inline void initModel4ParallelAAD(
    //  Inputs
    const Product<Number>&          prd,
    //  Cloned model, must have been allocated prior
    Model<Number>&                  clonedMdl,
    //  Path, also allocated prior
    Scenario<Number>&               path)
{
    //  Access tape
    Tape& tape = *Number::tape;

    //  Rewind tape
    tape.rewind();

    //  Put parameters on tape
    //      note this also initializes all adjoints
    clonedMdl.putParametersOnTape();

    //  Init the model
    //      not before the parameters are on tape:
    //      initialization uses parameters as arguments
    clonedMdl.init(prd.timeline(), prd.defline());

    //  Path
    initializePath(path);

    //  Mark the tape straight after parameters
    tape.mark();
    //
}
```

The template algorithm has the same signature as the single-threaded one and returns results in the same format:

```
#include "ThreadPool.h"

//  Parallel version of mcSimulAAD()
template<class F = decltype(defaultAggregator)>
inline AADSimulResults
mcParallelSimulAAD(
```

```
    const Product<Number>&  prd,
    const Model<Number>&    mdl,
    const RNG&              rng,
    const size_t            nPath,
    const F&                aggFun = defaultAggregator)
{
    const size_t nPay = prd.payoffLabels().size();
    const size_t nParam = mdl.numParams();

    //  Allocate results
    AADSimulResults results(nPath, nPay, nParam);

    //  Clear and initialise tape
    Number::tape->clear();
//  ...
```

and the entry-level function in main.h is already set up to fire it.

Next, we allocate working memory (without initializing anything quite yet), one mutable object for every thread (including the main thread, so number of worker threads + 1):

```
    //  ...

    //  We need one of all these for each thread
    //  0: main thread
    //  1 to n : worker threads

    ThreadPool *pool = ThreadPool::getInstance();
    const size_t nThread = pool->numThreads();

    //  Allocate workspace, do not initialize anything yet

    //  One model clone per thread
    //      models are mutable with AAD
    vector<unique_ptr<Model<Number>>> models(nThread + 1);
    for (auto& model : models)
    {
        model = mdl.clone();
        model->allocate(prd.timeline(), prd.defline());
    }

    //  One scenario per thread
    vector<Scenario<Number>> paths(nThread + 1);
    for (auto& path : paths)
    {
        allocatePath(prd.defline(), path);
    }

    //  One vector of payoffs per thread
    vector<vector<Number>> payoffs(nThread + 1, vector<Number>(nPay));

    //  ~workspace

    //  Tapes for the worker threads
    //  The main thread has one of its own
    vector<Tape> tapes(nThread);

```

```
37      //  Model initialized on this thread?
38      vector<int> mdlInit(nThread + 1, false);
39
40      //  ...
```

We also initialized a vector of tapes on line 35, one for each *worker* thread. The main thread already owns a tape: the static tape declared on the Number type instantiates of copy for the main thread when the application starts, as for any static object.

Finally, we have a vector of booleans on line 38, one for every thread, tracking what threads completed the initialization phase, starting with false for everyone.[6]

We must also initialize the RNGs and preallocate the Gaussian vector, but for that we need the simulation dimension D, and only an initialized model can provide this information. So the next step is initialize the main thread's model (the main thread always has index 0, the worker threads indices being 1 to nThread):

```
//  ...

//  Initialize main thread
initModel4ParallelAAD(prd, *models[0], paths[0]);

//  Mark main thread as initialized
mdlInit[0] = true;

//  Init the RNGs, one per thread
//  One RNG per thread
vector<unique_ptr<RNG>> rngs(nThread + 1);
for (auto& random : rngs)
{
    random = rng.clone();
    random->init(models[0]->simDim());
}

//  One Gaussian vector per thread
vector<vector<double>> gaussVecs
    (nThread + 1, vector<double>(models[0]->simDim()));

//  ...
```

Next, we dispatch the simulation loop for parallel processing to the thread pool, applying the same techniques as in Chapter 7:

```
//  ...

//  Reserve memory for futures
vector<TaskHandle> futures;
futures.reserve(nPath / BATCHSIZE + 1);

```

[6]We used a vector<int> in place of a vector<bool>. For legacy reasons, vector<bool> is not thread safe in standard C++ for concurrent writes into separate entries, and should be avoided.

```
    //  Start

    size_t firstPath = 0;
    size_t pathsLeft = nPath;
    while (pathsLeft > 0)
    {
        size_t pathsInTask = min<size_t>(pathsLeft, BATCHSIZE);

        futures.push_back(pool->spawnTask([&, firstPath, pathsInTask] ()
        {

    //  ...
```

The body of the lambda executes on different threads. This is the concurrent code. Here, the tape is the executing thread's tape and *pool–>threadNum*() returns the index of the executing thread. The first thing we do is note this index, and set the thread local tape accordingly:

```
    //  ...

            const size_t threadNum = pool->threadNum();

            //  Use this thread's tape
            //  Thread local magic: each thread has its own pointer
            //  Note main thread = 0 is not reset
            if (threadNum > 0) Number::tape = &tapes[threadNum - 1];

    //  ...
```

This is the core of parallel AAD right here. The tape is accessed through the *thread local* pointer *Number* : : *tape*. We set it *from the executing thread* to the tape we allocated for that thread in our vector of tapes. From there on, all the subsequent calculations coded in the lambda and executed on this thread are recorded on that tape. The main thread (0) already correctly refers to its global tape. It does not need resetting.

Only at this stage can we put the parameters on tape and initialize the model (other than allocation, we allocated all models earlier, on the main thread, knowing from Chapter 3 that we must avoid concurrent allocations):

```
    //  ...

            //  Initialize once on each thread
            if (!mdlInit[threadNum])
            {
                //  Initialize and put on the
                //      executing thread's tape
                initModel4ParallelAAD(
                    prd,
                    *models[threadNum],
                    paths[threadNum]);
```

```

            //  Mark as initialized
            mdlInit[threadNum] = true;
        }

    //  ...
```

Next, we pick this thread's RNG and skip it ahead, like in Chapter 7:

```
    //  ...

        // Get a RNG and position it correctly
        auto& random = rngs[threadNum];
        random->skipTo(firstPath);

    //...
```

Finally, the rest of the concurrent code executes the inner batch loop of Chapter 7 with the same contents as the serial differentiation code: rewind tape (to mark), generate random numbers, simulate scenario, compute payoffs, aggregate them, back-propagate adjoints (to mark), store results, repeat):

```
    //  ...

        //  And conduct the simulations, exactly same as sequential
        for (size_t i = 0; i < pathsInTask; i++)
        {
            //  Rewind tape to mark
            //  Notice : this is the tape for the executing thread

            Number::tape->rewindToMark();
            //  Next Gaussian vector, dimension D
            random->nextG(gaussVecs[threadNum]);
            //  Path
            models[threadNum]->generatePath(
                gaussVecs[threadNum],
                paths[threadNum]);
            //  Payoffs
            prd.payoffs(paths[threadNum], payoffs[threadNum]);

            //  Propagate adjoints
            Number result = aggFun(payoffs[threadNum]);
            result.propagateToMark();
            //  Store results for the path
            results.aggregated[firstPath + i] = double(result);
            convertCollection(
                payoffs[threadNum].begin(),
                payoffs[threadNum].end(),
                results.payoffs[firstPath + i].begin());
        }

        //  Remember tasks must return a bool
        return true;
    }));
```

```

        pathsLeft -= pathsInTask;
        firstPath += pathsInTask;
    }

    //  Wait and help
    for (auto& future : futures) pool->activeWait(future);

    //  ...
```

Like for serial differentiation, all adjoints are accumulated on the persistent piece of the tape at this stage, and we must propagate the initializers' adjoints back to the parameters' adjoints; see figure on page 416. The difference is that the adjoints are now spread across the multiple tapes belonging to multiple threads, but we are out of the concurrent code and back on the main thread. Never mind; this is one propagation conducted one time; we propagate all the threads' tapes from the main thread:

```
1   //  ...
2
3   //  Mark = limit between pre-calculations and path-wise operations
4   //  Operations above mark have been propagated and accumulated
5   //  We conduct one propagation mark to start
6   //  On the main thread's tape
7   Number::propagateMarkToStart();
8   //  And on the worker thread's tapes
9   Tape* mainThreadPtr = Number::tape;
10  for (size_t i = 0; i < nThread; ++i)
11  {
12      if (mdlInit[i + 1])
13      {
14          //  Set tape pointer
15          Number::tape = &tapes[i];
16          //  Propagate on that tape
17          Number::propagateMarkToStart();
18      }
19  }
20  //  Reset tape to main thread's
21  Number::tape = mainThreadPtr;
22
23  //  ...
```

We first propagate over the main thread's tape on line 7. Then, we sequentially propagate the other n tapes, from the main thread. In order to propagate the tape number i from the main thread, we temporarily set the main thread's tape as the tape i (line 15), and propagate (line 17).

In the end, we reset the main thread's tape to its static tape on line 21.[7]

[7]This code is not exception safe. If an exception occurs here, the main thread's tape is never reset. Ideally, we would code some RAII logic to reset the main thread's tape no matter what.

We are almost home safe. All the adjoints are correctly accumulated, but still across multiple tapes and multiple models. The tape of each thread contains the sum of the adjoints over the paths processed on that thread. The adjoints are accessible for every thread's copy of the model. We sum-up the adjoints across models, hence across tapes, normalize, pack, and return:

```
    //  ...

    //  Sum sensitivities over threads
    for (size_t j = 0; j < nParam; ++j)
    {
        results.risks[j] = 0.0;
        for (size_t i = 0; i < models.size(); ++i)
        {
            if (mdlInit[i]) results.risks[j] +=
                    models[i]->parameters()[j]->adjoint();
        }
        results.risks[j] /= nPath;
    }

    //  Clear the main thread's tape
    //  The other tapes are cleared when the vector of tapes exists scope
    Number::tape->clear();

    return results;
}
```

The code is found in mcBase.h.

12.6 PARALLEL RESULTS

We reproduce the example on page 424, this time in parallel: a 3y call 120 with a 150 barrier, in Dupire's model 1,800 local volatilities (of which 1,080 active) of 15% over 60 times and 30 spots, simulating 500,000 paths over 156 weekly steps with Sobol. With the serial implementation, it took around 8 seconds to compute the microbuckets (precisely 8,250 ms). With the improvements of Chapter 15, it takes less than 5 seconds (precisely 4,500 ms).

With the parallel implementation, we get the exact same results in exactly one second, with perfect parallel efficiency. With the improvements of Chapter 15, we get them in half a second (precisely 575 ms).

We produce more than 1,000 active sensitivities over 500,000 paths, 156 steps, in half a second.

With an efficient AAD library, coupled with an effective instrumentation of the simulation code, we computed derivatives over simulations with unthinkable speed. Not only can we produce lightning-fast risk reports on workstations and laptops, including for xVA, we can also conduct research that would have been impossible otherwise.

For instance, a parallel price with Sobol takes 350 ms in our example, that is, more than five minutes for a finite difference risk over the 1,080 active local volatilities (which perhaps may be improved to two to three minutes with substantial effort and invasive code). This is too slow. It is therefore common practice to reduce accuracy to produce risk reports in a manageable time. With 36 monthly time steps and 100,000 paths, a parallel price takes 20 ms, so a finite difference risk report could perhaps be produced in as little as ten seconds. The results, however, are not pretty, as seen in the charts below, first for the European, then for the barrier, to be compared to the charts on page 425.

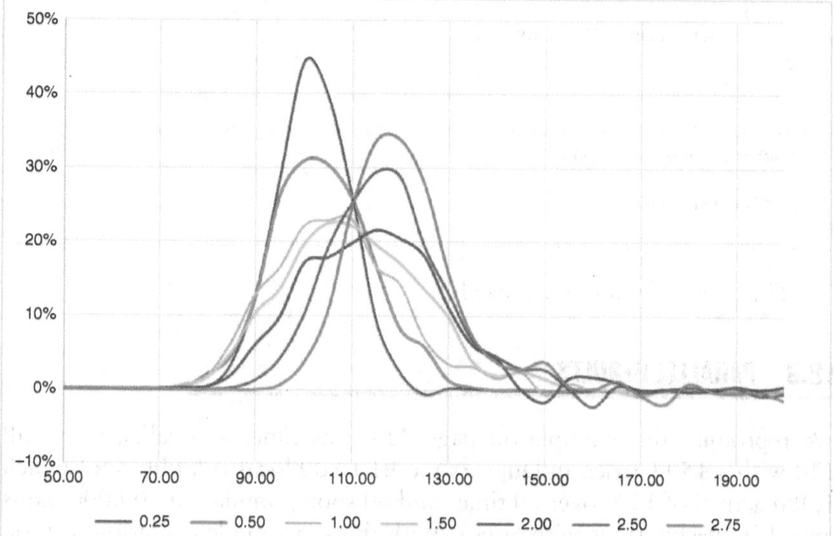

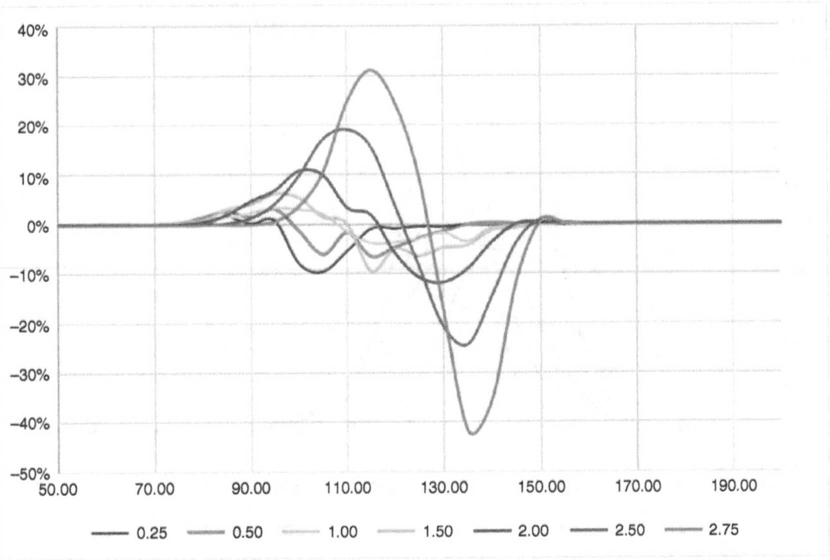

Slow financial software has the unfortunate consequence that users are encouraged to settle for approximate, inaccurate risk computed in reasonable time. With AAD, we produce the accurate charts on page 425 in half a second. We could run 1 M path to produce even better, smoother risk reports in a second.

Interestingly perhaps, one second is approximately the time it takes to produce a similar risk report with comparable accuracy, with FDM in place of Monte-Carlo, on a single core with finite differences. The combination of AAD and parallelism roughly brings FDM speed to Monte-Carlo simulations.

With 3M paths, we produce the much smoother charts below in around three seconds (that is three seconds *each*, although we produce them both in around three seconds in Chapter 14, and three seconds is with the improvements of Chapter 15):

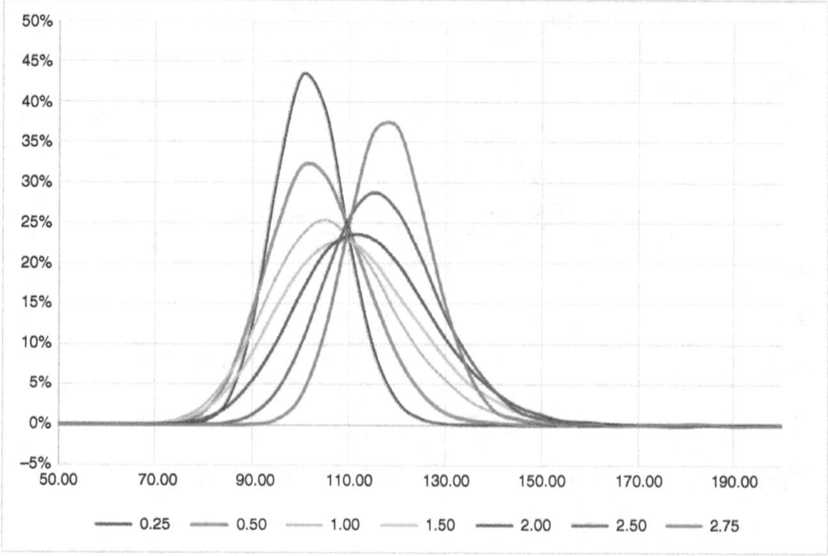

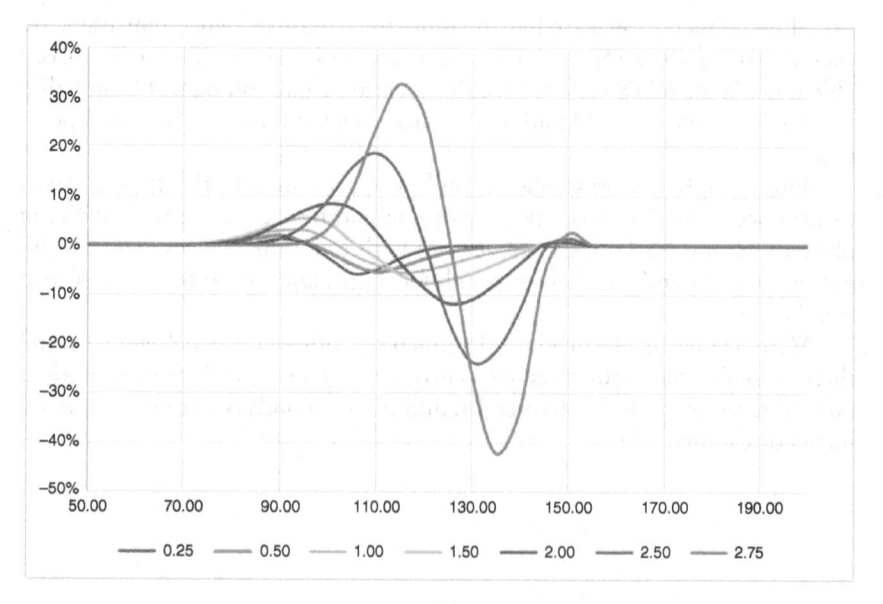

It would have taken at least half an hour with smart, multi-threaded bumping, at least four hours in serial mode. For all intents and purposes, to obtain such precise microbuckets with Monte-Carlo simulations is not feasible without AAD. AAD opens new research possibilities into financial risk management. One example is that it produces accurate microbuckets in a few seconds. For other, perhaps unexpected applications, see, for example, Reghai's [97] and Andreasen's [82].

We reiterate that the main purpose of derivatives systems is not to produce sensitivities to model parameters, but risks to market instruments. For this reason, this comparison between AAD and bumping is somewhat invidious. With bumping, sensitivities to hedge instruments could be produced faster by directly bumping a smaller number of market prices and repeating the whole valuation process, including the calibration of the model parameters. However, to produce market risks in this manner could lead to instabilities from differentiating through calibration. Further, the sensitivity to model parameters would not be computed, and that is valuable information in its own right. Finally, it would not offer the flexibility to compute risks with different instruments than those used to calibrate model parameters.

We have seen that AAD produces sensitivities to a large number of model parameters extremely fast. With the concern for costs of computing a large number of sensitivities out of the way, the better strategy for producing risks is to first compute sensitivities to model parameters, as explained in this chapter, and next turn them into sensitivities to hedge instruments, as explained in the next chapter, where we calibrate Dupire's model to market data and produce risks, not to local volatilities, but to implied volatilities, which directly reflect the market prices of European options.

CHAPTER 13

Check-Pointing and Calibration

In the previous chapter, we learned to differentiate simulations against model parameters in constant time, and produced *microbuckets* (sensitivities to local volatilities) in Dupire's model with remarkable speed. In this chapter, we present the key check-pointing algorithm, apply it to differentiation through calibrations to obtain market risks out of sensitivities to model parameters, and implement *superbucket* risk reports (sensitivities to implied volatilities) in Dupire's model.

13.1 CHECK-POINTING

Reducing RAM consumption with check-pointing

We pointed out in Chapter 11 that, due to RAM consumption, AAD cannot efficiently differentiate calculations taking more than 0.01 seconds on a core. Longer calculations, which include almost all cases of practical relevance, must be divided into pieces shorter than 0.01 seconds.core, and differentiated separately over each piece, wiping RAM in between and aggregating sensitivities in the end.

How exactly this is achieved depends on the instrumented algorithm. This is particularly simple in the case of path-wise simulations, where sensitivities are computed path by path and averaged in the end. But this solution is specific to simulations. This chapter discusses a more general solution called *check-pointing*. Check-pointing applies to many problems of practical relevance, in finance and elsewhere. Huge successfully applied it to the differentiation of multidimensional FDM in [90]. Huge and Savine applied it to efficiently differentiate the LSM algorithm and produce a complete xVA risk in [31]. We explain check-pointing in general mathematical and programmatic terms and apply it to the differentiation of calibration.

Formally, we differentiate a two-step algorithm, that is, a scalar function $F : \mathbb{R}^n \rightarrow \mathbb{R}$ that can be written as:

$$F(X) = H[G(X)]$$

where $G : \mathbb{R}^n \to \mathbb{R}^m$ is a vector-valued function and $H : \mathbb{R}^m \to \mathbb{R}$ is a scalar function, assumed differentiable in constant time. We denote $Y = G(X)$ and $z = F(X) = H[G(X)]$.

In the context of a calibrated financial simulation model, H is the value of some transaction in some model with m parameters Y. We learned to differentiate H in constant time in the previous chapter. G is an explicit[1] calibration that produces the m model parameters Y out of n market variables X, with $m \leq n$. F computes the value z of the transaction from the market variables X. The differentials of F are the hedge coefficients, also called *market risks*, that risk reports are designed to produce.

Importantly, there are many good reasons to split differentiation this way, besides minimizing the size of the tape. In the context of financial simulations, intermediate differentials provide useful information for research and risk management, the sensitivity of transactions to model parameters and the sensitivity of model parameters to market variables aggregated into market risks, but also constitute interesting information in their own right. Besides, to differentiate calibration is fine when it is explicit, but to differentiate a numerical calibration may result in unstable, inaccurate sensitivities. We present in Section 13.3 a specific algorithm for the differentiation of numerical calibrations. But we can only do that if we split differentiation between calibration and valuation.

Another example is how we differentiated the simulation algorithm in the previous chapter. We split the simulation algorithm F into an initialization phase G, which pre-calculates m deterministic results Y from the model parameters X, and a simulation phase H, which generates and evaluates paths and produces a value z out of X and Y.[2] We implemented H in a loop over paths, and differentiated each of its iterations separately, before propagating the resulting differentials over G. This is a direct application of check-pointing, although we didn't call it by its name at the time. If we hadn't split the differentiation of F into the differentiation of G and the differentiation of H, we could not have implemented path-wise differentiation that efficiently. More importantly, we could not have conducted the differentiation of H in parallel. In order to differentiate in parallel a parallel calculation, we must first extract the parallel piece and differentiate it separately from the rest, applying check-pointing to connect the resulting differentials.

As a final example, we could split the processing F of every path into the generation G of the m-dimensional scenario Y and the evaluation H of the payoff. We differentiated F altogether in the previous chapter, but for a

---

[1]Implicit (numerical) calibration is discussed later in this chapter when we present the Implicit Function Theorem. For now, G is an explicit function that expresses the model parameters out of market prices, like Dupire's formula [12].

[2]So in this case $F(X) = H[X, G(X)]$ not $H[G(X)]$, but this doesn't change anything, as will be demonstrated shortly.

product with a vast number of cash-flows valued over a high-dimensional path, like an xVA, we would separate the differentiation of the two to limit RAM consumption.

In all these cases, H can be differentiated in constant time with AAD because it is a scalar function. In theory, F is also differentiable in constant time because it is also a scalar function. But we discussed some of the many reasons why it may be desirable to split its differentiation. Hence, the exercise is to split the differentiation of F into a differentiation of G and a differentiation of H while preserving the constant time property. The problem is that G is *not* a scalar function, hence it *cannot* be differentiated in constant time with straightforward AAD. To achieve this, we need additional logic, and it is this additional logic that is called check-pointing.

Formally, from the chain rule:

$$\frac{\partial F}{\partial X} = \frac{\partial G}{\partial X}\frac{\partial H}{\partial Y}$$

and our assumption is that we have a constant time computation for $\frac{\partial H}{\partial Y}$. But AAD cannot compute the Jacobian $\frac{\partial G}{\partial X}$ in constant time. We have seen in Chapter 11 that Jacobians take linear time in the number of results m. With bumping, it takes linear time in n. In any case, it cannot be computed in constant time. Furthermore, $\frac{\partial G}{\partial X}\frac{\partial H}{\partial Y}$ is the product of the m vector $\frac{\partial H}{\partial Y}$ by the $m \times n$ matrix $\frac{\partial G}{\partial X}$, linear in nm.

Check-pointing applies adjoint calculus to compute $\frac{\partial F}{\partial X}$ in constant time, in a sequence of steps where the adjoints of F and G are propagated separately, without ever computing a Jacobian or performing a matrix product.

If, hypothetically, we did differentiate F altogether with a single application of AAD, what would the tape look like?

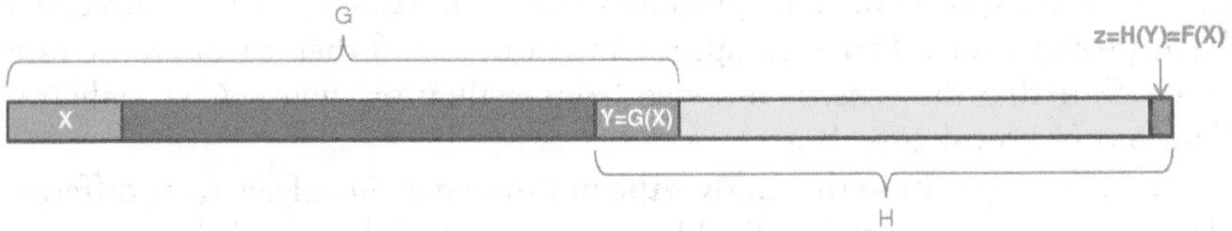

It must be this way, because G is computed before H, and H only depends on the results of G.

The part of the tape that belongs to H is self-contained for adjoint propagation, because H only depends on Y, and not on the details of its internal calculations within G.[3] So the arguments to all calculations within H must

[3]We assume that G and H are functional in the sense that they produce outputs out of inputs without modification of hidden state.

be located on H's part of the tape, including Y; they cannot belong to G's part of the tape.

The section of the tape that belongs to G (inclusive of inputs X and outputs Y) is also self-contained. Evidently, the calculations within G cannot depend on the calculations in H, which is evaluated *later*. And we have seen that the calculations within H cannot directly reference those of G, except through its outputs Y.

Hence, the tape for F is *separable* for the purpose of adjoint propagation: it can be split into two self-contained tapes, with a common section Y as the output to G's tape and the input to H's tape.

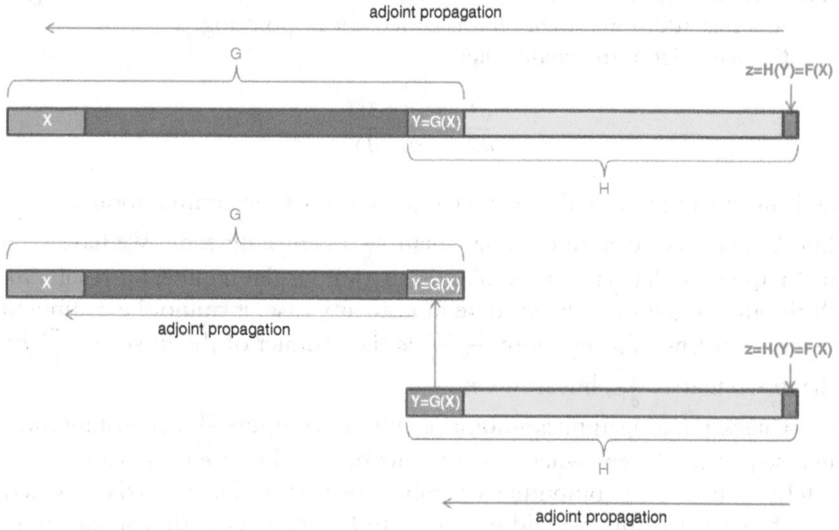

It should be clear that an overall back-propagation through the entire tape of F is equivalent, and produces the same results, as two successive back-propagations, first through the tape of H, and then through the tape of G. Note that the order is reversed from evaluation, where G is evaluated first and H is evaluated last.

It is this separation that allows the multiple benefits of separate differentiations, including a smaller RAM consumption. Only one of the two tapes of G and H is active at a time in memory, and the differentials of $z = F(X)$ are accumulated through adjoint propagation alone, hence, in constant time. The check-pointing algorithm is articulated below:

1. Starting with a clean tape, compute and store $Y = G(X)$ without AAD instrumentation. The only purpose here is to compute Y. Put Y on tape.
2. Compute the final result $z = H(Y) = F(X)$ with an *instrumented* evaluation of H. This builds the tape for H.

3. Back-propagate from z to Y, producing the adjoints of Y: $\frac{\partial H}{\partial Y}$. Store this result.
4. Wipe H's tape. It is no longer needed.
5. Put X on tape.
6. Recompute $Y = G(X)$ as in step 1, this time with AAD instrumentation. This builds the tape for G.
7. Seed that tape with the adjoints of Y, that is the $\frac{\partial H}{\partial Y}$, known from step 3. This is the defining step in the check-pointing algorithm. Instead of seeding the tape with 1 for the end result (and 0 everywhere else), seed it with the known adjoints for all the components of the vector Y.
8. Conduct back-propagation through the tape of G, from the known adjoints of Y to those, unknown, of X.
9. The adjoints of X are the final, desired result: $\frac{\partial F}{\partial X} = \frac{\partial G}{\partial X}\frac{\partial H}{\partial Y}$.
10. Wipe the tape.

The following figure shows the state of the tape after each step:

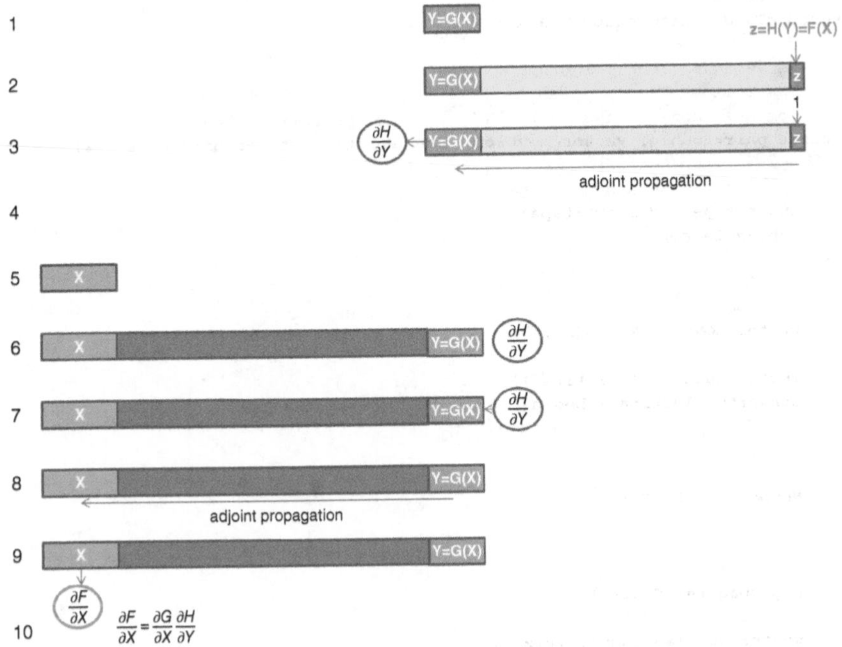

It should be clear that this algorithm guarantees all of the following:

Constant time computation since only adjoint propagations are involved. Note that the Jacobian of G is *never* computed. We don't know it at the term of the computation, and we don't need it to produce the end result. Also note that successive functions are propagated in the reverse order to their evaluation. Check-pointing is a sort of "macro-level" AAD where the nodes are not mathematical operations but steps in an algorithm.

Correct adjoint accumulation since it should be clear that these computations produce the exact same results as a full adjoint propagation throughout the entire tape for F. It is actually the same propagations that are executed, but through pieces of tape at a time.

Reduced RAM consumption since only the one tape for G or H lives in memory at a time.

In code, check-pointing goes as follows:

```
template<class T>
T H(const vector<T>& Y);

template<class T>
vector<T> G(const vector<T>& X);

//  Implements check-pointing
//  Takes input X
//  Computes and returns F(X) = H[G(X)] and its derivatives
inline pair<double, vector<double>> checkPoint(const vector<double>& X)
{
    //  Start with a clean tape
    auto* tape = Number::tape;
    tape->clear();

    //  1
    //  Compute Y
    vector<double> Y = G(X);
    //  Convert to numbers
    vector<Number> iY(Y.size());
    convertCollection(Y.begin(), Y.end(), iY.begin());
    //  Note that also puts iY on tape

    //  2
    Number z = H(iY);

    //  3
    //  Propagate
    z.propagateToStart();
    //  Store derivatives
    vector<double> dhdy(iY.size());
    transform(iY.begin(), iY.end(), dhdy.begin(),
        [](const Number& y) {return y.adjoint(); });

    //  4
    tape->clear();

```

```
    // 5
    vector<Number> iX(X.size());
    convertCollection(X.begin(), X.end(), iX.begin());

    // 6
    vector<Number> oY = G(iX);

    // 7
    for (size_t i = 0; i < oY.size(); ++i)
    {
        oY[i].adjoint() = dhdy[i];
    }

    // 8
    Number::propagateAdjoints(prev(tape->end()), tape->begin());

    // 9
    pair<double, vector<double>> results;
    results.first = z.value();
    results.second.resize(iX.size());
    transform(iX.begin(), iX.end(), results.second.begin(),
        [](const Number& x) {return x.adjoint(); });

    // 10
    tape->clear();

    return results;
}
```

We could write a generic higher-order function to encapsulate check-pointing logic. But we refrain from doing so. It is best left to client code to implement check-pointing at best in different situations. The AAD library provides all the basic constructs to implement check-pointing easily and in a flexible manner.

Note that the algorithm also works in the more general context where:

$$F(X) = H[X, G(X)]$$

because, then:

$$\frac{\partial F}{\partial X} = \frac{\partial H}{\partial X} + \frac{\partial H}{\partial Y}\frac{\partial G}{\partial X}$$

The left-hand side $\frac{\partial H}{\partial X}$ is computed in constant time by differentiation of H (we can do that: it has been our working hypothesis all along). The right-hand side $\frac{\partial H}{\partial Y}\frac{\partial G}{\partial X}$ is computed by check-pointing.

Alternatively, we may redefine G to return the n coordinates of X in addition to its result Y, and we are back to the initial case where $F(X) = H[G(X)]$.

This concludes our general discussion of check-pointing. Check-pointing applies in vast number of contexts, to the point that every nontrivial AAD instrumentation involves some form of check-pointing, including the instrumentation of our simulation library in the previous chapter, as pointed out earlier. In the next section, we apply check-pointing to calibration and the production of market risks. In the meantime, we quickly discuss application to black box code.

Check-pointing black box code

AAD instrumentation cannot be partial. The entire calculation code must be instrumented, and all the active code must be templated. A partial instrumentation would break the chain rule and prevent adjoints to correctly propagate through non-instrumented active calculations, resulting in wrong differentials. It follows that all the source code implementing a calculation must be available, and modifiable, so it may be instantiated with the Number type.

In a real-world production environment, this is not always the case. It often happens that part of the calculation code is a black box. We can call this code to conduct some intermediate calculations, but we cannot easily see or modify the source code. The routine may be part of third-party software with signatures in headers and binary libraries, but no source code. Or, the source code may be written in a different language. Or, the source is available but cannot be modified, for technical, policy, or legal reasons.

Or maybe we could instrument the code but we don't want to. An intensive calculation code with a low number of active inputs may be best differentiated either analytically or with finite differences. Or, as we will see in the case of a numerical calibration, some code must be differentiated in a specific manner, a blind differentiation, either with finite differences or AAD, resulting in wrong or unstable derivatives. This applies to many iterative algorithms, like eigenvalue decomposition, Cholesky decomposition, or SVD regression, as noted by Huge in [93].

In all these cases, we have an intermediate calculation that remains non-instrumented, and differentiated in its own specific way, which may or may not be finite differences.[4] Check-pointing allows to consistently connect this piece in the context of a larger differentiation, the rest of the calculation being differentiated in constant time with AAD.

For example, consider the differentiation of a calculation F that is evaluated in three steps, $G : \mathbb{R}^n \to \mathbb{R}^m$, $BB : \mathbb{R}^m \to \mathbb{R}^p$, and $H : \mathbb{R}^p \to \mathbb{R}$:

$$F(X) = H\{BB[G(X)]\}$$

where BB is the black box. It is not instrumented, and its Jacobian $\partial BB/\partial G$ is computed by specific means, perhaps finite differences. The problem is to

[4] A third-party black box can always be differentiated with finite differences.

conduct the rest of the differentiation in constant time with AAD, and connect the Jacobian of the black box without breaking the derivatives chain. We have discussed a walkaround in Chapter 8 in the context of manual adjoint code. In the context of automatic adjoint differentiation, we have two choices.

We could *overload* BB and make it a building block in the AAD library, like we did for the Gaussian functions in Chapter 10. This solution invades and grows the AAD library. It is recommended when BB is a low-level, general-purpose algorithm, called from many places in the software.

In most situations, however, BB would be a necessary intermediate calculation in a specific context, which doesn't justify an invasion of the AAD library. All we need is a walkaround in the instrumentation of F, along the lines of Chapter 8, but with automatic adjoint propagation. We can implement such walkaround with check-pointing. Denote:

$$Y = G(X) \quad , \quad Z = BB(Y) \quad , \quad z = H(Z)$$

then, by the chain rule:

$$\frac{\partial F}{\partial X} = \frac{\partial H}{\partial Z}\frac{\partial BB}{\partial Y}\frac{\partial G}{\partial X}$$

We start with a non-instrumented calculation of $Y = G(X)$, as is customary with check-pointing. Next, we compute the value $Z = BB(Y)$ of BB, as well as its Jacobian $\partial BB/\partial Y$, computed, as discussed, by specific means.

Knowing Z, we compute the gradient $\partial H/\partial Z$, a row vector in dimension p, of H, in constant time, with AAD instrumentation. We multiply it on the right by $\partial BB/\partial Y$ to find:

$$\frac{\partial F}{\partial Y} = \frac{\partial H}{\partial Z}\frac{\partial BB}{\partial Y}$$

This row vector in dimension m is by definition the adjoint of Y in the calculation of F. We can therefore apply the check-pointing algorithm. Execute an instrumented instance of:

$$Y = G(X)$$

which builds the tape of G, seed the adjoints of the results Y with the known $\partial F/\partial Y$, and back-propagate to find the desired adjoints of X, that is $\partial F/\partial X$.

We successfully applied check-pointing to connect the specific differentiation of BB with the rest of the differentiated calculation. The differentiation of BB takes the times it takes, and a matrix-by-vector product is necessary for the connection, but the rest of the differentiation proceeds with AAD in constant time.

13.2 EXPLICIT CALIBRATION

Dupire's formula

We now turn toward financial applications of the check-pointing algorithm, more precisely, the important matter of the production of market risks.

So far in this book, we implemented Dupire's model with a given local volatility surface $\sigma(S, t)$, represented in practice by a bilinearly interpolated matrix. Its differentiation produced the derivatives of some transaction's value in the model with respect to this local volatility matrix.

But this is not the application Dupire meant for his model. Traders are not interested in risks to a theoretical, abstract local volatility. Dupire's model is meant to be *calibrated* to the market prices of European calls and puts, or, equivalently, market-implied Black and Scholes volatilities, such that its values are consistent with the market prices of European options, and its risk sensitivities are derivatives to *implied* volatilities, which represent the market prices of concrete instruments that traders may buy and sell to hedge the sensitivities of their transactions.

Dupire's model is unique in that its calibration is *explicit*. The calibrated local volatility is expressed directly as a function of the market prices of European calls by Dupire's famous formula [12]:

$$\sigma^2(S, t) = \frac{2C_T(S, t)}{S^2 C_{KK}(S, t)}$$

where $C(K, T)$ is today's price of the European call of strike K and maturity T, and subscripts denote partial derivatives.

Dupire's formula may be elegantly demonstrated in a couple of lines with Laurent Schwartz's generalized derivatives and Tanaka's formula (essentially an extension of Ito's lemma in the sense of distributions), following the footsteps of Savine, 2001 [44].

By application of Tanaka's formula to the function $f(x) = (x - K)^+$ under Dupire's dynamics $\frac{dS_t}{S_t} = \sigma(S_t, t)dW_t$, we find:

$$d(S_t - K)^+ = 1_{\{S_t > K\}} dS_t + \frac{1}{2}\delta(S_t - K)S_t^2\sigma^2(S_t, t)dt$$

where δ is the Dirac mass. Taking (risk-neutral) expectations on both sides:

$$dE(S_t - K)^+ = 0 + \frac{1}{2}\varphi_t(K)K^2\sigma^2(K, t)dt$$

where $\varphi_t(K) = C_{KK}(K,t)$ is the (risk-neutral) density of S_t in K, and since $E(S_t - K)^+ = C(K,t)$, we have Dupire's result:

$$\frac{dC(K,t)}{dt} = \frac{1}{2}\frac{d^2C(K,t)}{dK^2}K^2\sigma^2(K,t)$$

Similar formulas are found with this methodology in extensions of Dupire's model with rates, dividends, stochastic volatility, and jumps; see [44].

The Implied Volatility Surface (IVS)

Dupire's formula refers to today's prices of European calls of all strikes K and maturities T, or, equivalently, the continuous surface of Black and Scholes's market-implied volatilities $\hat{\sigma}(K,T)$. In Chapter 4, we pointed out that this is also equivalent to marginal risk-neutral densities for all maturities, and called this continuous surface of market prices an *Implied Volatility Surface* or IVS.

The IVS must satisfy some fundamental properties to feed Dupire's formula: it must be continuous, differentiable in T, and twice differentiable in K. C_{KK} must be strictly positive, meaning call prices must be convex in strike. We also require that $C_T > 0$, meaning call prices are increasing in maturity. $C_{KK} < 0$ or $C_T < 0$ would allow a static arbitrage (see for instance [46]) so any non-arbitrageable IVS guarantees that $C_{KK} \geq 0, C_T \geq 0$. But this is not enough. We need strict positiveness as well as continuity and differentiability.

We pointed out in Chapter 4 that the market typically provides prices for a discrete number of European options, and that to interpolate a complete, continuous, differentiable IVS out of these prices was not a trivial exercise. The implementation of the accepted solutions in the industry, including Gatheral's SVI ([40]) and Andreasen and Huge's LVI ([41]), are out of our scope here.

We circumvent this difficulty by *defining* an IVS from Merton's jump-diffusion model of 1976 [98]. Merton's model is an extension of Black and Scholes where the underlying asset price is not only subject to a diffusion, but also random discontinuities, or jumps, occurring at random times and driven by a Poisson process:

$$\frac{dS_t}{S_t} = \sigma dW + J_t dN_t - comp \cdot dt$$

where N is a Poisson process with intensity λ and the Js are a collection IID random variables such that $\log(1 + J_t) \approx J_t \to N(m, v)$. The Poisson process and the jumps are independent from each other and independent from the Brownian motion. Jumps are roughly Gaussian with mean m and variance v. $comp = \lambda \left[\exp\left(m + \frac{v}{2}\right) - 1\right] \approx \lambda m$ guarantees that S satisfies the martingale property so the model remains non-arbitrageable.

Merton demonstrated that the price of a European call in this model can be expressed explicitly as a weighted average of Black and Scholes prices:

$$C(K, T) = \sum_{n=0}^{\infty} \frac{\exp(-\lambda T)}{n!} (\lambda T)^n BS$$
$$\times \left(S_0 \exp\left[n\left(m + \frac{v}{2}\right) - comp \cdot T \right], \sqrt{\sigma^2 + \frac{nv}{T}}, K, T \right)$$

where $BS(S, \hat{\sigma}, K, T)$ is Black and Scholes's formula. The model is purposely written so the distribution of S_T, *conditional to the number n of jumps*, is log-normal with known mean and variance. The conditional expectation of the payoff is therefore given by Black and Scholes's formula, with a different forward and variance depending on the number of jumps. It follows that the price, the *unconditional* expectation, is the average of the conditional expectations, weighted by the distribution of the number of jumps. The distribution of a Poisson process is well known, and Merton's formula follows.

The term in the infinite sum dies quickly with the factorial, so it is safe, in practice, to limit the sum to its first 5–10 terms. The formula is implemented in analytics.h, along with the Black and Scholes's formula.

We are using a continuous-time, arbitrage-free model to define the IVS; therefore, the properties necessary to feed Dupire's formula are guaranteed. In addition, Merton's model is known to produce realistic IVS with a shape similar to major equity derivatives markets.

We are using a fictitious Merton market in place of the "real" market so as to get around some technical difficulties unrelated to the purpose of this document. This is evidently for illustration purposes only and not for production.

We declare the IVS as a polymorphic class that provides a Black and Scholes market-implied volatility for all strikes and maturities. The implementation is simplified in that it ignores rates or dividends. The following code is found in ivs.h:

```
class IVS
{
    //  To avoid reference to a linear market
    double mySpot;
```

```
public:

    IVS(const double spot) : mySpot(spot) {}

    //  Read access to spot
    double spot() const
    {
        return mySpot;
    }

    //  Raw implied vol
    virtual double impliedVol(const double strike, const Time mat)
                const = 0;

    //  ...
```

where the concrete IVS derives *impliedVol*() to provide a volatility surface. The IVS also provides a method for the pricing of European calls:

```
    //  ...

    //  Call price
    template<class T = double>
    T call(
        const double strike,
        const Time mat) const
    {
        //  blackScholes is defined in analytics.h, templated
        return blackScholes<T>(
            mySpot,
            strike,
            impliedVol(strike, mat),
            mat);
    }

    //  ...
```

where the function *blackScholes* is implemented in analytics.h, templated. By application of Dupire's formula, the IVS also provides the local volatility for a given spot and time:

```
    //  ...

    //  Local vol, dupire's formula
    template<class T = double>
    T localVol(
        const double strike,
        const double mat) const
    {
        //  Derivative to time
        const T c00 = call(strike, mat, risk);
        const T c01 = call(strike, mat - 1.0e-04, risk);
        const T c02 = call(strike, mat + 1.0e-04, risk);
        const T ct = (c02 - c01) * 0.5e04;
```

```
        //  Second derivative to strike = density
        const T c10 = call(strike - 1.0e-04, mat, risk);
        const T c20 = call(strike + 1.0e-04, mat, risk);
        const T ckk = (c10 + c20 - 2.0 * c00) * 1.0e08;

        //  Dupire's formula
        return sqrt(2.0 * ct / ckk) / strike;
    }

    //  Virtual destructor needed for polymorphic class
    virtual ~IVS() {}
};
```

As discussed, we define a concrete IVS from Merton's model. All a concrete IVS must do is derive *impliedVol*():

```
class MertonIVS : public IVS
{
    double myVol;
    double myIntensity, myAverageJmp, myJmpStd;

public:

    MertonIVS(const double spot, const double vol,
        const double intens, const double aveJmp, const double stdJmp)
        : IVS(spot),
        myVol(vol),
        myIntensity(intens),
        myAverageJmp(aveJmp),
        myJmpStd(stdJmp)
    {}

    double impliedVol(const double strike, const Time mat) const override
    {
        //  Merton's formula is defined in analytics.h
        const double call
            = merton(
                spot(),
                strike,
                myVol,
                mat,
                myIntensity,
                myAverageJmp,
                myJmpStd);

        //  Implied volatility from price, also in analytics.h
        return blackScholesIvol(spot(), strike, call, mat);
    }
};
```

where *merton*() is an implementation of Merton's formula, and *blackScholesIvol*() implements a numerical procedure to find an implied volatility from an option price. Both are implemented in analytics.h.

We implemented a generic framework for IVS. Although we only implemented one concrete IVS, and a particularly simple one that defines the market from Merton's model, we could implement any other concrete IVS, including:

- Hagan's SABR [36] with parameters interpolated in maturity and underlying, as is market practice for interest rate options,
- Heston's stochastic volatility model [42] with parameters interpolated in maturity, as is market practice for foreign exchange options.[5]
- Gatheral's SVI implied volatility interpolation [40], as is market standard for equity derivatives, or
- Andreasen and Huge's recent award-winning LVI [41] argitrage-free interpolation.

Any concrete IVS implementation must only override the *impliedVol*() method to provide a Black and Scholes implied volatility for any strike and maturity. The rest, in particular the computation of Dupire's local volatility, is on the base IVS.

Calibration of Dupire's model

It is easy to calibrate Dupire's model to an IVS; all it takes is an implementation of Dupire's formula. The formula guarantees that the resulting local volatility surface in Dupire's model matches the option prices in the IVS. We write a free calibration function in mcMdlDupire.h. It accepts a target IVS, a grid of spots and times, and returns a local volatility matrix, calibrated sequentially in time:

```
#include "ivs.h"

#define ONE_HOUR 0.000114469

//  Returns a struct with spots, times and lVols
template<class T = double>
inline auto dupireCalib(
        //  The IVS we calibrate to
        const IVS& ivs,
        //  The local vol grid
        //  The spots to include
        const vector<double>& inclSpots,
        //  Maximum space between spots
```

[5]Some companies still use an approximation based on second-order sensitivities to volatility, a so-called "vega-volga-vanna" interpolation that vaguely mimics a stochastic volatility model, although it is hard to see any advantage over a correct implementation of Heston.

```
        const double maxDs,
        //  The times to include, note NOT 0
        const vector<Time>& inclTimes,
        //  Maximum space between times
        const double maxDt)
{
    //  Results
    struct
    {
        vector<double> spots;
        vector<Time> times;
        matrix<T> lVols;
    } results;

    //  Spots and times
    results.spots = fillData(inclSpots, maxDs, 0.01); //  min space = 0.01
    results.times = fillData(inclTimes, maxDt,
        ONE_HOUR,                   //  min space = 1 hour
        &maxDt, &maxDt + 1          //  hack to include maxDt as first time
    );

    //  Allocate local vols, transposed maturity first
    matrix<T> lVolsT(results.times.size(), results.spots.size());

    //  Maturity by maturity
    const size_t n = results.times.size();
    for (size_t j = 0; j < n; ++j)
    {
        dupireCalibMaturity(
            ivs,
            results.times[j],
            results.spots.begin(),
            results.spots.end(),
            lVolsT[j]);
    }

    //  transpose is defined in matrix.h
    results.lVols = transpose(lVolsT);

    return results;
}
```

It is convenient to conduct the calibration sequentially in time, although our Dupire stores local volatility in spot major. For this reason, we calibrate a temporary volatility matrix in time major, and return its transpose (defined in matrix.h). We calibrate each time slice independently with the free function dupireCalibMaturity() defined in mcMdlDupire.h:

```
//  Calibrates one maturity
//  Main calibration function below
template <class IT, class OT, class T = double>
inline void dupireCalibMaturity(
    //  IVS we calibrate to
    const IVS& ivs,
    //  Maturity to calibrate
    const Time maturity,
```

```
    //  Spots for local vol
    IT spotsBegin,
    IT spotsEnd,
    //  Results, by spot
    //  With (random access) iterator, STL style
    OT  lVolsBegin)
{
    //  Number of spots
    IT spots = spotsBegin;
    const size_t nSpots = distance(spotsBegin, spotsEnd);

    //  Estimate ATM so we cut the grid 2 stdevs away to avoid instabilities
    const double atmCall = double(ivs.call(ivs.spot(), maturity));
    //  Standard deviation, approx. atm call * sqrt(2pi)
    const double std = atmCall * 2.506628274631;

    //  Skip spots below and above 2.5 std
    int il = 0;
    while (il < nSpots && spots[il] < ivs.spot() - 2.5 * std) ++il;
    int ih = nSpots - 1;
    while (ih >= 0 && spots[ih] > ivs.spot() + 2.5 * std) --ih;

    //  Loop on spots
    for (int i = il; i <= ih; ++i)
    {
        //  Dupire's formula
        lVolsBegin[i] = ivs.localVol(spots[i], maturity);
    }

    //  Extrapolate flat outside std
    for (int i = 0; i < il; ++i)
        lVolsBegin[i] = lVolsBegin[il];
    for (int i = ih + 1; i < nSpots; ++i)
        lVolsBegin[i] = lVolsBegin[ih];
}
```

Finally, we have the following higher-level function in main.h for our application:

```
//  Returns spots, times and lVols in a struct
inline auto
dupireCalib(
    //  The local vol grid
    //  The spots to include
    const vector<double>& inclSpots,
    //  Maximum space between spots
    const double maxDs,
    //  The times to include, note NOT 0
    const vector<Time>& inclTimes,
    //  Maximum space between times
    const double maxDt,
    //  The IVS we calibrate to
    //  'B'achelier, Black'S'choles or 'M'erton
    const double spot,
    const double vol,
    const double jmpIntens = 0.0,
    const double jmpAverage = 0.0,
```

```
    const double jmpStd = 0.0)
{
    // Create IVS
    MertonIVS ivs(spot, vol, jmpIntens, jmpAverage, jmpStd);

    // Go
    return dupireCalib(ivs, inclSpots, maxDs, inclTimes, maxDt);
}
```

It takes around 50 milliseconds to calibrate a local volatility grid of 30 spots between 50 and 200 and 60 times between now and 5 years, to a Merton IVS with spot 100, volatility 15, jump intensity 5, mean −15 and standard deviation 10. With 150 spots and 260 times, it takes 400 ms. Calibration is embarrassingly parallel and trivially multi-threadable across maturities. This is left as an exercise.

We can easily test the quality of the calibration. Initialize Dupire's model with the result of the calibration. Price a set of European options of different strikes and maturities (developed as a single product with multiple payoffs on page 238) by simulation in this model, and compare with Merton's price as implemented in the *merton*() function in analytics.h in closed-form. In our tests, Dupire and Merton prices match within a couple of basis points over a wide range of strikes and maturities (with 500,000 paths, weekly time steps, where a parallel Sobol pricing of 20 European calls with maturities up to three years takes 400 milliseconds).

Risk views

The process calibration + simulation produces the value of a transaction *out of market-implied volatilities*. Its differentials are sensitivities to market-traded variables, more relevant for trading and hedging than sensitivities to model parameters:

$$\hat{\sigma}(K,T) \xrightarrow[G]{\text{calibration}} \sigma(S,t) \xrightarrow[H]{\text{simulation}} V_0$$

Model parameters are obtained from market variables with a prior calibration step. Model sensitivities are obtained with AAD as explained and developed in Chapter 12. We can therefore obtain the market sensitivities by *check-pointing the model sensitivities into calibration.*

We developed, in the previous chapter, functionality to obtain the *micro-bucket* $\frac{\partial V_0}{\partial \sigma(S,t)}$ in constant time. We check-point this result into calibration to obtain $\frac{\partial V_0}{\partial \hat{\sigma}(K,T)}$, what Dupire calls a *superbucket*.

We are missing one piece of functionality: our IVS $\hat{\sigma}(K,T)$ is defined in derived IVS classes, from a set of parameters, which nature depends on the concrete IVS. For instance, the Merton IVS is parameterized with a continuous volatility, jump intensity, and the mean and standard deviation of jumps. The desired derivatives are not to the parameters of the concrete IVS, but to a discrete set of implied Black and Scholes market-implied volatilities, *irrespective* of how these volatilities are produced or interpolated.

To achieve this result, we are going to use a neat technique that professional financial system developers typically apply in this situation: we are going to define a *risk surface*:

$$s(K, T)$$

such that if we denote $\hat{\sigma}(K, T)$ the implied volatilities given by the concrete IVS, our calculations will not use these original implied volatilities, but implied volatilities *shifted by the risk surface*:

$$\sum(K, T) = \hat{\sigma}(K, T) + s(K, T)$$

Further, we interpolate the risk surface $s(K, T)$ from a discrete set of knots:

$$s_{ij} = s(K_i, T_j)$$

that we call the *risk view*. All the knots are set to 0, so:

$$\sum(K, T) = \hat{\sigma}(K, T)$$

so the results of all calculations remain evidently unchanged by shifting implied volatilities by zero, but in terms of risk, we get:

$$\frac{\partial}{\sigma(K, T)} = \frac{\partial}{\partial s(K, T)}$$

The risk view does not affect the value, and its derivatives exactly correspond to derivatives to implied volatilities, irrespective of how these implied volatilities are computed.

We compute sensitivities to implied volatilities as sensitivities to the risk view:

$$\frac{\partial V_0}{\partial s_{ij}}$$

Risk views apply to bumping as well as AAD and are extremely useful, in many contexts, to aggregate risks over selected market instruments.

In the context of Dupire's model, we apply a risk view over an IVS fed to Dupire's formula. Dupire's formula depends on the first- and second-order derivatives of call prices, so the risk view must be differentiable. (Bi-)linear interpolation is not an option. We must implement a smooth interpolation.

A vast amount of smooth interpolations exist in literature, but what we need is a *localized* one, otherwise the resulting risk spills over the volatility surface. For these reasons, we implement a well-known, simple, localized and efficient smooth interpolation algorithm called *smoothstep*, presented in many places, including Wikipedia's "Smoothstep" article. Like linear interpolation, smoothstep interpolation finds x_i such that $x_i < x_0 \leq x_{i+1}$, and, unlike linear interpolation, which returns:

$$y_0 = y_i + (y_{i+1} - y_i)t$$

where $t = (x_0 - x_i)/(x_{i+1} - x_i)$, smoothstep returns:

$$y_0 = y_i + (y_{i+1} - y_i)t^2(3 - 2t)$$

Practically, we upgrade the *interp*() function of Chapter 6 to implement either linear or smoothstep interpolation. We also produce a two-dimensional variant:

```
//  interp.h

#include <algorithm>
using namespace std;

//  Utility for interpolation
//  Interpolates the vector y against knots x in value x0
//  Interpolation is linear or smooth, extrapolation is flat
template <bool smoothStep=false, class ITX, class ITY, class T>
inline auto interp(
    //  sorted on xs
    ITX                                 xBegin,
    ITX                                 xEnd,
    //  corresponding ys
    ITY                                 yBegin,
    ITY                                 yEnd,
    //  interpolate for point x0
    const T&                            x0)
    ->remove_reference_t<decltype(*yBegin)>
{
    //  STL binary search, returns iterator on 1st no less than x0
    //  upper_bound guarantees logarithmic search
    auto it = upper_bound(xBegin, xEnd, x0);

    //  Extrapolation?
    if (it == xEnd) return *(yEnd - 1);
    if (it == xBegin) return *yBegin;

    //  Interpolation
    size_t n = distance(xBegin, it) - 1;
    auto x1 = xBegin[n];
    auto y1 = yBegin[n];
    auto x2 = xBegin[n + 1];
    auto y2 = yBegin[n + 1];
```

```
    auto t = (x0 - x1) / (x2 - x1);

    //  Note constexpr if
    if constexpr (smoothStep)
    {
        //  smoothstep
        return y1 + (y2 - y1) * t * t * (3.0 - 2 * t);
    }

    else
    {
        //  linear
        return y1 + (y2 - y1) * t;
    }
}

//  2D
template <bool smoothStep=false, class T, class U, class V, class W, class X>
inline V interp2D(
    //  sorted on xs
    const vector<T>&            x,
    //  sorted on ys
    const vector<U>&            y,
    //  zs in a matrix
    const matrix<V>&            z,
    //  interpolate for point (x0,y0)
    const W&                    x0,
    const X&                    y0)
{
    const size_t n = x.size();
    const size_t m = y.size();

    //  STL binary search, returns iterator on 1st no less than x0
    //  upper_boung guarantees logarithmic search
    auto it = upper_bound(x.begin(), x.end(), x0);
    const size_t n2 = distance(x.begin(), it);

    //  Extrapolation in x?
    if (n2 == n)
        return interp<smoothStep>(
            y.begin(),
            y.end(),
            z[n2 - 1],
            z[n2 - 1] + m,
            y0);
    if (n2 == 0)
        return interp<smoothStep>(
            y.begin(),
            y.end(),
            z[0],
            z[0] + m,
            y0);

    //  Interpolation in x
    const size_t n1 = n2 - 1;
    auto x1 = x[n1];
```

```
    auto x2 = x[n2];
    auto z1 = interp<smoothStep>(
                y.begin(),
                y.end(),
                z[n1],
                z[n1] + m,
                y0);
    auto z2 = interp<smoothStep>(
                y.begin(),
                y.end(),
                z[n2],
                z[n2] + m,
                y0);

    auto t = (x0 - x1) / (x2 - x1);
    if constexpr (smoothStep)
    {
        // Smooth step
        return z1 + (z2 - z1) * t * t * (3.0 - 2 * t);
    }
    else
    {
        // linear
        return z1 + (z2 - z1) * t;;
    }
}
```

Armed with smooth interpolation, we can define the RiskView object in ivs.h. Note that (contrarily to the IVS), the risk view is templated since we will be computing derivatives to its knots, and we want to do that with AAD:

```
// ivs.h

#include "interp.h"

// Risk view
template <class T>
class RiskView
{
    bool              myEmpty;

    vector<double>    myStrikes;
    vector<Time>      myMats;
    matrix<T>         mySpreads;

public:

    // Default constructor, empty view
    RiskView() : myEmpty(true) {}

    // Intializes risk view AND put on tape
    // Sets all spreads to 0
    RiskView(const vector<double>& strikes, const vector<Time>& mats) :
        myEmpty(false),
        myStrikes(strikes),
        myMats(mats),
```

```
        mySpreads(strikes.size(), mats.size())
    {
        for (auto& spr : mySpreads) spr = T(0.0);
    }

    //  Get spread
    T spread(const double strike, const Time mat) const
    {
        return myEmpty
            ? T(0.0)
            : interp2D<true>(myStrikes, myMats, mySpreads, strike, mat);
    }

    //  Accessors by const ref
    bool empty() const { return myEmpty; }
    size_t rows() const { return myStrikes.size(); }
    size_t cols() const { return myMats.size(); }
    const vector<double>& strikes() const { return myStrikes; }
    const vector<Time>& mats() const { return myMats; }
    const matrix<T>& risks() const { return mySpreads; }

    //  Iterators
    typedef typename matrix<T>::iterator iterator;
    typedef typename matrix<T>::const_iterator const_iterator;
    iterator begin() { return mySpreads.begin(); }
    iterator end() { return mySpreads.end(); }
    const_iterator begin() const { return mySpreads.begin(); }
    const_iterator end() const { return mySpreads.end(); }

    //  For bump risk
    void bump(const size_t i, const size_t j, const double bumpBy)
    {
        mySpreads[i][j] += bumpBy;
    }
};
```

This code should be self-explanatory; the risk view is nothing more than a two-dimensional interpolation object with convenient accessors and iterators and knots set to zero. The method *bump*() modifies one knot by a small amount *bumpBy*, from zero to *bumpBy*, so we can apply bump risk.

The next step is to effectively incorporate the risk view in the calculation. We extend the methods *call*() and *localVol*() on the base IVS so they may be called with a risk view.

```
//  IVS base class
//  ...

    //  Call price
    template<class T = double>
    T call(
        const double strike,
        const Time mat,
        const RiskView<T>* risk = nullptr) const
    {
        //  blackScholes is defined in analytics.h, templated
```

```
        return blackScholes<T>(
            mySpot,
            strike,
            impliedVol(strike, mat)
                + (risk ? risk->spread(strike, mat) : T(0.0)),
            mat);
    }

    // Local vol, dupire's formula
    template<class T = double>
    T localVol(
        const double strike,
        const double mat,
        const RiskView<T>* risk = nullptr) const
    {
        // Derivative to time
        const T c00 = call(strike, mat, risk);
        const T c01 = call(strike, mat - 1.0e-04, risk);
        const T c02 = call(strike, mat + 1.0e-04, risk);
        const T ct = (c02 - c01) * 0.5e04;

        // Second derivative to strike = density
        const T c10 = call(strike - 1.0e-04, mat, risk);
        const T c20 = call(strike + 1.0e-04, mat, risk);
        const T ckk = (c10 + c20 - 2.0 * c00) * 1.0e08;

        // Dupire's formula
        return sqrt(2.0 * ct / ckk) / strike;
    }

// ...
```

The modification is minor. A shift, interpolated from the risk view, is added to the implied volatility for the computation of call prices, hence local volatilities. Since the risk view is set to zero, this doesn't modify results, but it does produce risk with respect to the risk view's knots.

Finally, we apply the same minor modification to the calibration functions in McMdlDupire.h so they accept an optional risk view:

```
// mcMdlDupire.h

// Calibrates one maturity
// Main calibration function below
template <class IT, class OT, class T = double>
inline void dupireCalibMaturity(
    // IVS we calibrate to
    const IVS& ivs,
    // Maturity to calibrate
    const Time maturity,
    // Spots for local vol
    IT spotsBegin,
    IT spotsEnd,
    // Results, by spot
    // With (random access) iterator, STL style
    OT lVolsBegin,
```

```
    //  Risk view
    const RiskView<T>& riskView = RiskView<double>())
{
    //  Number of spots
    IT spots = spotsBegin;
    const size_t nSpots = distance(spotsBegin, spotsEnd);

    //  Estimate ATM so we cut the grid 2 stdevs away to avoid instabilities
    const double atmCall = double(ivs.call(ivs.spot(), maturity));
    //  Standard deviation, approx. atm call * sqrt(2pi)
    const double std = atmCall * 2.506628274631;

    //  Skip spots below and above 2.5 std
    int il = 0;
    while (il < nSpots && spots[il] < ivs.spot() - 2.5 * std) ++il;
    int ih = nSpots - 1;
    while (ih >= 0 && spots[ih] > ivs.spot() + 2.5 * std) --ih;

    //  Loop on spots
    for (int i = il; i <= ih; ++i)
    {
        //  Dupire's formula
        lVolsBegin[i] = ivs.localVol(spots[i], maturity, &riskView);
    }

    //  Extrapolate flat outside std
    for (int i = 0; i < il; ++i)
        lVolsBegin[i] = lVolsBegin[il];
    for (int i = ih + 1; i < nSpots; ++i)
        lVolsBegin[i] = lVolsBegin[ih];
}

#define ONE_HOUR 0.000114469

//  Returns a struct with spots, times and lVols
template<class T = double>
inline auto dupireCalib(
        //  The IVS we calibrate to
        const IVS& ivs,
        //  The local vol grid
        //  The spots to include
        const vector<double>& inclSpots,
        //  Maximum space between spots
        const double maxDs,
        //  The times to include, note NOT 0
        const vector<Time>& inclTimes,
        //  Maximum space between times
        const double maxDt,
        //  Risk view if required
        //  omitted: T = double , no risk view
        const RiskView<T>& riskView = RiskView<double>())
{
    //  Results
    struct
    {
        vector<double> spots;
        vector<Time> times;
```

```
        matrix<T> lVols;
    } results;

    //  Spots and times
    results.spots = fillData(inclSpots, maxDs, 0.01); //  min space = 0.01
    results.times = fillData(inclTimes, maxDt,
        ONE_HOUR,                   //  min space = 1 hour
        &maxDt, &maxDt + 1          //  dirty trick to include maxDt
    );

    //  Allocate local vols, transposed maturity first
    matrix<T> lVolsT(results.times.size(), results.spots.size());

    //  Maturity by maturity
    const size_t n = results.times.size();
    for (size_t j = 0; j < n; ++j)
    {
        dupireCalibMaturity(
            ivs,
            results.times[j],
            results.spots.begin(),
            results.spots.end(),
            lVolsT[j],
            riskView);
    }

    //  transpose is defined in matrix.h
    results.lVols = transpose(lVolsT);

    return results;
}
```

Superbuckets

We now have all the pieces to compute superbuckets by check-pointing. We build a higher-level function in main.h that executes the steps of our check-pointing algorithm:

```
struct SuperbucketResults
{
    double value;
    double delta;
    vector<double> strikes;
    vector<Time> mats;
    matrix<double> vega;
};

//  Returns value, delta, strikes, maturities
//      and vega = derivatives to implied vols = superbucket
inline auto
    dupireSuperbucket(
    //  Model parameters that are not calibrated
    const double            spot,
    const double            maxDt,
```

```
    //  Product
    const string&            productId,
    const map<string, double>&   notionals,
    //  The local vol grid
    //  The spots to include
    const vector<double>&    inclSpots,
    //  Maximum space between spots
    const double             maxDs,
    //  The times to include, note NOT 0
    const vector<Time>&      inclTimes,
    //  Maximum space between times
    const double             maxDtVol,
    //  The IVS we calibrate to
    //  Risk view
    const vector<double>&    strikes,
    const vector<Time>&      mats,
    //  Merton params
    const double             vol,
    const double             jmpIntens,
    const double             jmpAverage,
    const double             jmpStd,
    //  Numerical parameters
    const NumericalParam&    num)
{
    //  Results
    SuperbucketResults results;

    //  ...
```

The first check-pointing step is compute local volatilities by calibration (after initialization of the tape) and store the calibrated model in memory.

```
    //  ...

    //  Start with a clean tape
    auto* tape = Number::tape;
    tape->rewind();

    //  Calibrate the model
    auto params = dupireCalib(
        inclSpots,
        maxDs,
        inclTimes,
        maxDtVol,
        spot,
        vol,
        jmpIntens,
        jmpAverage,
        jmpStd);
    const vector<double>& spots = params.spots;
    const vector<Time>& times = params.times;
    const matrix<double>& lvols = params.lVols;

    //  Put in memory
    putDupire(spot, spots, times, lvols, maxDt, "superbucket");

    //  ...
```

Next, we compute the microbucket:

```
    // ...

    // Find delta and microbucket
    auto mdlDerivs = dupireAADRisk(
        "superbucket",
        productId,
        notionals,
        num);
    results.value = mdlDerivs.value;
    results.delta = mdlDerivs.delta;
    const matrix<double>& microbucket = mdlDerivs.vega;

    // ...
```

where *dupireAADRisk*() is essentially a wrapper around *AADrisk-Aggregate*() from the previous chapter, with an interface specific to Dupire that returns a delta and a microbucket matrix:

```
// Returns a struct with price, delta and vega matrix
inline auto dupireAADRisk(
    // model id
    const string&           modelId,
    // product id
    const string&           productId,
    const map<string, double>&   notionals,
    // numerical parameters
    const NumericalParam&   num)
{
    // Check that the model is a Dupire
    const Model<Number>* model = getModel<Number>(modelId);
    if (!model)
    {
        throw runtime_error("dupireAADRisk() : Model not found");
    }
    const Dupire<Number>* dupire
        = dynamic_cast<const Dupire<Number>*>(model);
    if (!dupire)
    {
        throw runtime_error("dupireAADRisk() : Model not a Dupire");
    }

    // Results
    struct
    {
        double value;
        double delta;
        matrix<double> vega;
    } results;

    // Go
    auto simulResults = AADriskAggregate(
                            modelId,
                            productId,
```

```
                        notionals,
                        num);

    //  Find results

    //  Value
    results.value = simulResults.riskPayoffValue;

    //  Delta

    results.delta = simulResults.risks[0];

    //  Vegas

    results.vega.resize(dupire->spots().size(), dupire->times().size());
    copy(
        next(simulResults.risks.begin()),
        simulResults.risks.end(),
        results.vega.begin());

    return results;
}
```

The next part is the crucial one for the check-pointing algorithm: we clean the tape, build the risk view (which puts it on tape), and conduct calibration again, this time in instrumented mode:

```
    //  dupireSuperbucket()
    //  ...

    //  Clear tape
    //  tape->rewind();
    tape->clear();

    //  Convert market inputs to numbers, put on tape

    //  Create IVS
    MertonIVS ivs(spot, vol, jmpIntens, jmpAverage, jmpStd);

    //  Risk view --> that is the AAD input
    //  Note: that puts the view on tape
    RiskView<Number> riskView(strikes, mats);

    //  Calibrate again, in AAD mode, make tape
    auto nParams = dupireCalib(
        ivs,
        inclSpots,
        maxDs,
        inclTimes,
        maxDtVol,
        riskView);
    matrix<Number>& nLvols = nParams.lVols;

    //  ...
```

We seed the calibration tape with the microbucket obtained earlier:

```
//  ...

//  Seed local vol adjoints on tape with microbucket results
for (size_t i = 0; i < microbucket.rows(); ++i)
{
    for (size_t j = 0; j < microbucket.cols(); ++j)
    {
        nLvols[i][j].adjoint() = microbucket[i][j];
    }
}

//  ...
```

and propagate back to the risk view:

```
//  ...

//  Propagate
Number::propagateAdjoints(prev(tape->end()), tape->begin());

//  ...
```

This completes the computation. We pick the desired derivatives as the adjoints of the knots in the risk view, clean the tape, and return the results:

```
    //  ...

    //  Results: superbucket = risk view

    //  Copy results
    results.strikes = strikes;
    results.mats = mats;
    results.vega.resize(riskView.rows(), riskView.cols());
    transform(riskView.begin(), riskView.end(), results.vega.begin(),
        [](const Number& n)
    {
        return n.adjoint();
    });

    //  Clear tape
    tape->clear();

    //  Return results
    return results;
}
```

We illustrated two powerful and general techniques for the production of financial risk sensitivities: the risk view, which allows to aggregate risks over instruments selected by the user, irrespective of calibration or the definition of the market; and check-pointing, which separates the differentiation of simulations from the differentiation of calibration, so that the differentiation of simulations may be performed with path-wise AAD, with limited RAM consumption, in parallel and in constant time.

Finally, note that we left delta unchanged through calibration. We are returning the Dupire delta, the sensitivity to the spot with *local* volatility unchanged. With the superbucket information, we could easily adjust delta for any smile dynamics assumption (sticky, sliding, . . .). There is no strong consensus within the trading and research community as to what the "correct" delta is, but most convincing research points to the Dupire delta as the correct delta within a wide range of models; see, for instance, [99].

Finite difference superbucket risk

As a reference, and in order to test the results of the check-pointing algorithm, we implement a superbucket bump risk, where we differentiate, with finite differences, the whole process calibration + simulation in a trivial manner in main.h:

```
//  Returns value, delta, strikes, maturities
//      and vega = derivatives to implied vols = superbucket
inline auto
    dupireSuperbucketBump(
        //  Model parameters that are not calibrated
        const double            spot,
        const double            maxDt,
        //  Product
        const string&           productId,
        const map<string, double>&   notionals,
        //  The local vol grid
        //  The spots to include
        const vector<double>&   inclSpots,
        //  Maximum space between spots
        const double            maxDs,
        //  The times to include, note NOT 0
        const vector<Time>&     inclTimes,
        //  Maximum space between times
        const double            maxDtVol,
        //  The IVS we calibrate to
        //  Risk view
        const vector<double>&   strikes,
        const vector<Time>&     mats,
        //  Merton params
        const double            vol,
        const double            jmpIntens,
        const double            jmpAverage,
        const double            jmpStd,
        //  Numerical parameters
        const NumericalParam&   num)
{
    //  Results
    SuperbucketResults results;

    //  Calibrate the model
    auto params = dupireCalib(
        inclSpots,
        maxDs,
        inclTimes,
```

```
        maxDtVol,
        spot,
        vol,
        jmpIntens,
        jmpAverage,
        jmpStd);
    const vector<double>& spots = params.spots;
    const vector<Time>& times = params.times;
    const matrix<double>& lvols = params.lVols;

    //  Create model
    Dupire<double> model(spot, spots, times, lvols, maxDt);

    //  Get product
    const Product<double>* product = getProduct<double>(productId);

    //  Base price
    auto baseVals = value(model, *product, num);

    //  Vector of notionals
    const vector<string>& allPayoffs = baseVals.identifiers;
    vector<double> vnots(allPayoffs.size(), 0.0);
    for (const auto& notional : notionals)
    {
        auto it = find(
                    allPayoffs.begin(),
                    allPayoffs.end(),
                    notional.first);
        if (it == allPayoffs.end())
        {
            throw runtime_error(
                "dupireSuperbucketBump() : payoff not found");
        }
        vnots[distance(allPayoffs.begin(), it)] = notional.second;
    }

    //  Base book value
    results.value = inner_product(
                        vnots.begin(),
                        vnots.end(),
                        baseVals.values.begin(),
                        0.0);

    //  Create IVS
    MertonIVS ivs(spot, vol, jmpIntens, jmpAverage, jmpStd);

    //  Create risk view
    RiskView<double> riskView(strikes, mats);

    //  Bumps

    //  bump, recalibrate, reset model, reprice, pick value, unbump

    //  Delta

    //  Recreate model
    Dupire<double> bumpedModel(
                        spot + 1.0e-08,
                        spots,
```

```
                        times,
                        lvols,
                        maxDt);
    //  Reprice
    auto bumpedVals = value(bumpedModel, *product, num);
    //  Pick results and differentiate
    results.delta = (
        inner_product(
            vnots.begin(),
            vnots.end(),
            bumpedVals.values.begin(),
            0.0)
        - results.value) * 1.0e+08;

    // Vega

    const size_t n = riskView.rows(), m = riskView.cols();
    results.vega.resize(n, m);
    for (size_t i = 0; i < n; ++i) for (size_t j = 0; j < m; ++j)
    {
        //  Bump
        riskView.bump(i, j, 1.0e-05);
        //  Recalibrate
        auto bumpedCalib = dupireCalib(
                            ivs,
                            inclSpots,
                            maxDs,
                            inclTimes,
                            maxDtVol,
                            riskView);
        //  Recreate model
        Dupire<double> bumpedModel(
                            spot,
                            bumpedCalib.spots,
                            bumpedCalib.times,
                            bumpedCalib.lVols,
                            maxDt);
        //  Reprice
        auto bumpedVals = value(bumpedModel, *product, num);
        //  Pick results and differentiate
        results.vega[i][j] = (
            inner_product(
                    vnots.begin(),
                    vnots.end(),
                    bumpedVals.values.begin(),
                    0.0)
            - results.value) * 1.0e+05;
        //  Unbump
        riskView.bump(i, j, -1.0e-05);
    }

    //  Copy results and strikes
    results.strikes = strikes;
    results.mats = mats;

    //  Return results
    return results;
}
```

Results

We start with the superbucket of a European call of maturity 3 years, strike 120, over a risk view with 14 knot strikes, every 10 points between 50 and 180, and 5 maturities every year between 1y and 5y.

We define the European option market as a Merton market with volatility 15, jump intensity 5, mean jump −15 and jump standard deviation 10. We calibrate a local volatility matrix with 150 spots between 50 and 200, and 60 times between now and 5 years.[6] We simulate with 300,000 Sobol points in parallel over 312 (biweekly) times steps.

The Merton price is 4.25. Dupire's price is off two basis points at 4.23. The corresponding Black and Scholes implied volatility is 15.35. The Black and Scholes vega is 59. We should expect a superbucket with 59 on the strike 120, maturity 3 years, and zero everywhere else. The superbucket is obtained in around two seconds. Almost all of it is simulation. Calibration and check-pointing are virtually free.

With the improvements of Chapter 15, the superbucket is produced in one second.

The resulting superbucket is displayed on the chart below.

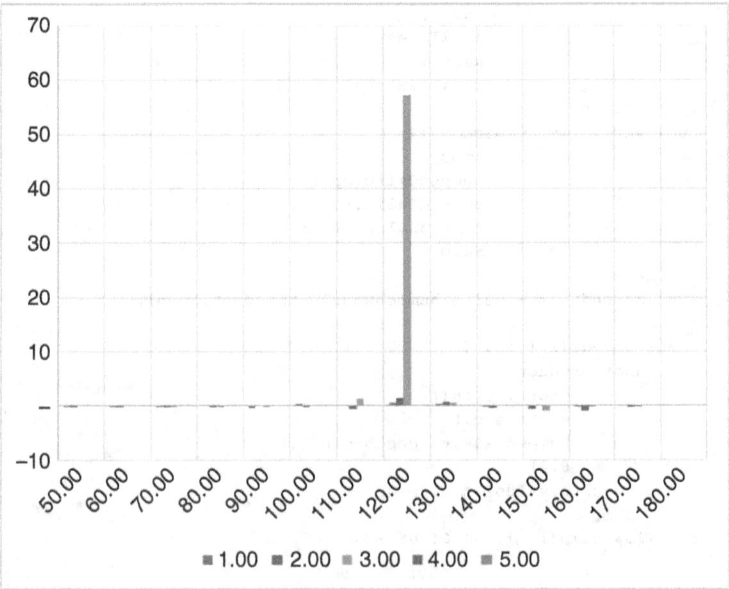

[6]To obtain a stable superbucket takes a fine-grained microbucket and simulation timeline, many paths, and a sparse risk view.

This is a good-quality superbucket, especially given it was obtained with simulations in just over a second. Superbuckets traditionally obtained with FDM are of similar quality and also take around a second to compute. Once again, we notice that AAD and parallelism bring FDM performance to Monte-Carlo simulations.

With the same settings (10 points spacing on the risk view), we compute the superbucket for a 2 years 105 call (first chart) and a 2.5 years 85 call (second chart). The results are displayed below. The calculation is proportionally faster for lower maturities, linearly in the total number of time steps. We see that vega is correctly interpolated over the risk view (with limited spilling that eventually disappears as we increase the number of paths and time steps).

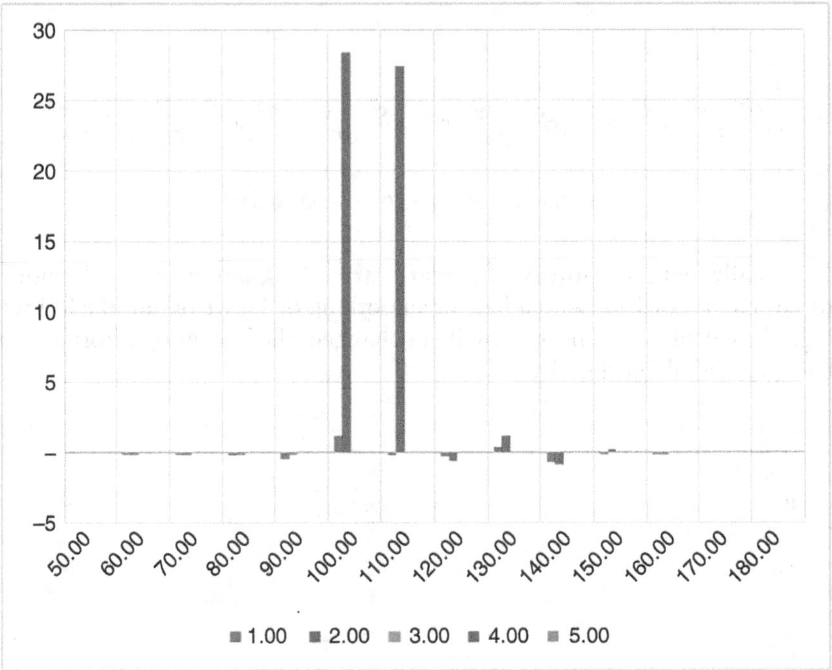

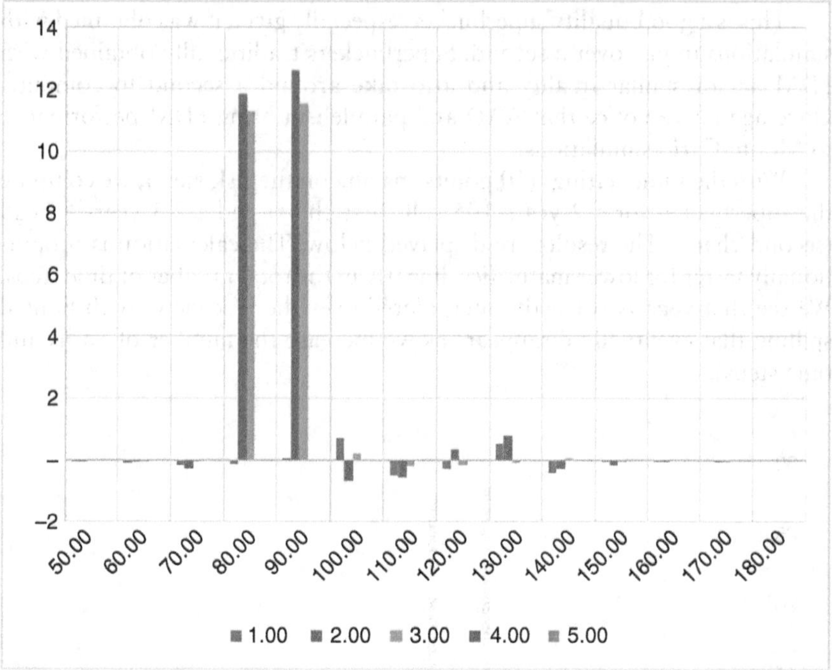

Finally, with a maturity of 3 years, strike 120, and a (biweekly monitored) barrier of 150 (with a barrier smoothing of 1), we obtain the following. The calculation time is virtually unchanged, the barrier monitoring cost being essentially negligible.

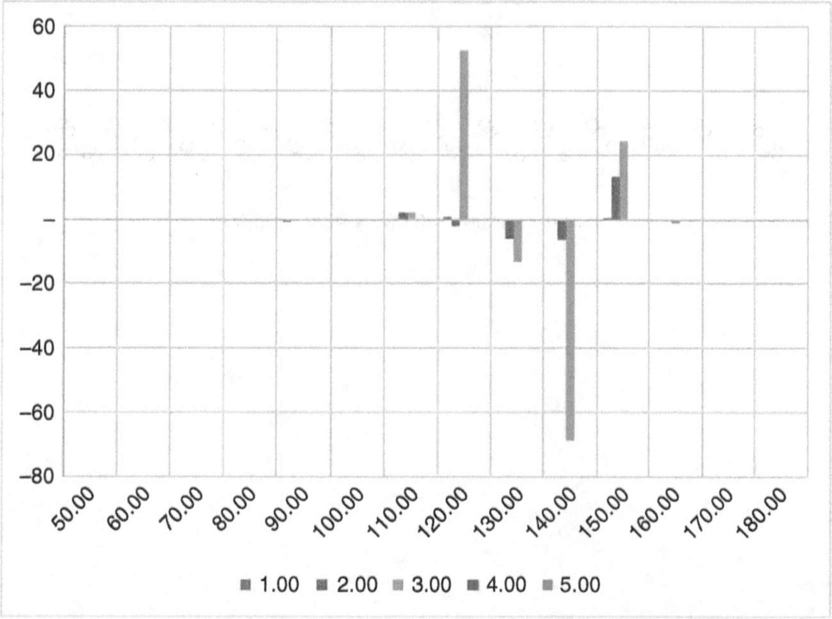

This barrier superbucket has the typical, expected shape for an up-and-out call: positive vega concentrated at maturity on the strike, negative vega, also concentrated at maturity (with some spilling over the preceding maturity on the risk view from the interpolation), *below* the barrier, partly unwound by (perhaps counterintuitive but a systematic observation nonetheless) *positive* vega on the barrier.

Comparing with a bump risk for performance and correctness, we find that finite differences produce a very similar risk report in 45 around seconds. For a 3y maturity, it could be reduced to 30 seconds by only bumping active volatilities. We are computing "only" 42 risk sensitivities (14 strikes and 3 maturities up to 3y on the risk view), so AAD acceleration is less impressive here: times 30, probably down to times 20 with a smarter implementation of the bump risk.

However, in this particular case, AAD risk is also much more stable. The results of the bumped superbucket depend on the size of the bumps and the spacing of the local volatility and the risk view, in an unstable, explosive manner. It frequently produces results in the thousands in random cells where vega is expected in the tens. AAD superbuckets are resilient and stable, because derivatives are computed analytically, without ever actually changing the initial conditions.

Finally, the quality of the superbucket is dependent on how sparse is the risk view, and rapidly deteriorates when more strikes and maturities are added to it. This is a problem with superbuckets known in the industry: to obtain decent superbuckets over a thinly spaced risk view forces to increase the time steps at the expense of speed. It helps to implement simulation schemes more sophisticated than Euler's.

13.3 IMPLICIT CALIBRATION

Iterative calibration

We mentioned that explicit, analytic calibration was an exceptional feature of Dupire's model. Other models are typically calibrated numerically. In general terms, calibration proceeds as follows. To calibrate the m parameters Y of a model (say the m local volatilities in a simulation model a la Dupire) to a market (say an IVS with a risk view X), pick $N \geq m$ instruments (say European calls of different strikes and maturities) and find the model parameters so that the model price of these instruments matches their market price.

Denote $f_i(X)$ the market price of the instrument i and $g_i(Y)$ its model price with parameters Y. The functions f and g are always explicit and generally analytic or quasi-analytic, depending on the nature of the market and model.

We call calibration error for instrument i the quantity:

$$e_i \equiv w_i[g_i(Y) - f_i(X)]$$

where w_i is the weight of instrument i in the calibration. Since f and g are explicit, it follows that e is also explicit.

Calibration consists in finding the Y that minimizes the norm of e. The optimal Y is a function of the market X:

$$Y^* = \arg\min_Y \left(\|e\|^2 = \sum_{i=1}^{n} w_i^2 [g_i(Y) - f_i(X)]^2 \right) = G(X)$$

The minimization is conducted numerically, with an iterative procedure. Numerical Recipes [20] provides the code and explanations for many common optimization routines. The most commonly used in the financial industry is Levenberg and Marquardt's algorithm from their chapter 15.

It follows that Y is a function G of X, just like in the explicit case, but here this function is defined implicitly as the result of an iterative procedure.

Differentiation through calibration

As before, we compute a price in the calibrated model:

$$V_0 = H(Y) = F(X)$$

and assume that we effectively differentiated the valuation step so we know:

$$\frac{\partial V_0}{\partial Y} = \frac{\partial H}{\partial Y}$$

Our goal, as before, is to compute the market risks:

$$\frac{\partial V_0}{\partial X} = \frac{\partial G}{\partial X}\frac{\partial H}{\partial Y}$$

The difference is that G is now an implicit function that involves a numerical fit, and it is not advisable to differentiate it directly, with AAD or otherwise.

First, numerical minimization is likely to take several seconds, which would not be RAM or cache efficient with AAD. Second, numerical minimization always involves control flow. AAD cannot differentiate control flow, and bumping through control flow is unstable. Finally, in case $N > m$, the result is a best fit, not a perfect fit, and to blindly differentiate it could produce unstable sensitivities.

It follows that we compute the Jacobian of G without actually differentiating G, with a variant of the ancient Implicit Function Theorem, as described next. It also follows that check-pointing is no longer an option.

We must compute the Jacobian of G $\frac{\partial G}{\partial X}$ and calculate a matrix product to find $\frac{\partial V_0}{\partial X} = \frac{\partial G}{\partial X}\frac{\partial H}{\partial Y}$. The matrix product must be calculated efficiently; we refer to our Chapter 1.

Note that although we cannot implement check-pointing in this case, we still want to separate the differentiation of H from the differentiation of G, because then we can compute $\frac{\partial H}{\partial Y}$ efficiently with parallel, path-wise AAD

The Implicit Function Theorem

We want to compute the Jacobian $\frac{\partial G}{\partial X}$ without actually differentiating G. Remember that:

$$Y^* = G(X) = \arg\min_Y \|e\|^2$$

and that:

$$e_i(X, Y) = w_i[g_i(Y) - f_i(X)]$$

is an explicit function that may be safely differentiated with AAD or otherwise.

To find $\frac{\partial G}{\partial X}$ as a function of the differentials of e, we demonstrate a variant of the Implicit Function Theorem, or IFT.

The result of the calibration $Y^* = G(X)$ that realizes the minimum satisfies:

$$\frac{\partial \|e(X, Y^*)\|^2}{\partial Y} = 2\left[\frac{\partial e(X, Y^*)}{\partial Y}\right]^t e(X, Y^*) = 0_m$$

It follows that:

$$\left\{\frac{\partial e[X, G(X)]}{\partial G}\right\}^t e[X, G(X)] = 0_m$$

Differentiating with respect to X, we get:

$$\frac{\partial}{\partial X}\left\{\frac{\partial e[X, G(X)]}{\partial G}\right\}^t e[X, G(X)] + \left\{\frac{\partial e[X, G(X)]}{\partial G}\right\}^t \frac{\partial e[X, G(X)]}{\partial X} = 0_{mn}$$

Assuming that the fit is "decent," the errors at the optimum are negligible compared to the derivatives so we can drop the left term and get:

$$\left\{\frac{\partial e[X, G(X)]}{\partial G}\right\}^t \frac{\partial e[X, G(X)]}{\partial X} \approx 0_{mn}$$

For clarity, we denote $\partial_1 e$ the derivative of e with respect to its first variable X, and $\partial_2 e$ its derivative to the second variable Y. Then, dropping the approximation in the equality:

$$\partial_2 e[X, G(X)]^t \left\{ \partial_1 e[X, G(X)] + \partial_2 e[X, G(X)]^t \frac{\partial G(X)}{\partial X} \right\} = 0_{mn}$$

And it follows that:

$$\frac{\partial G}{\partial X} = -\left[\partial_2 e(X, Y^*)^t \partial_2 e(X, Y^*)\right]^{-1} \partial_2 e(X, Y^*)^t \partial_1 e(X, Y^*)$$

The Jacobian of the calibration, that is, the matrix of the derivatives of the calibrated model parameters to the market parameters, is the product of the pseudo-inverse of the differentials $\partial_2 e$ of the calibration errors to the model parameters, by the differentials $\partial_1 e$ of the errors to the market parameters.

The differentials $\partial_1 e$ and $\partial_2 e$ are computed by explicit differentiation, and the Jacobian of G is computed with linear algebra without ever actually differentiating through the minimization. Finally, we have an expression for the market risks:

$$\begin{aligned}\frac{\partial V_0}{\partial X} &= \frac{\partial G}{\partial X}\frac{\partial H}{\partial Y} \\ &= -\left[\partial_2 e(X, Y^*)^t \partial_2 e(X, Y^*)\right]^{-1} \partial_2 e(X, Y^*)^t \partial_1 e(X, Y^*) \frac{\partial H(Y^*)}{\partial Y}\end{aligned}$$

We call this formula the IFT. We note that this is the same formula as the well-known normal equation for a multidimensional linear regression:

$$\beta = (X^t X)^{-1} X^t Y$$

The market risks correspond to the regression of the model risks of the transaction onto the market risks of the calibration instruments. When $\partial_2 e$ is square of full rank (the calibration is a unique perfect fit), that expression simplifies into:

$$\frac{\partial V_0}{\partial X} = -\partial_2 e(X, Y^*)^{-1} \partial_1 e(X, Y^*) \frac{\partial H(Y^*)}{\partial Y}$$

In addition, in the case of a perfect fit, calibration weights are superfluous, hence:

$$\partial_1 e(X, Y) = -\left[\frac{\partial f_i}{\partial X}\right], \partial_2 e(X, Y) = \left[\frac{\partial g_i}{\partial Y}\right]$$

and finally:

$$\frac{\partial V_0}{\partial X} = \frac{\partial g}{\partial Y}^{-1} \frac{\partial f}{\partial X}\frac{\partial H}{\partial Y}$$

This formula further illustrates that we are projecting the model risks of the transaction onto the market risks of the calibration instruments. Note that in addition to dealing with weights, the complete IFT formula simply replaces the inverse in the simplified formula by a pseudo-inverse. It is best to use the complete IFT formula in all cases, the pseudo-inverse stabilizing results when $\frac{\partial g}{\partial Y}$ approaches singularity.

We refer to [94] for an independent discussion of differentiation through calibration and application of the IFT.

We have derived two very different means of computing market risks out of model risks in this chapter: with the extremely efficient check-pointing technique when model parameters are explicitly derived from market parameters, and with the IFT formula when model parameters are calibrated to the market prices of a chosen set of instruments. These two methodologies combine to provide the means to separate the differentiation of the valuation, from the propagation of risks to market parameters. This separation is crucial, because it allows to differentiate valuation efficiently with parallel AAD in isolation, and *then* propagate the resulting derivative sensitivities to market variables with check-pointing when possible, or IFT otherwise.

Model hierarchies and risk propagation

We mentioned in Chapter 4 that linear markets, IVS, and dynamic models really form a hierarchy of models, where the parameters of a parent model are derived, or calibrated, from the parameters of its child models.

For instance, a one-factor Libor Market Model (LMM, [29]) is parameterized by a set of initial forward rates $F(0, T_i)$ and a volatility matrix for these forward rates $\sigma(t_j, T_i)$.

Neglecting Libor basis in the interest of simplicity, the initial rates $F(0, T_i)$ are derived from the continuous discount curve $DF(0, T)$ on the linear rate market (LRM) by the explicit formula:

$$F(0, T_i) = \frac{DF(0, T_i) - \delta DF(0, T_i + \delta)}{\delta DF(0, T_i + \delta)}$$

where δ is the duration of the forward rates, typically 3 or 6 months. The LRM constructs the discount curve by interpolation, which is a form of calibration, to a discrete set of par swap rates and other market-traded instruments. The reality of modern LRMs is much more complicated due to basis curves and collateral discounting, see [39], but they are still constructed by fitting parameters to a set of market instruments, hence, by calibration.

The LMM volatility surface $\sigma(t_j, T_i)$ is calibrated to an IVS $\hat{\sigma}(K, T_1, T_2)$, which, in this case, represents swaptions of strike K, maturity T_1 into a swap of lifetime T_2. This IVS itself typically calibrates to a discrete set of swaption-implied volatilities, where the corresponding forward swap rates are computed from the LRM's discount curve through the formula:

$$S(T_1, T_2) = \frac{\sum_i \delta F(0, T_i) DF(0, T_i + \delta)}{\sum_i \delta DF(0, T_i + \delta)}$$

After the model is fully specified and calibrated, it may be applied to compute the value V of a (perhaps exotic) transaction through (parallel) simulations.

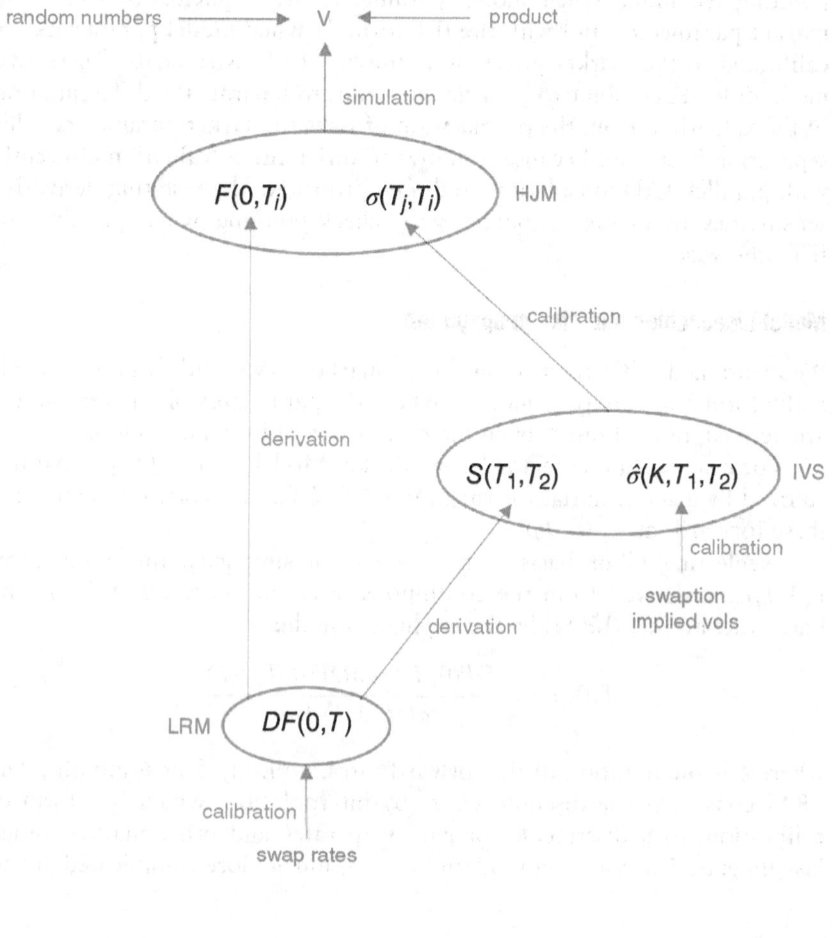

This figure above illustrates the entire process for the valuation of a transaction, including calibrations. This figure is reminiscent of the DAGs of Chapter 9. The nodes here do not represent elementary mathematical operations, but successive transformations of parameters culminating into a valuation on the top node. Derivative sensitivities are propagated through this DAG, from model risks to market risks, in the exact same way that adjoints are propagated through a calculation DAG: in the reverse order from evaluation. AAD over (parallel) simulations results in derivatives to model parameters. From there, sensitivities to parameters propagate toward the leaves, applying IFT for traversing calibrations and check-pointing for traversing derivations. The risk propagation results in the desired market risks: derivatives to swap rates and swaption-implied volatilities.

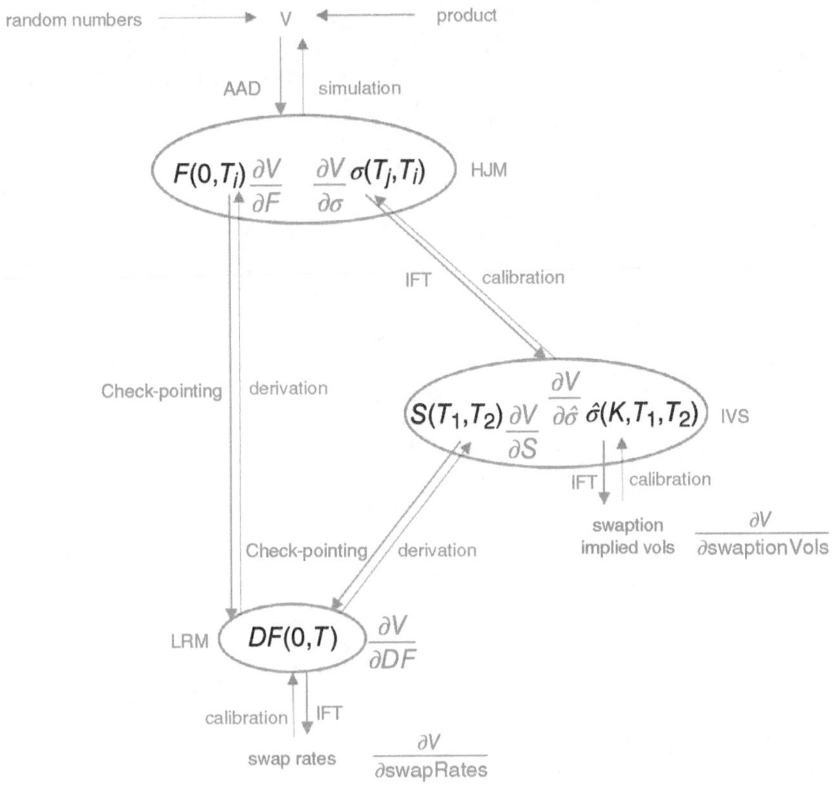

Although we are not developing the code for the general risk propagation process (this is thousands of lines of code and the potential topic of a dedicated publication), we point out that this algorithm is at the core of a well-designed derivatives risk management system.

CHAPTER 14

Multiple Differentiation in *Almost* Constant Time

14.1 MULTIDIMENSIONAL DIFFERENTIATION

In this chapter, we discuss the efficient differentiation of multidimensional functions. In general terms, we have a function:

$$F : \mathbb{R}^n \to \mathbb{R}^m$$

that produces m results $(F_k)_{1 \le k \le m}$ out of n inputs $(x_i)_{1 \le i \le n}$. The function F is implemented in code and we want to compute its $m \times n$ Jacobian matrix:

$$\frac{\partial F_k}{\partial x_i}$$

as efficiently as possible.

We have seen that AAD run time is constant in n, but linear in m, and that this is the case even when the *evaluation* of F is constant in m, because the adjoint propagation step is always linear in m. Finite differences, and most other differentiation algorithms, on the other hand, are constant in m and linear in n. As a result, AAD offers an order of magnitude acceleration in cases where the number of inputs n is significantly larger than the number of results m, as is typical in financial applications.

We have covered so far the case where $m = 1$, and AAD computes a large number n of differentials very efficiently in constant time. When $m > 1$, linear time is inescapable. The time spent in the computation of the Jacobian matrix is necessarily:

$$\Delta T = \alpha + \beta m$$

The theoretical, asymptotic differentiation time will remain linear irrespective of our efforts. In situations of practical relevance, however, when m is limited compared to n, and the marginal cost β is many times smaller than the fixed cost α, the Jacobian computes in "almost" constant time. This is the best result we can hope to achieve.

In the context of financial simulations, we have efficiently differentiated *one* result in Chapter 12. We pointed out that this is not limited to the value of one transaction. Our library differentiates, for example, the value of a portfolio of many European options, defined as a single product on page 238. We can differentiate the value of a portfolio. What we cannot do is differentiate separately every transaction in the portfolio, and obtain an *itemized* risk report, one per transaction, as opposed to one aggregated risk report for the entire portfolio. The techniques in this chapter permit to compute itemized risk reports in "almost" the same time it takes to compute an aggregated one.

The reason why we deferred this discussion to the penultimate chapter is not difficulty. What we cover here is not especially hard if the previous chapters are well understood. We opted to explain the core mechanics of AAD in the context of a one-dimensional differentiation, in a focused manner and without the distraction of the notions and constructs necessary in the multidimensional case. In addition, one-dimensional differentiation, including the aggregated risk of a portfolio of transactions, constitutes by far the most common application in finance. For this reason, we take special care that the additional developments conducted here should not slow down the one-dimensional case.

14.2 TRADITIONAL MULTIDIMENSIONAL AAD

Traditional AAD does not provide specific support for multiple results in the AAD library, and leaves it to the client code to compute the Jacobian of F in the following, rather self-evident, manner:

1. Evaluate an instrumented instance of F, which builds its tape once and for all.
2. For each result F_k:
   (a) *Reseed the tape*, setting the adjoint of F_k to one and all other adjoints on tape to zero.
   (b) Back-propagate. Read the kth row of the Jacobian $\partial F_k / \partial x_i$ in the adjoints of the n inputs x_i. Store the results. Repeat.

Our code contains all the necessary pieces to implement this algorithm. The method:

resetAdjoints()

on the Tape and Number classes reset the value of all adjoints on tape to zero. The method overloads:

propagateAdjoints()

on the Number class set the caller Number's adjoint to one prior to propagation.

The complexity of this algorithm is visibly linear in m, back-propagation being repeated m times. It may be efficient or not depending on the instrumented code. In the context of our simulation code, it is *extremely* inefficient.

To see this, remember that we optimized the simulation code of Chapter 6 so that as much calculation as possible happens once on initialization and as little as possible during the repeated simulations. The resulting tape, illustrated on page 414, is mainly populated with the model parameters and the initial computations, the simulation part occupying a minor fraction of the space. Consequently, we implemented in Chapter 12 a strong optimization whereby adjoints are back-propagated through the minor simulation piece repeatedly during simulations, and only once after all simulations completed, through the initialization phase.

It is this optimization, along with selective instrumentation, that allowed to differentiate simulations with remarkable speed, both against model parameters in Chapter 12 and against market variables in Chapter 13. But this is not scalable for a multidimensional differentiation. With the naive multidimensional AAD algorithm explained above, the adjoints of F_k are reset after they are computed, so the adjoints of F_{k+1} populate the tape next. We can no longer accumulate the adjoints of the parameters and initializers across simulations, because adjoints are overwritten by those of other results every time. To implement the traditional multidimensional algorithm would force us to repeatedly propagate adjoints, in every simulation, *back to the beginning of the tape*, resulting in a slower differentiation by orders of magnitude.

Traditional multidimensional AAD may be sufficient in some situations, but simulation is not one of them. We must find another way.

14.3 MULTIDIMENSIONAL ADJOINTS

After implementing and testing a number of possible solutions, we decided on a particularly efficient and rather simple one, although it cannot be implemented in client code without specific support from the AAD library: we work with not one, but *multiple adjoints*, one per result.

With the definitions and notations of Section 9.3, we define the instrumented calculation as a sequence f_i, $1 \le i \le N$ of operations f_i, that produce the intermediate results y_i out of arguments y_j where $j \in C_i$, C_i being the set of indices of f_i's arguments in the sequence:

$$y_i = f_i(y_j, j \in C_i)$$

The first ys are the inputs to the calculation: for $1 \leq i \leq n$, $y_i = x_i$, and some ys, typically late in the sequence, are the results:

$$y_{N_k} = F_k$$

Like in Section 9.3, we define the adjoints A_i of all the intermediate results y_i, except that adjoints are no longer scalars, but m-dimensional vectors:

$$A_i = (a_i^1, \ldots, a_i^m)$$

with $a_i^k \equiv \frac{\partial F_k}{\partial y_i}$.

To every operation number i involved in the calculation, correspond not one adjoint a_i, the derivative of the final result to the intermediate result y_i, but m adjoints, the derivatives of the m final results F_k to the intermediate result y_i. This definition can be written in a compact, vector form:

$$A_i \equiv \frac{\partial F}{\partial y_i}$$

It immediately follows this definition that:

$$\frac{\partial F}{\partial x_i} = A_i$$

so the desired Jacobian, the differentials of results to inputs, are still the input's adjoints. It also follows that for every component k of F:

$$A_{N_k} = (1_{\{N_k = k'\}}, 1 \leq k' \leq m)$$

which specifies the boundary condition for adjoint propagation. The adjoint equation itself immediately follows from the chain rule:

$$a_j^k = \sum_{i/j \in C_i} \frac{\partial y_i}{\partial y_j} a_i^k$$

or, in compact vector form:

$$A_j = \sum_{i/j \in C_i} \frac{\partial y_i}{\partial y_j} A_i$$

and the back-propagation equation from node i to nodes $j \in C_i$ follows:

$$a_j^k += \frac{\partial y_i}{\partial y_j} a_i^k$$

or in compact form:

$$A_j += \frac{\partial y_i}{\partial y_j} A_i$$

Unsurprisingly, this is the exact same equation as previously, only it is now a vector equation in dimension m, no longer a scalar equation in dimension 1. The m adjoints Aj are propagated together from the m adjoints A_i, weighted by the same local derivative $\partial y_i / \partial y_j$. It follows that back-propagation remains, as expected, linear in m, but the rest is unchanged. For instance, in the case of simulations, we can propagate through a limited piece of the tape repeatedly during simulations, and through the initialization part, only once in the end. Therefore, the *marginal* cost β of additional differentiated results remains limited, resulting in an "almost" constant time differentiation, as will be verified shortly with numerical results. First, we implement these equations in code.

14.4 AAD LIBRARY SUPPORT

The manipulations above cannot be directly implemented in client code. We must develop support in the AAD library for the representation, storage, and propagation of multidimensional adjoints. Those manipulations necessarily carry overhead, and we don't want to adversely affect the speed of one-dimensional differentiations. For this reason, and at the risk of code duplication, we keep the one-dimensional and multidimensional codes separate.

Static multidimensional indicators

First, we add some static indicators on the Node and Tape classes, that set the context to either single or multidimensional and specify its dimension. On the Node class, we add a static data member *numAdj* $(= m)$ that stores the dimension of adjoints, also the number of differentiated results. We keep it private, although it must be accessible from the Tape and Number classes, so we make these classes friends of the Node:

```
//  Node class in AADNode.h
class Node
{
    friend class Tape;
    friend class Number;

    //  Number of adjoints (results) to propagate, usually 1
    static size_t   numAdj;

//  ...
```

Similarly, we put a static boolean indicator *multi* on the Tape class. It is somewhat redundant with *Node* : : *numAdj*, although it is faster to test a boolean than compare a size_t to one. Since we test every time we record an operation, the performance gain justifies the duplication:

```
//  Tape class in AADTape.h
class Tape
{
    friend class Number;

    //      Working with multiple results / adjoints?
    static bool                             multi;

//  ...
```

We also provide a function *setNumResultsForAAD*() in AAD.h so client code may set the context prior to a differentiation:

```
inline auto setNumResultsForAAD(
    const bool multi = false,
    const size_t numResults = 1)
{
    Tape::multi = multi;
    Node::numAdj = numResults;
    return make_unique<numResultsResetterForAAD>();
}
```

This function returns a (unique) pointer on an RAII object that resets dimension to 1 on destruction:

```
//  Also in AAD.h
struct numResultsResetterForAAD
{
    ~numResultsResetterForAAD()
    {
        Tape::multi = false;
        Node::numAdj = 1;
    }
};
```

Client code sets the context by a call like:

```
    //  Set context for differentiation in dimension 50
    auto resetter = setNumResultsForAAD(true, 50);
```

When *resetter* exits scope for whatever reason, its destructor is invoked and the static context is reset to dimension one. This guarantees that the static context is always reset, even in the case of an exception, and that client code doesn't need to concern itself with the restoration of the context after completion.

The function and the RAII class must be declared friends to the Node and the Tape class to access the private indicators.

The Node class

The Node class is modified as follows:

```
class Node
{
    friend class Tape;
    friend class Number;
    friend auto setNumResultsForAAD(const bool, const size_t);
    friend struct numResultsResetterForAAD;

    //  Number of adjoints (results) to propagate, usually 1
    static size_t   numAdj;

    //  Number of childs (arguments)
    const size_t n;

    //  The adjoint(s)
    //  in single case, self held
    double          mAdjoint = 0;
    //  in multi case, held separately and accessed by pointer
    double*         pAdjoints;

    //  Data lives in separate memory

    //  the n derivatives to arguments,
    double*         pDerivatives;

    //  the n pointers to the adjoints of arguments
    double**        pAdjPtrs;

public:

    Node(const size_t N = 0) : n(N) {}

    //  Access to adjoint(s)
    //  single
    double& adjoint() { return mAdjoint; }
    //  multi
    double& adjoint(const size_t n) { return pAdjoints[n]; }

    //  Back-propagate adjoints to arguments adjoints

    //  Single case
    void propagateOne()
    {
```

```
43         // Nothing to propagate
44         if (!n || !mAdjoint) return;
45 
46         for (size_t i = 0; i < n; ++i)
47         {
48             *(pAdjPtrs[i]) += pDerivatives[i] * mAdjoint;
49         }
50     }
51 
52     // Multi case
53     void propagateAll()
54     {
55         // No adjoint to propagate
56         if (!n || all_of(pAdjoints, pAdjoints + numAdj,
57             [](const double& x) { return !x; }))
58             return;
59 
60         for (size_t i = 0; i < n; ++i)
61         {
62             double *adjPtrs = pAdjPtrs[i], ders = pDerivatives[i];
63 
64             // Vectorized!
65             for (size_t j = 0; j < numAdj; ++j)
66             {
67                 adjPtrs[j] += ders * pAdjoints[j];
68             }
69         }
70     }
71 };
```

In addition to the static indicator and friends discussed earlier, we have three modifications:

1. The storage of adjoints on line 14. We keep a scalar adjoint *mAdjoint* on the node for the one-dimensional case. In the multidimensional case, the $m = nAdj$ adjoints live in separate memory on tape, like the derivatives and child adjoint pointers of Chapter 10, and are accessed with the pointer *pAdjoints*.
2. Consequently, we must modify how client code accesses adjoints on line 32. In the one-dimensional case, the single adjoint is accessed, as previously, with the *adjoint*() accessor without arguments. In the multidimensional case, adjoints are accessed with an overload that takes the index of the adjoint between 0 and $m - 1$ as an argument.
3. Finally, propagation on line 52 is unchanged in the one-dimensional case,[1] and we code another propagation function *propagateAll*() for the multidimensional case. This function back-propagates the m adjoints altogether to the child adjoints, multiplied by the same local derivatives.

[1]And we understand now why we called the function "propagateOne()".

Visual Studio 2017 vectorizes the innermost loop, another step on the way of an almost constant time differentiation.

The Tape class

The Tape class is modified in a similar manner:

```
#include "blocklist.h"
#include "AADNode.h"

constexpr size_t BLOCKSIZE = 16384;        //  Number of nodes
constexpr size_t ADJSIZE   = 32768;        //  Number of adjoints
constexpr size_t DATASIZE  = 65536;        //  Data in bytes

class Tape
{
    //  Working with multiple results / adjoints?
    static bool                        multi;

    //  Storage for adjoints
    blocklist<double, ADJSIZE>         myAdjointsMulti;

    //  Storage for the nodes
    blocklist<Node, BLOCKSIZE>         myNodes;

    //  Storage for derivatives and child adjoint pointers
    blocklist<double, DATASIZE>        myDers;
    blocklist<double*, DATASIZE>       myArgPtrs;

    //  Padding so tapes in a vector don't interfere
    char                               myPad[64];

    friend auto setNumResultsForAAD(const bool, const size_t);
    friend struct numResultsResetterForAAD;
    friend class Number;

public:

    //  Build note in place and return a pointer
    //  N : number of childs (arguments)
    template <size_t N>
    Node* recordNode()
    {
        //  Construct the node in place on tape
        Node* node = myNodes.emplace_back(N);

        //  Store and zero the adjoint(s)
        if (multi)
        {
           node->pAdjoints = myAdjointsMulti.emplace_back_multi(Node::numAdj);
           fill(node->pAdjoints, node->pAdjoints + Node::numAdj, 0.0);
        }

        //  Store the derivatives and child adjoint pointers unless leaf
        if constexpr(N)
```

```
        {
            node->pDerivatives = myDers.emplace_back_multi<N>();
            node->pAdjPtrs = myArgPtrs.emplace_back_multi<N>();

        }

        return node;
    }

    void resetAdjoints()
    {
        if (multi)
        {
            myAdjointsMulti.memset(0);
        }
        else
        {
            for (Node& node : myNodes)
            {
                node.mAdjoint = 0;
            }
        }
    }

    //  Clear
    void clear()
    {
        myAdjointsMulti.clear();
        myDers.clear();
        myArgPtrs.clear();
        myNodes.clear();
    }

    //  Rewind
    void rewind()
    {
        if (multi)
        {
            myAdjointsMulti.rewind();
        }
        myDers.rewind();
        myArgPtrs.rewind();
        myNodes.rewind();
    }

    //  Set mark
    void mark()
    {
        if (multi)
        {
            myAdjointsMulti.setmark();
        }
        myDers.setmark();
        myArgPtrs.setmark();
        myNodes.setmark();
    }
```

```
106     //  Rewind to mark
107     void rewindToMark()
108     {
109         if (multi)
110         {
111             myAdjointsMulti.rewind_to_mark();
112         }
113         myDers.rewind_to_mark();
114         myArgPtrs.rewind_to_mark();
115         myNodes.rewind_to_mark();
116     }
117
118     //  Iterators, etc. unmodified
119     //  ...
120 };
```

1. We have an additional blocklist for the multidimensional adjoints on line 14.
2. The method *recordNode*() performs, in the multidimensional case, the additional work following line 41: initialize the m adjoints to zero, store them on the adjoint blocklist, and set the Node's pointer *pAdjoints* to access their memory space in the blocklist, like local derivatives and child adjoint pointers.
3. The method *resetAdjoints*() is upgraded to handle the multidimensional case with a global memset in the adjoint blocklist, line 60.
4. Finally, all the methods that clean, rewind, or mark the tape, like *clear*() on line 76 or *rewind*() on line 85, are upgraded to also clean, rewind, or mark the adjoint blocklist so it is synchronized with the rest of the data on tape.

The Number class

Finally, we need some minor support in the Number class. First, we need some accessors for the multidimensional adjoints so client code may seed them before propagation and read them to pick differentials after propagation:

```
//  Number class in AADNumber.h
//  ...

    //  Accessors: value and adjoint

    double& value()
    {
        return myValue;
    }
    double value() const
    {
        return myValue;
    }
```

```
    //  1D case
    double& adjoint()
    {
        return myNode->adjoint();
    }
    double adjoint() const
    {
        return myNode->adjoint();
    }

    //  m-dimensional case
    double& adjoint(const size_t n)
    {
        return myNode->adjoint(n);
    }
    double adjoint(const size_t n) const
    {
        return myNode->adjoint(n);
    }

//  ...
```

We also coded some methods to facilitate back-propagation on the Number type in Chapter 10. For multidimensional propagation, we only provide a general static method, leaving it to the client code to seed the tape and provide the correct start and end propagation points:

```
//  Number class in AADNumber.h
//  ...

    //  Propagate adjoints
    //      from and to both INCLUSIVE
    static void propagateAdjointsMulti(
    Tape::iterator propagateFrom,
    Tape::iterator propagateTo)
    {
        auto it = propagateFrom;
        while (it != propagateTo)
        {
            it->propagateAll();
            --it;
        }
        it->propagateAll();
    }

//  ...
```

14.5 INSTRUMENTATION OF SIMULATION ALGORITHMS

With the correct instrumentation of the AAD library, we can easily develop the multidimensional counterparts of the template algorithms of Chapter 12 in mcBase.h. The code is mostly identical to the one-dimensional code, so the template codes could have been merged. We opted for duplication in the interest of clarity.

The nature of the results is different: we no longer return the *vector* of the risk sensitivities of one aggregate of the product's payoffs to the model parameters. We return the *matrix* of the risk sensitivities of all the product's payoffs to the model parameters:

```
struct AADMultiSimulResults
{
    AADMultiSimulResults(
        const size_t nPath,
        const size_t nPay,
        const size_t nParam) :
            payoffs(nPath, vector<double>(nPay)),
            risks(nParam, nPay)
    {}

    //  matrix(0..nPath - 1, 0..nPay - 1) of payoffs, same as mcSimul()
    vector<vector<double>>  payoffs;

    //  matrix(0..nParam - 1, 0..nPay - 1) of risk sensitivities
    //      of all payoffs, averaged over paths
    matrix<double>      risks;
};
```

The signature of the single-threaded and multi-threaded template algorithms is otherwise unchanged, with the exception that they no longer require a payoff aggregator from the client code, since they produce an itemized risk for all the payoffs. They are listed in mcBase.h, like the one-dimensional counterparts. The single-threaded template algorithm is listed below with differences to the one-dimensional version highlighted and commented:

```
inline AADMultiSimulResults
mcSimulAADMulti(
        const Product<Number>&  prd,
        const Model<Number>&    mdl,
        const RNG&              rng,
        const size_t            nPath)
{
    auto cMdl = mdl.clone();
    auto cRng = rng.clone();

    Scenario<Number> path;}
    allocatePath(prd.defline(), path);
    cMdl->allocate(prd.timeline(), prd.defline());

    const size_t nPay = prd.payoffLabels().size();
    const vector<Number*>& params = cMdl->parameters();
    const size_t nParam = params.size();

    Tape& tape = *Number::tape;
    tape.clear();

```

```
    //  Set the AAD environment to multi-dimensional with dimension nPay
    //  Reset to 1D is automatic when resetter exits scope
        auto resetter = setNumResultsForAAD(true, nPay);

    cMdl->putParametersOnTape();
    cMdl->init(prd.timeline(), prd.defline());
    initializePath(path);
    tape.mark();

    cRng->init(cMdl->simDim());

    vector<Number> nPayoffs(nPay);
    vector<double> gaussVec(cMdl->simDim());

    //  Allocate multi-dimensional results
    //      including a matrix(0..nParam - 1, 0..nPay - 1)
    //      of risk sensitivities
        AADMultiSimulResults results(nPath, nPay, nParam);

    for (size_t i = 0; i<nPath; i++)
    {
        tape.rewindToMark();

        cRng->nextG(gaussVec);
        cMdl->generatePath(gaussVec, path);
        prd.payoffs(path, nPayoffs);

        //  Multi-dimensional propagation
        //      client code seeds the tape
        //      with the correct boundary conditions
          for (size_t j = 0; j < nPay; ++j) nPayoffs[j].adjoint(j) = 1.0;
        //      multi-dimensional propagation over simulation,
        //      end to mark
          Number::propagateAdjointsMulti(prev(tape.end()), tape.markIt());

        convertCollection(
            nPayoffs.begin(),
            nPayoffs.end(),
            results.payoffs[i].begin());
    }

    //  Multi-dimensional propagation over initialization, mark to start
      Number::propagateAdjointsMulti(tape.markIt(), tape.begin());

    //  Pack results
      for (size_t i = 0; i < nParam; ++i)
          for (size_t j = 0; j < nPay; ++j)
              results.risks[i][j] = params[i]->adjoint(j) / nPath;

    tape.clear();

    return results;
}
```

The same modifications apply to the multi-threaded code, which is therefore listed below without highlighting or comments:

```
inline AADMultiSimulResults
mcParallelSimulAADMulti(
    const Product<Number>&  prd,
    const Model<Number>&    mdl,
    const RNG& rng,
    const size_t            nPath)
{
    const size_t nPay = prd.payoffLabels().size();
    const size_t nParam = mdl.numParams();

    Number::tape->clear();
    auto resetter = setNumResultsForAAD(true, nPay);

    ThreadPool *pool = ThreadPool::getInstance();
    const size_t nThread = pool->numThreads();

    vector<unique_ptr<Model<Number>>> models(nThread + 1);
    for (auto& model : models)
    {
        model = mdl.clone();
        model->allocate(prd.timeline(), prd.defline());
    }

    vector<Scenario<Number>> paths(nThread + 1);
    for (auto& path : paths)
    {
        allocatePath(prd.defline(), path);
    }

    vector<vector<Number>> payoffs(nThread + 1, vector<Number>(nPay));

    vector<Tape> tapes(nThread);

    vector<int> mdlInit(nThread + 1, false);

    initModel4ParallelAAD(prd, *models[0], paths[0]);

    mdlInit[0] = true;

    vector<unique_ptr<RNG>> rngs(nThread + 1);
    for (auto& random : rngs)
    {
        random = rng.clone();
        random->init(models[0]->simDim());
    }

    vector<vector<double>> gaussVecs
    (nThread + 1, vector<double>(models[0]->simDim()));

    AADMultiSimulResults results(nPath, nPay, nParam);

```

```
    vector<TaskHandle> futures;
    futures.reserve(nPath / BATCHSIZE + 1);

    size_t firstPath = 0;
    size_t pathsLeft = nPath;
    while (pathsLeft > 0)
    {
        size_t pathsInTask = min<size_t>(pathsLeft, BATCHSIZE);

        futures.push_back(pool->spawnTask([&, firstPath, pathsInTask]()
        {
            const size_t threadNum = pool->threadNum();

            if (threadNum > 0) Number::tape = &tapes[threadNum - 1];

            if (!mdlInit[threadNum])
            {
                initModel4ParallelAAD(
                    prd,
                    *models[threadNum],
                    paths[threadNum]);

                mdlInit[threadNum] = true;
            }

            auto& random = rngs[threadNum];
            random->skipTo(firstPath);

            for (size_t i = 0; i < pathsInTask; i++)
            {

                Number::tape->rewindToMark();
                random->nextG(gaussVecs[threadNum]);
                models[threadNum]->generatePath(
                    gaussVecs[threadNum],
                    paths[threadNum]);
                prd.payoffs(paths[threadNum], payoffs[threadNum]);

                const size_t n = payoffs[threadNum].size();
                for (size_t j = 0; j < n; ++j)
                {
                    payoffs[threadNum][j].adjoint(j) = 1.0;
                }
                Number::propagateAdjointsMulti(
                    prev(Number::tape->end()),
                    Number::tape->markIt());

                convertCollection(
                    payoffs[threadNum].begin(),
                    payoffs[threadNum].end(),
                    results.payoffs[firstPath + i].begin());
            }

            return true;
        }));

```

```
        pathsLeft -= pathsInTask;
        firstPath += pathsInTask;
    }

    for (auto& future : futures) pool->activeWait(future);

    Number::propagateAdjointsMulti(
        Number::tape->markIt(),
        Number::tape->begin());

    for (size_t i = 0; i < nThread; ++i)
    {
        if (mdlInit[i + 1])
        {
            Number::propagateAdjointsMulti(
                        tapes[i].markIt(),
                        tapes[i].begin());
        }
    }

    for (size_t j = 0; j < nParam; ++j) for (size_t k = 0; k < nPay; ++k)
    {
        results.risks[j][k] = 0.0;
        for (size_t i = 0; i < models.size(); ++i)
        {
            if (mdlInit[i])
                results.risks[j][k] += models[i]->parameters()[j]->adjoint(k);
        }
        results.risks[j][k] /= nPath;
    }

    Number::tape->clear();

    return results;
}
```

14.6 RESULTS

We need an entry-level function *AADriskMulti*() in main.h, almost identical to the *AADriskOne*() and *AADriskAggregate*() of Chapter 12:

```
inline RiskReports AADriskMulti(
    const string&           modelId,
    const string&           productId,
    const NumericalParam&   num)
{
    const Model<Number>* model = getModel<Number>(modelId);
    const Product<Number>* product = getProduct<Number>(productId);

    if (!model || !product)
    {
        throw runtime_error(
            "AADrisk() : Could not retrieve model and product");
    }
```

```

    RiskReports results;

    //  Random Number Generator
    unique_ptr<RNG> rng;
    if (num.useSobol) rng = make_unique<Sobol>();
    else rng = make_unique<mrg32k3a>(num.seed1, num.seed2);

    //  Simulate
    const auto simulResults = num.parallel
        ? mcParallelSimulAADMulti(*product, *model, *rng, num.numPath)
        : mcSimulAADMulti(*product, *model, *rng, num.numPath);

    results.params = model->parameterLabels();
    results.payoffs = product->payoffLabels();
    results.risks = move(simulResults.risks);

    //  Average values across paths
    const size_t nPayoffs = product->payoffLabels().size();
    results.values.resize(nPayoffs);
    for (size_t i = 0; i < nPayoffs; ++i)
    {
        results.values[i] = accumulate(
            simulResults.payoffs.begin(),
            simulResults.payoffs.end(),
            0.0,
            [i](const double acc, const vector<double>& v)
                {
                    return acc + v[i];
                }
        ) / num.numPath;
    }

    return results;
}
```

so we can produce an itemized risk report. To produce numerical results, we set Dupire's model in the same way as for the numerical results of Chapter 12, where it took a second (half a second with the improvements in the next chapter) to produce the accurate microbucket of a barrier option of maturity 3y (with Sobol numbers, 500,000 paths over 156 weekly steps, in parallel over 8 cores). With a European call, without barrier monitoring, it takes around 850 ms (450 ms with expression templates).

To obtain the exact same microbucket for a single European call, but in a portfolio of one single transaction, with the multidimensional algorithm, takes 600 ms with expression templates. Irrespective of the number of itemized risk reports we are producing, multidimensional differentiation carries a measurable additional fixed overhead of around 30%. This is why we kept two separate cases instead of programming the one-dimensional case as multidimensional with dimension one.

To produce 50 microbuckets for 50 European options, with different strikes and maturity 3y or less, takes only 4,350 ms with expression templates, around seven times one microbucket. The following chart shows the times, in milliseconds, to produce itemized risk reports, with expression templates, as a function of the number of microbuckets:

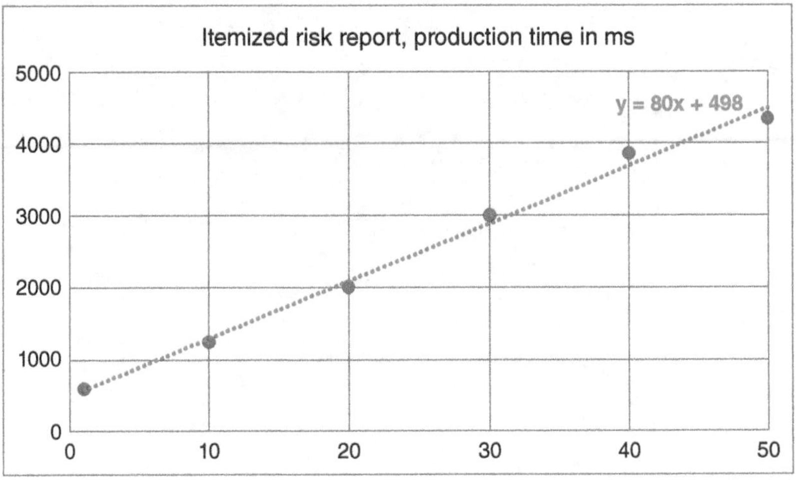

As promised, the time is obviously linear in the number of transactions, but almost constant in the sense that the marginal time to produce an additional microbucket is only 80 ms, compared to the 600 ms for the production of one microbucket.

We produced 50 microbuckets in less than 5 seconds, but this is with the improvements of the next chapter. With the traditional AAD library, it takes around 15 seconds. The expression template optimization is particularly effective in this example, with an acceleration by a factor of three. Expression templates are introduced in the next and final chapter.

CHAPTER 15

Acceleration with Expression Templates

In this final chapter, we rewrite the AAD library of Chapter 10 in modern C++ with the *expression template* technology. The application of expression templates to AAD was initially introduced in [89] as a means to accelerate AAD and approach the speed of manual AD (see Chapter 8) with the convenience of an automatic differentiation library.

The result is an overall acceleration *by a factor two to three* in cases of practical relevance, as measured in the numerical results of Chapters 12, 13, and 14. The acceleration applies to both single-threaded and multi-threaded instrumented code. It is also worth noting that we implemented a selective instrumentation in our simulation code of Chapter 12. It obviously follows that only the instrumented part of the code is accelerated with expression templates. It also follows that the instrumented part is accelerated by a very substantial factor, resulting in an overall acceleration by a factor two to three.

The resulting sensitivities are identical in all cases. The exact same mathematical operations are conducted, and the complexity is the same as before. The difference is a more efficient memory access resulting from smaller tapes, and the delegation of some *administrative* (not mathematical) work to *compile time*. The benefits are computational, not mathematical or algorithmic. Like in the matrix product of Chapter 1, the acceleration is nonetheless substantial.

The new AAD library exposes the exact same interface as the library of Chapter 10. It follows that there is no modification whatsoever in the instrumented code. The AAD library consists of three files: AADNode.h, AADTape.h, and AADNumber.h. We only rewrite AADNumber.h, in a new file AADExpr.h. AADNode.h and AADTape.h are unchanged. We put a convenient pragma in AAD.h so we can switch between the traditional implementation of Chapter 10 and the implementation of this chapter by changing a single definition:

```
//  AAD with expression templates
#define AADET   true

#if AADET

#include "AADExpr.h"

#else

#include "AADNumber.h"

#endif
```

although this is only intended for testing and debugging. The expression template implementation is superior by all metrics, so we never revert to the traditional implementation in a production context.

15.1 EXPRESSION NODES

At the core of AADET (AAD with Expression Templates) lies the notion that nodes no longer represent elementary mathematical operations like +, *, *log*() or *sqrt*, but entire expressions like our example of Chapter 9:

$$y = (y_1 + x_3 y_2)(y_1 + y_2), y_1 = x_2(5x_0 + x_1), y_2 = \log(y_1)$$

or, in (templated) code:

```
template <class T>
T f(T x[5])
{
    auto y1 = x[2] * (5.0 * x[0] + x[1]);
    auto y2 = log(y1);
    auto y = (y1 + x[3] * y2) * (y1 + y2);
    return y;
}
```

Recall that the DAG of this expression is:

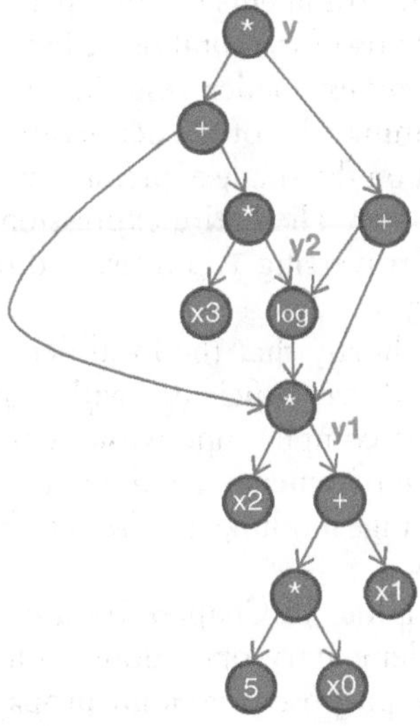

We had 12 nodes on tape for the expression, 6 binaries, 2 unaries, and 4 leaves (a binary that takes a constant on one side is a unary, and a constant does not produce a node, by virtue of an optimization we put in the code in Chapter 10).

The size of a Node in memory is given by *sizeof*(*Node*), which readers may verify easily, is 24 bytes in a 32-bit build. This is 288 bytes for the expression. Unary and binary nodes store additional data on tape: the local derivatives and child adjoint pointers, one double (8 bytes) and one pointer (4 bytes) for unaries (12 bytes), two of each (24 bytes) for binaries. For our expression, this is another 168 bytes of memory, for a total of 456 bytes.[1] In addition, the data is scattered in memory across different nodes.

The techniques in this chapter permit to represent the whole expression *in a single node*. This node does not represent an elementary mathematical

[1]Or more in the multidimensional case of Chapter 14, which requires additional memory for the storage of m adjoints.

operation, but the entire expression, and it has 4 arguments: the 4 leaves x_0, x_1, x_2, and x_3. The size of the node is still 24 bytes. It stores on tape the four local derivatives to its arguments (32 bytes) and the four pointers to the arguments adjoints (16 bytes) for a total of 72 bytes. The memory footprint for the expression is *six times* smaller, resulting in shorter tapes that better fit in CPU caches, explaining part of the acceleration.

In addition, the data on the single expression node is coalescent in memory so propagation is faster. The entire expression propagates on a single node, saving the cost of traversing 12 nodes, and the cost of working with data scattered in memory.

Finally, we will see shortly that the local derivatives of the expression to its inputs are produced very efficiently with expression templates, with costly work performed at compile time without run-time overhead. Those three *computational* improvements combine to substantially accelerate AAD, although they change nothing to the number or the nature of the mathematical operations.

With the operation nodes of Chapter 10, a calculation is decomposed in a large sequence of elementary operations with one or two arguments, recorded on tape for the purpose of adjoint propagation. With expression nodes, the calculation is decomposed in a smaller sequence of larger expressions with an arbitrary number of arguments. Adjoint mathematics apply in the exact same manner. Our Node class from Chapter 10 is able to represent expressions with an arbitrary number of arguments, even though the traditional AAD code only built unary or binary operator nodes.

This is all very encouraging, but we must somehow build the expression nodes. We must figure out, from an expression written in C++, the number of its arguments, the location of the adjoints of these arguments, and the local derivatives of the expression to its arguments, so we can store all of these on the expression's node on tape. We know since Chapter 9 that this information is given by the expression's DAG, and we learned to apply operator overloading to build the DAG in memory.

However, the expression DAGs of Chapter 9 were built at run time, in dynamic memory and with dynamic polymorphism, at the cost of a substantial run-time overhead. To do it in the same way here would defeat the benefits of expression nodes and probably result in a *slower* differentiation overall. Expression templates allow to build the DAG of expressions *at compile time*, in static memory (the so-called stack memory, the application's working memory for the execution of its functions), so their run-time traversal to compute derivatives is extremely efficient.

The derivatives of an expression to its inputs generally depend on the value of the inputs, which is only known at run time. It follows that the derivatives cannot be computed at compile time. But the DAG only depends on the expression, and, to state the obvious, the expression, written in C++

code, is very much known at compile time. It follows that there must exist a way to build the DAG at compile time, saving the cost of DAG creation at run time, as long as the programming language provides the necessary constructs. It so happens that this can be achieved, very practically, with an application of the template meta-programming facilities of standard C++. The only work that remains at run time is to propagate adjoints to compute derivatives, in the reverse order, through the DAG, and this is very quick because compile-time DAGs are constructed on the stack memory, and the adjoint propagation sequence is generally inlined and further optimized by the compiler.

In our example, the value and the partial derivatives of y depend on the array x, which is unknown at compile time, and may depend on user input. The value and derivative of the expression may only be computed at run time. But we do have valuable information at compile time: we know the number of inputs (five, out of which one, $x[4]$, is inactive) and we know *the sequence of operations* that produce y out of x, so we can produce its DAG, illustrated above, while the code is being compiled.

The technology that implements these notions is known as *expression templates*. This is a well-known idiom in advanced C++ that has been successfully applied to accelerate mathematical code in many contexts, like high-performance linear algebra libraries.

It follows that the differentials of the expressions that constitute a calculation are still recorded on tape, and propagated in reverse order, to accumulate the derivatives of the calculation to its inputs in constant time. But the tape doesn't record the individual operations that constitute the expressions; it records the expressions themselves, which results in a shorter tape storing a lower number of records. What is new is that the differentials of the expressions themselves, also computed automatically and in constant time with a reverse adjoint propagation over their constituent operations, are accumulated without a tape and without the related overhead, with the help of expression templates. The next section explains this technology and its application to a tape-less, constant time differentiation of the expressions.

15.2 EXPRESSION TEMPLATES

Template meta-programming

C++ templates offer vastly more than their common use as a placeholder for types in generic classes and methods. C++ templates may be applied to meta-programming, that is, very literally, code that generates code. Templates are resolved into template-free code *before compilation*, when all templates are *instantiated*, so it is the instantiated code which is ultimately translated into machine language by the compiler. Because the translation from the templated code into the instantiated code occurs at compile time, that transformation is free from any run-time overhead. This application of C++ templates to write general code that is transformed, at instantiation time, into specific code to be compiled into machine code, is called *template meta-programming*. Well-designed meta-programming can produce elegant, readable code where the necessary boilerplate is generated on instantiation. It can also produce extremely efficient code, where part of the work is conducted on instantiation, hence, at compile time.

The classic example is the compile-time factorial:

```
template <unsigned n>
struct factorial
{
    enum { value = n * factorial<n - 1>::value };
};

template <>
struct factorial<0>
{
    enum { value = 1 };
};

int main()
{
    cout << factorial<5>::value << endl;    // 120
}
```

The factorial is computed at compile time. The instantiated code that is compiled and executed is simply:

```
int main()
{
    cout << 120 << endl;    // 120
}
```

This traditional C++ code uses the fact that enums are resolved at compile time, like templates, sizeof(), and functions marked constexpr. We will use these to figure the number of the active arguments of an expression at compile time.

Template meta-programming is probably the most advanced form of C++ programming. It is a fascinating area, only available in C++. The most complete reference in this field is Vandevoorde et al.'s C++ Templates – The Complete Guide, recently updated in modern C++ in [76]. Motivated readers may find extensive information on template meta-programming, including expression templates, in this publication. Our introduction barely scratches the surface.

Static polymorphism and CRTP

Expression templates implement the CRTP (curiously recursive template pattern) idiom, so to understand expression templates, we must introduce CRTP first.

Classical object-oriented polymorphism offers a neat separation between interface and implementation, at the cost of run-time overhead, a textbook example being:

```
class Animal
{
public:

    virtual void makeNoise() const = 0;
    virtual ~Animal() {}
};

class Duck : public Animal
{
public:

    void makeNoise()  const override
    {
        cout << "Quack!" << endl;
    }
};

class Dog : public Animal
{
public:

    void makeNoise() const override
    {
        cout << "Woof!" << endl;
    }
};

inline void speakTo(const Animal& animal)
{
    animal.makeNoise();
}

int main()
{
    unique_ptr<Animal> donald = make_unique<Duck>();
    unique_ptr<Animal> goofy = make_unique<Dog>();

    speakTo(*donald);    // Quack
    speakTo(*goofy);     // Woof

}
```

We applied this pattern to build DAGs with polymorphic nodes in Chapter 9. Edges were represented by base node pointers, and concrete nodes represented the operators that combine nodes to produce an expression: +, $*$, *log*(), *sqrt*(), etc. Concrete nodes overrode methods to evaluate or differentiate operators, providing the means of recursively evaluating or differentiating the whole expression. To apply polymorphism to graphs in this manner is a well-known design pattern, applicable in many contexts and identified in GOF [25] as the *composite pattern*.

What is perhaps less well known is that polymorphism can be also be achieved at compile time, saving run-time allocation, creation, and resolution overhead, as long as the concrete type is known at compile time, as opposed to, say, dependent on user input:

```
template <class A>
class Animal
{
public:
    void makeNoise() const
    {
        //  Downcast, we know it's a A
        const A& a = static_cast<const A&>(*this);
        //  Call makeNoise on class A
        a.makeNoise();
    }
};

//  CRTP
class Duck : public Animal<Duck>
{
public:
    void makeNoise() const
    {
        cout << "Couac!" << endl;
    }
};

class Dog : public Animal<Dog>
{
public:
    void makeNoise() const
    {
        cout << "Waf!" << endl;
    }
};

template <class A>
inline void speakTo(const Animal<A>& animal)
{
    animal.makeNoise();
}

```

```
39 int main()
40 {
41     Duck donald;
42     Dog goofy;
43
44     speakTo(donald);      // Couac
45     speakTo(goofy);       //  Waf
46 }
```

We have the same results as before, without vtable pointers or allocations. Nothing is virtual in the code above; execution is conducted on the stack with resolution at compile time. The base class must know the derived class so it can downcast itself in the "virtual" function call on line 8 and call the correct function defined on the "concrete" class, not with run-time *overriding*, but with compile time *overloading*. It follows that the base class must be templated on the derived class, hence the curious syntax on line 15 where a class derives another class templated on itself. This syntax is what permits compile-time polymorphism, and it may appear awkward at first sight, hence the name "curiously recursive template pattern," or CRTP.

Static polymorphism can achieve many behaviors of classical run-time polymorphism, without the associated overhead. Our animals bark and quack all the same, only faster: the identification of animals as ducks or dogs (resolution) occurs at compile time, and they live on the stack (static memory), not on the heap (dynamic memory), so they are typically accessed faster. A compiler will generally inline the call on lines 44 and 45, whereas it is rarely able to inline true virtual calls. This being said, static polymorphism cannot do everything dynamic polymorphism does. If it were the case, and since static polymorphism is always faster, language support for dynamic polymorphism would be unnecessary. In particular, resolution may only occur at compile time if the concrete type is known at compile time. Our simulation library of Chapter 6 implemented virtual polymorphism so users can mix and match models, products, and RNGs at run time. This cannot be achieved with CRTP. To apply compile time polymorphism, the concrete type of the objects must be static and cannot depend on user input.

As far as C++ expressions are concerned, concrete expression types are compile time constants. For example, in the expression code on page 504, the operation on line 5 is a *log*(). The sequence of operations on line 4 is $*, +, *$. The concrete types of the operations is known at compile time, so we can construct the DAG with CRTP in place of virtual mechanisms. As a counterexample, in our publication [11], we build and evaluate DAGs that determine a product's cash-flows from a user-supplied script. In this case, the sequence of the concrete operations is not known at compile time, so CRTP cannot be applied.

Finally, CRTP is type safe, like run-time polymorphism, in the sense that *speakTo*() only accepts Animals as arguments. CRTP is different from simple templates, like:

```
template <class T>
inline void speakTo(const T& t)
{
    t.makeNoise();
}
```

This template function catches any argument and compiles as long as its type implements a method *makeNoise*(). This may be convenient in specific cases, but this is not CRTP. This is actually not any kind of polymorphism. And this is not what we need for expression templates. In order to build a compile-time DAG, like the run-time DAGs of Chapter 9, we need operator overloads that catch all kinds of expressions, *but only expressions and nothing else*. This cannot be achieved with simple templates; it requires a proper type hierarchy, either virtual or CRTP.

Building expressions

Expression templates are a modern counterpart of the composite pattern. They represent expressions in a CRTP hierarchy, which, combined with operator overloading, turns expressions into DAGs at compile time.

An expression is an essentially recursive thing: by definition, an expression consists in one or multiple expressions combined with an operator. A number is an elementary expression. For example, in the last line of the code on page 504:

$$y = (y_1 + x_3 y_2)(y_1 + y_2)$$

y_1, y_2, and x_3 are (leaf) expressions, $x_3 y_2$ is the expression that applies the operator $*$ to the expressions x_3 and y_2. The left-hand-side factor is the expression that applies the operator $+$ to the sub-expressions y_1 and $x_3 y_2$, and y is the expression that applies $*$ to the sub-expressions $y_1 + x_3 y_2$ and $y_1 + y_2$, the latter being an expression that applies $+$ to the (leaf) expressions y_1 and y_2.

We used this recursive definition in Chapter 9 to develop a classical object-oriented hierarchy of expressions (which we called nodes). We can do the same here with CRTP:

```
template <class E>
class Expression
{};

//  Times
```

```

// CRTP
template <class LHS, class RHS>
class ExprTimes : public Expression<ExprTimes<LHS, RHS>>
{
    LHS lhs;
    RHS rhs;

public:

    // Constructor
    explicit ExprTimes
        (const Expression<LHS>& l, const Expression<RHS>& r)
        : lhs(static_cast<const LHS&>(l)),
        rhs(static_cast<const RHS&>(r)) {}

    double value() const
    {
        return lhs.value() * rhs.value();
    }
};

// Operator overload for expressions
template <class LHS, class RHS>
inline ExprTimes<LHS, RHS> operator*(
    const Expression<LHS>& lhs, const Expression<RHS>& rhs)
{
    return ExprTimes<LHS, RHS>(lhs, rhs);
}
```

The base class is empty. We need it to build a CRTP hierarchy, so the overloaded operators catch all expressions and nothing else. The concrete *ExprTimes* expression represents the multiplication of two expressions. It is constructed from two expressions, storing them in accordance with their concrete type after a static cast in the constructor. The overloaded operator $*$ takes two expressions (all types of expressions but nothing other than expressions) and constructs the corresponding *ExprTimes* expression.

We must code expression types and operator overloads for all standard mathematical functions. We will not do this in this section, where the code only demonstrates and explains the expression template technology. Our actual AADET code does implement all standard functions, including the Gaussian density and cumulative distribution, as discussed in Chapter 12. For now, we only consider multiplication and logarithm:

```
// Log

template <class ARG>
class ExprLog : public Expression<ExprLog<ARG>>
{
    ARG arg;

```

```
public:

    //  Constructor
    explicit ExprLog(const Expression<ARG>& a)
        : arg(static_cast<const ARG&>(a)) {}

    double value() const
    {
        return log(arg.value());
    }
};

//  Operator overload for expressions
template <class ARG>
inline ExprLog<ARG> log(const Expression<ARG>& arg)
{
    return ExprLog<ARG>(arg);
}
```

Finally, we have our custom number type, which is also an expression, although a special one because it is a leaf in the DAG:

```
//  Number type, also an expression
class Number : public Expression<Number>
{
    double val;

public:

    //  Constructor
    explicit Number(const double v) : val(v) {}

    double value() const
    {
        return val;
    }
};
```

We can test our toy code and implement a lazy evaluation:

```
template <class T>
auto calculate(const T t1, const T t2)
{
    return t1 * log(t2);
}

int main()
{
    Number x1(2.0), x2(3.0);

    auto e = calculate(x1, x2);

    cout << e.value() << endl;  // 2.19722 = 2 * log(3)
}
```

We achieved the exact same result as in Chapter 9. The call to *calculate*() did not calculate anything. It built the DAG for the calculation. The code in *calculate*(), instantiated with the Number type, is:

```
auto calculate(const Number t1, const Number t2)
{
    return t1 * log(t2);
}
```

which is nothing else than syntactic sugar for:

```
ExprTimes<Number,ExprLog<Number>> calculate(
    const Number t1,
    const Number t2)
{
    return ExprTimes<Number,ExprLog<Number>>(t1, ExprLog<Number>(t2));
}
```

The reason is that *t*2 is a Number, hence an Expression. Therefore, the resolution of *log*(*t*2) is caught in the *log*() overload for expressions, producing *ExprLog* < *Number* > (*t*2), which is itself an Expression, as is t_1, so the multiplication is also caught in the overload for expressions, resulting in a type *ExprTimes* < *Number*, *ExprLog* < *Number* >>, constructed with the left-hand-side t_1 and the right-hand-side *ExprLog* < *Number* > (*t*2).

Provided the compiler correctly inlines the calls to the overloaded operators,[2] the code that is effectively compiled is:

```
1  int main()
2  {
3      Number x1(2.0), x2(3.0);
4
5      //  auto e = calculate(x1, x2);
6      ExprTimes<Number,ExprLog<Number>>
7          e  =  ExprTimes<Number,ExprLog<Number>>(
8                x1,
9                ExprLog<Number>(x2));
10
11     //  cout << e.value() << endl;
12     cout << 2.0 * log(3.0) << endl;
13 }
```

where it is apparent that line 5, which originally said "e = calculate(x1, x2)," does not calculate anything, but builds an *expression tree*, which is another name for the expression's DAG. Furthermore, as evident in the instantiated code above, and provided that the compiler performs the correct inlining, this DAG is built at compile time and stored on the stack. Finally, the compiler always applies operators, overloaded or not, in the

[2]Which Visual Studio 2017 always does; for avoidance of doubt, we set the project property C/C++ / Optimization / Inline Function Expansion to Any Suitable.

correct order, respecting conventional precedence, as well as parentheses, so the resulting DAG is automatically correct.

The (lazy) evaluation is conducted on line 11, from the DAG. The call to *value*() on the top node evaluates the DAG in postorder, as in Chapter 9. This evaluation happens at run time, but over a DAG prebuilt at compile time, and stored on the stack. No time is wasted constructing the DAG at run time, and evaluation on the stack is typically faster than on the heap. In addition, the compiler should inline and optimize the nested sequence of calls to *value*() over the expression tree, something the compiler could not do in Chapter 9, where the DAG was not available at compile time.

We applied expression templates to build an expression DAG, like in Chapter 9, but at compile time, on the stack, without allocation or resolution overhead. We effectively achieved the same result while cutting most of the overhead involved, hence the performance gain.

Traversing expressions

The DAG may have been built at compile time on the stack; it still does everything the run time DAG of Chapter 9 did. We already demonstrated lazy evaluation. We can also count the active numbers in an expression, at compile time, using enums and constexpr functions (functions evaluated at compile time):

```
//  class ExprTimes
    enum { numNumbers = LHS::numNumbers + RHS::numNumbers };

//  class Log
    enum { numNumbers = ARG::numNumbers };

//  class Number
    enum { numNumbers = 1 };

//  constexpr function evaluates at compile time
template <class E>
constexpr auto countNumbersIn(const Expression<E>&)
{
    return E::numNumbers;
}

int main()
{
    Number x1(2.0), x2(3.0);

    auto e = calculate(x1, x2);
```

```

    cout << e.value() << endl;  // 2.19722 = 2 * log(3)

    cout << countNumbersIn(e) << endl;     // 2
}
```

The active Numbers are counted at compile time in the exact same way that we computed a compile-time factorial earlier. This information gives us, at compile time, the number of arguments we need for the expression's node on tape.

We can also reverse engineer the original program from the DAG, like we did in Chapter 9:

```

//  class ExprTimes
    string writeProgram(
        //  On input, the number of nodes processed so far
        //  On return the total number of nodes processed on exit
        size_t& processed)
    {
        //  Process the left sub-DAG
        const string ls = lhs.writeProgram(processed);
        const size_t ln = processed - 1;

        //  Process the right sub_DAG
        const string rs = rhs.writeProgram(processed);
        const size_t rn = processed - 1;

        //  Process this node
        const string thisString = ls + rs +
            "y" + to_string(processed)
            + " = y" + to_string(ln) + " * y" + to_string(rn) + "\n";

        ++processed;

        return thisString;
    }

//  class ExprLog
    string writeProgram(
        //  On input, the number of nodes processed so far
        //  On return the total number of nodes processed on exit
        size_t& processed)
    {
        //  Process the arg sub-DAG
        const string s = arg.writeProgram(processed);
        const size_t n = processed - 1;

```

```
            //  Process this node
            const string thisString = s +
                "y" + to_string(processed)
                + " = log(y" + to_string(n) + ")\n";

            ++processed;

            return thisString;
        }

//  class Number
        string writeProgram(
            //  On input, the number of nodes processed so far
            //  On return the total number of nodes processed on exit
            size_t& processed)
        {
            const string thisString = "y" + to_string(processed)
                + " = " + to_string(val) + "\n";

            ++processed;

            return thisString;
        }


int main()
{
        Number x1(2.0), x2(3.0);

        auto e = calculate(x1, x2);

        cout << e.value() << endl;   // 2.19722 = 2 * log(3)

        cout << numNumbersIn(e) << endl;     // 2

        size_t processed = 0;
        cout << e.writeProgram(processed);
}
```

We get:

y0 = 2.000000
y1 = 3.000000
y2 = log(y1)
y3 = y0 * y2

Differentiating expressions in constant time

More importantly, we can apply adjoint propagation through the compile-time DAG to compute in constant time the derivatives of an expression to its active inputs. The mathematics are identical to classical adjoint propagation. Adjoints still accumulate in reverse order over the operations that constitute the expression. The difference is that this reverse adjoint propagation is conducted over the expression's DAG built at compile time, saving the cost of building and traversing a tape in dynamic memory. It is only the resulting expression derivatives that are stored on tape, so that the adjoints of the entire calculation, consisting of a sequence of expressions, may be accumulated over a shorter tape in reverse order at a later stage.

At this point, we must examine the expression's DAG in further detail. This DAG is actually different in nature to the DAGs of Chapter 9. Our DAG here is a *tree*, in the sense that every node has only one parent. When an expression uses the same number multiple times, as in:

```
Number x, y;
//  ...
Number z = x * x * y;
```

each instance (in the sense that there are two instances of x in the expression above) is a different node in the expression tree, although both point to the same node on tape. When we compute the expression's derivatives, we produce separate derivatives for the two xs. But on tape, the two child adjoint pointers refer to the same adjoint. Therefore, at propagation time, the two are summed up into the adjoint on x's node, so in the end the adjoint for x accumulates correctly.

The expression's DAG being a tree makes it easier to "push" adjoints top down, applying basic adjoint mathematics as in the previous chapters, from the top node, whose adjoint is one, to the leaves (Numbers), which store the adjoints propagated through the tree.

```
//  class ExprTimes
    //  Input: accumulated adjoint for this node or 1 if top node
    void pushAdjoint(const double adjoint)
    {
        lhs.pushAdjoint(adjoint * rhs.value());
        rhs.pushAdjoint(adjoint * lhs.value());
    }

//  class Log
    //  Input: accumulated adjoint for this node or 1 if top node
    void pushAdjoint(const double adjoint)
    {
        arg.pushAdjoint(adjoint / arg.value());
    }


//  class Number
class Number : public Expression<Number>
{
    double val;
    double adj;

```

```
public:

    // Constructor
    explicit Number(const double v) : val(v), adj(0.0) {}

    double value() const
    {
        return val;
    }

    double adjoint() const
    {
        return adj;
    }

//  ...

    void pushAdjoint(const double adjoint)
    {
        adj = adjoint;
    }

int main()
{
    Number x1(2.0), x2(3.0);

    auto e = calculate(x1, x2);

    cout << e.value() << endl;  // 2.19722 = 2 * log(3)

    cout << numNumbersIn(e) << endl;     // 2

    size_t processed = 0;
    cout << e.writeProgram(processed);

    e.pushAdjoint(1.0);
    cout << "x1 adjoint = " << x1.adjoint() << endl;     //  1.09861 = log(3)
    cout << "x2 adjoint = " << x2.adjoint() << endl;     //  0.666667 = 2/3
}
```

Note that *pushAdjoint*() is *overloaded* for different expression types. Contrarily to virtual overriding, overloading is resolved at compile time, without run-time overhead. This function accepts the node's adjoint as argument, implements the adjoint equation to compute the adjoints of the child nodes, and recursively pushes them down the child expression sub-trees by calling *pushAdjoint*() on the child expressions. The recursion starts on the top node of the expression tree, the result of the expression, with argument 1, and stops on Numbers, leaf nodes without children, which register their adjoint, given as argument, in their internal data member. The actual AADET code implements the same pattern, where Number leaves register adjoints on the expression's node on tape.

Flattening expressions

This tutorial demonstrated how expression templates work, how they are applied to produce compile-time expression trees, and how these trees are traversed at run time to compute the derivatives of an expression to its active numbers so we can construct the expression's node on tape.

We know how to efficiently compute the derivatives of an *expression*, but a *calculation* is a *sequence* of expressions. All expressions are still recorded on tape, and their adjoints are still back-propagated through the tape to compute the differentials of the whole calculation.

To work with expression trees is faster than working with the tape: DAGs are constructed at compile time, saving run-time overhead; computations are conducted in faster, static memory on the stack, and optimized by the compiler because they are known at compile time. In principle, it is best working as much as possible with expression trees and as little as possible with the tape. In practice, it is generally impossible to represent a whole calculation as a single expression. Variables of expression types cannot be overwritten, since different expressions are instances of different types. Expressions cannot grow through control flow like loops. In general, an expression is contained in one line of code. For instance, the example we used earlier (here instantiated with the Number type):

```
Number f(Number x[5])
{
    Number y1 = x[2] * (5.0 * x[0] + x[1]);
    Number y2 = log(y1);
    Number y = (y1 + x[3] * y2) * (y1 + y2);
    return y;
}
```

consists in three expressions. The three lines assign expressions of different types on the right-hand side, into Numbers on the left-hand side. From the moment an expression is assigned to a Number (which is itself a leaf expression), the expression is *flattened* into a single Number and no longer exists as a compound expression. This is when the expression and its derivatives are evaluated, and where the expression is recorded on tape, on a node that is assigned to the left-hand-side Number's *myNode* pointer. The flattening code will be discussed shortly, with the AADET library. It is implemented in the Number's assignment operator with an Expression argument:

```
//  On the Number class
    template <class E>
    Number& operator=
        (const Expression<E>& e)
    {
        myNode = evaluateFlattenAndRecord(e);
        return *this;
    }
```

AADET in general, and flattening in particular, are illustrated in the figure below:

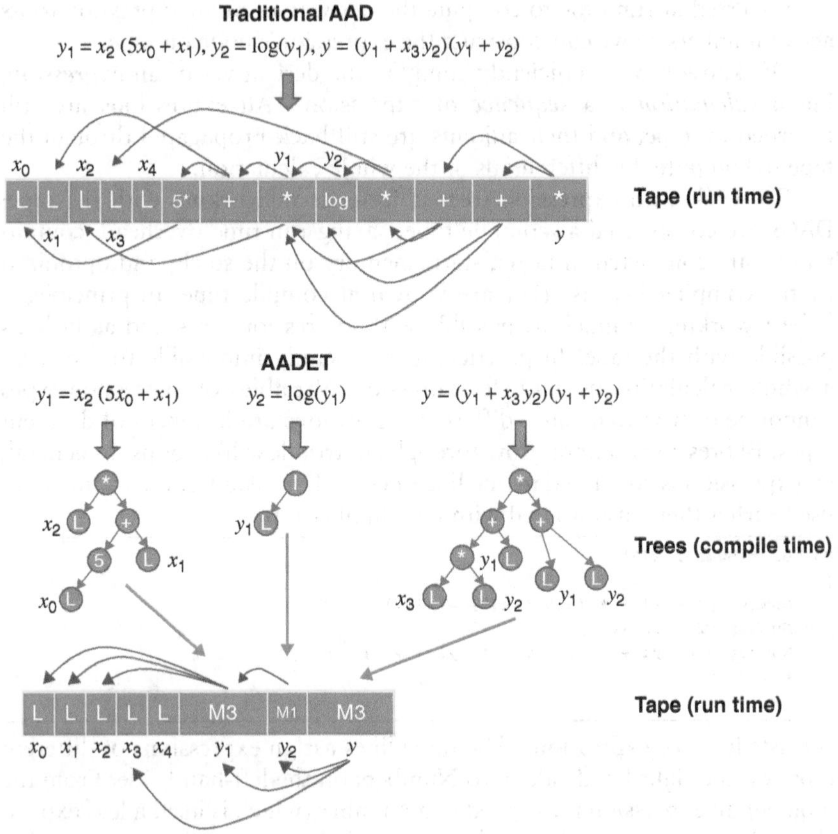

This graphical representation should clarify how AADET works and why it produces such a considerable acceleration. Expressions adjoints are back-propagated over trees that live on the stack and are constructed at compile time, so only their results live on tape. A large part of the adjoint propagation occurs on faster grounds. We want as much propagation as possible to be performed over expression trees and as little as possible on tape.

Hence, we want expressions to grow. In order to keep the expression growing as much as possible, the instrumented code should use *auto* as the result type for expressions, so the resulting variable is of type Expression, not Number, no flattening occurs, and the expression keeps growing. It is only on assignment *into a Number* that the expression is flattened and a multinomial node is created on tape. The same code rewritten with auto:

```
Number f(Number x[5])
{
    auto y1 = x[2] * (5.0 * x[0] + x[1]);
    auto y2 = log(y1);
    auto y = (y1 + x[3] * y2) * (y1 + y2);
    return y;
}
```

creates only one node on tape (with four arguments) and therefore differentiates faster. Flattening occurs on return when the expression y is assigned into a temporary Number returned by $f()$. Expression templates offer an additional optimization opportunity to instrumented code by systematically using *auto* as the type to hold expression results. Importantly, calls to functions with templated arguments and auto return type also result in expressions, so yet another version of $f()$:

```
auto f(Number x[5])
{
    auto y1 = x[2] * (5.0 * x[0] + x[1]);
    auto y2 = log(y1);
    auto y = (y1 + x[3] * y2) * (y1 + y2);
    return y;
}
```

is even more efficient because $f()$ returns an expression that may further grow outside of $f()$, in the caller code.

What we cannot do is traverse control flow without flattening expressions, because the overwritten variables would be of different expression types, as mentioned earlier. For instance, the following code would not compile:

```
Number f(Number x[5])
{
    auto y = x[2] * (5.0 * x[0] + x[1]);
    y = log(y1);                          //  error: different type
    y = (y1 + x[3] * y2) * (y1 + y2);     //  error: different type
    return y;
}
```

We cannot overwrite variables of expression types with different expression types. Every auto variable must be a new variable. In particular, variables overwritten in a loop must be of Number type, which means we cannot grow expressions through loops:

```
Number f(Number x[5])
{
    /*  Error

    auto s = x[0] + x[1];    //  type = +(Number, Number)
```

```

    for (size_t i = 2; i < 5; ++i)
    {
        s += x[i]; //  on 1st iter type = +((+Number,Number),Number) etc.
    }

    */


    //  Correct

    Number s = x[0] + x[1];     //  flattened into Number

    for (size_t i = 2; i < 5; ++i)
    {
        s += x[i]; //  flattened into Number on every iteration
    }

    return y;
}
```

The flattening code in *evaluateFlattenAndRecord*() (which is not actually called that) is given, along with the rest of the AADET code, in the next section.

15.3 EXPRESSION TEMPLATED AAD CODE

In this final section, we deliver and discuss the AADET code in AADExpr.h. It is hoped that the extended tutorial of the previous section helps readers read and understand this code. Template meta-programming in general, and expression templates in particular, are advanced C++ idioms, notoriously hard to understand or explain. It is our hope that we have given enough background to clarify what follows.

We start with a base expression class that implements CRTP for evaluation:

```
//  Base CRTP class so operators catch expressions
template <class E>
struct Expression
{
    double value() const
    {
        return static_cast<const E*>(this)->value();
    }

    explicit operator double() const
    {
        return value();
    }
};
```

Next, we have a single CRTP expression class to handle *all binary operations*:

```
//  Binary expression
//  LHS : the expression on the left
//  RHS : the expression on the right
//  OP : the binary function
template <class LHS, class RHS, class OP>
class BinaryExpression
    //  CRTP
    : public Expression<BinaryExpression<LHS, RHS, OP>>
{
    const double myValue;

    const LHS lhs;
    const RHS rhs;

public:

    //  Constructor out of 2 expressions
    //  Note : eager evaluation on construction
    explicit BinaryExpression(
        const Expression<LHS>& l,
        const Expression<RHS>& r)
        : myValue(OP::eval(l.value(), r.value())),
        lhs(static_cast<const LHS&>(l)),
        rhs(static_cast<const RHS&>(r))
    {}

    //  Value accessors
    double value() const { return myValue; }

    //  Expression template magic
    //  Expressions know
    //      AT COMPILE TIME
    //      the count of Numbers in their sub-DAG
    enum { numNumbers = LHS::numNumbers + RHS::numNumbers };

    //  Push adjoint down the DAG
    //  And write information into the node for the expression
    //  N : total number of numbers in the expression
    //  n : numbers already processed
    template <size_t N, size_t n>
    void pushAdjoint(
        //  Node for the complete expression being processed
        Node&        exprNode,
        //  Adjoint accumulated for this binary node
        const double    adjoint)
        const
    {
        //  Push on the left, if numbers there
        if (LHS::numNumbers > 0)
        {
            lhs.pushAdjoint<N, n>(
                exprNode,
                adjoint * OP::leftDerivative(
```

```
                                 lhs.value(),
                                 rhs.value(),
                                 value()));
        }

        //  Push on the right
        if (RHS::numNumbers > 0)
        {
            //  Note left push processed LHS::numNumbers numbers
            //  So the next number to be processed is n + LHS::numNumbers
            rhs.pushAdjoint<N, n + LHS::numNumbers>(
                exprNode,
                adjoint * OP::rightDerivative(
                                     lhs.value(),
                                     rhs.value(),
                                     value()));
        }
    }
};
```

This single class handles all binary operations, and delegates the evaluation of the value and derivatives of specific binary operators to its template parameter OP. This is an example of the so-called *policy design*, promoted by Andrei Alexandrescu in [100]. This is a very neat idiom that allows to minimize code duplication and gather the common parts to all binary expressions in one place. It is a modern counterpart to GOF's template pattern in [25].

The CRTP design was discussed earlier. The implementation of *pushAdjoint*() is more complicated than earlier, and deserves an explanation. It is templated on two numbers: N and n. N is the total count of active Numbers in the expression, and n is the count of these active Numbers processed so far. Hence, we start by pushing adjoints onto the left sub-tree with the same n, increase n by the number of active Numbers on the left, and process the right sub-tree.

The method takes as an input a reference to the current expression's node on the tape, where the resulting derivatives and pointers are stored when the traversal of the expression is complete. Concretely, the traversal stops on leaves, which are Numbers. The implementation of *pushAdjoint*() on the Number class, which has nothing to "push," having no children, stores the resulting derivatives on the expression's Node.

Next, we define the different OPs for different binary operations. Their only responsibility is to provide values and derivatives for specific binary operations:

```
//  "Concrete" binaries, we only need to define operations and derivatives
struct OPMult
{
    static const double eval(const double l, const double r)
    {
        return l * r;
    }
```

```

    static const double leftDerivative
        (const double l, const double r, const double v)
    {
        return r;
    }

    static const double rightDerivative
        (const double l, const double r, const double v)
    {
        return l;
    }
};

struct OPAdd
{
    static const double eval(const double l, const double r)
    {
        return l + r;
    }

    static const double leftDerivative
        (const double l, const double r, const double v)
    {
        return 1.0;
    }

    static const double rightDerivative
        (const double l, const double r, const double v)
    {
        return 1.0;
    }
};

//  ...
```

Subtraction, division, power, maximum, and minimum are defined identically. Next, we have the binary operator overloads, which build the binary expressions and are no different from our toy code earlier in this chapter:

```
//  Operator overloading for binary expressions
//  So DAG is built on the stack at compile time
//  And traversed at run time for evaluation and propagation

template <class LHS, class RHS>
 BinaryExpression<LHS, RHS, OPMult> operator*(
    const Expression<LHS>& lhs, const Expression<RHS>& rhs)
{
    return BinaryExpression<LHS, RHS, OPMult>(lhs, rhs);
}

template <class LHS, class RHS>
 BinaryExpression<LHS, RHS, OPAdd> operator+(
    const Expression<LHS>& lhs, const Expression<RHS>& rhs)
```

```
{
    return BinaryExpression<LHS, RHS, OPAdd>(lhs, rhs);
}

//  ...
```

We also have −, /, *pow*(), *max*(), and *min*().

Next, with the exact same design, we have the unary expressions:

```
//  Unary expressions : Same logic with one argument

//  The CRTP class
template <class ARG, class OP>
class UnaryExpression
    //  CRTP
    : public Expression<UnaryExpression<ARG, OP>>
{
    const double myValue;

    const ARG arg;
    //  For binary operators with a double on one side
    const double dArg = 0.0;

public:

    //  Constructor
    //  Note : eager evaluation on construction
    explicit UnaryExpression(
        const Expression<ARG>& a) :
            myValue(OP::eval(a.value(), 0.0)),
            arg(static_cast<const ARG&>(a))
        {}

    //  Constructor for binary expressions with a double on one side
    explicit UnaryExpression(
        const Expression<ARG>& a,
        const double b) :
            myValue(OP::eval(a.value(), b)),
            arg(static_cast<const ARG&>(a)), dArg(b)
        {}

    //  Value accessors
    double value() const { return myValue; }

    //  Expression template magic
    enum { numNumbers = ARG::numNumbers };

    //  Push adjoint down the expression DAG
    template <size_t N, size_t n>
    void pushAdjoint(
        //  Node for the complete expression being processed
        Node&           exprNode,
        const double    adjoint)        //  Adjoint accumulated on the node
        const
    {
        //  Push to argument, if numbers there
        if (ARG::numNumbers > 0)
```

```
        {
            arg.pushAdjoint<N, n>(
                exprNode,
                adjoint * OP::derivative(arg.value(), value(), dArg));
        }
    }
};

//  The unary operators

struct OPExp
{
    static const double eval(const double r, const double d)
    {
        return exp(r);
    }

    static const double derivative
        (const double r, const double v, const double d)
    {
        return v;
    }
};

struct OPLog
{
    static const double eval(const double r, const double d)
    {
        return log(r);
    }

    static const double derivative
        (const double r, const double v, const double d)
    {
        return 1.0 / r;
    }
};

//  OPSqrt and OPFabs are defined in the same manner

//  And overloading

template <class ARG>
 UnaryExpression<ARG, OPExp> exp(const Expression<ARG>& arg)
{
    return UnaryExpression<ARG, OPExp>(arg);
}

template <class ARG>
 UnaryExpression<ARG, OPLog> log(const Expression<ARG>& arg)
{
    return UnaryExpression<ARG, OPLog>(arg);
}

//  sqrt() and fabs() are overloaded in the same manner
```

We also implement the same optimization as in Chapter 10: we consider a binary expression with a constant (double) on one side as a unary expression: if d is a double (not a Number, so a "constant" as far as AAD is concerned), then xd or dx are not binary $*$ but a unary $(* d)$. Hence, the unary expression class stores that constant, and we have additional OPs and operator overloads for these expressions:

```
//  Binary operators with a double on one side

//  * double or double *
struct OPMultD
{
    static const double eval(const double r, const double d)
    {
        return r * d;
    }

    static const double derivative
    (const double r, const double v, const double d)
    {
        return d;
    }
};

//  double -
struct OPSubDL
{
    static const double eval(const double r, const double d)
    {
        return d - r;
    }

    static const double derivative
    (const double r, const double v, const double d)
    {
        return -1.0;
    }
};

//  - double
struct OPSubDR
{
    static const double eval(const double r, const double d)
    {
        return r - d;
    }

    static const double derivative
    (const double r, const double v, const double d)
    {
        return 1.0;
    }
};

//  We have (many) identical structs for all the binary operators

//  And overloading
```

```
//  Binary operators with a double on one side

template <class ARG>
 UnaryExpression<ARG, OPMultD> operator*(
     const double d, const Expression<ARG>& rhs)
{
     return UnaryExpression<ARG, OPMultD>(rhs, d);
}

template <class ARG>
 UnaryExpression<ARG, OPMultD> operator*(
     const Expression<ARG>& lhs, const double d)
{
     return UnaryExpression<ARG, OPMultD>(lhs, d);
}

template <class ARG>
 UnaryExpression<ARG, OPSubDL> operator-(
     const double d, const Expression<ARG>& rhs)
{
     return UnaryExpression<ARG, OPSubDL>(rhs, d);
}

template <class ARG>
 UnaryExpression<ARG, OPSubDR> operator-(
     const Expression<ARG>& lhs, const double d)
{
     return UnaryExpression<ARG, OPSubDR>(lhs, d);
}

//  We have (many) identical overloads for all the binary operators
```

It may happen that we want to include another function or operator to be considered as an elementary building block. For instance, the Normal density and cumulative distribution functions from gaussians.h should be considered as elementary building blocks, even though they are not standard C++ mathematical functions, as discussed in Chapter 10. We pointed out that it is a fair optimization to overload all those frequently called, low-level functions, and consider them as elementary building blocks, rather than instrument them and record their expressions on tape. We saw that in the case of the cumulative normal distribution, in particular, the results are both faster and more accurate. This is how we proceed:

1. Code the corresponding unary or binary operator, with the evaluation and differentiation methods. For the Gaussian functions, operators are unary:

```
struct OPNormalDens
{
    static const double eval(const double r, const double d)
    {
        return normalDens(r);
    }

```

```
        static const double derivative
        (const double r, const double v, const double d)
        {
            return - r * v;
        }
    };

    struct OPNormalCdf
    {
        static const double eval(const double r, const double d)
        {
            return normalCdf(r);
        }

        static const double derivative
        (const double r, const double v, const double d)
        {
            return normalDens(r);
        }
    };

```

2. Code the overload for the function, which builds a unary or binary expression with the correct operator:

```

    template <class ARG>
    UnaryExpression<ARG, OPNormalDens>
        normalDens(const Expression<ARG>& arg)
    {
        return UnaryExpression<ARG, OPNormalDens>(arg);
    }

    template <class ARG>
    UnaryExpression<ARG, OPNormalCdf>
        normalCdf(const Expression<ARG>& arg)
    {
        return UnaryExpression<ARG, OPNormalCdf>(arg);
    }

```

This is it. The two snippets of code above effectively made *normalDens*() and *normalCdf*() AADET building blocks. There is no need to template the definitions of the original functions (no harm, either; the expression overloads have precedence, being the more specific).

Next, we have comparison operators for expressions:

```
//  Comparison, as normal

template<class E, class F>
 bool operator==(const Expression<E>& lhs, const Expression<F>& rhs)
{
    return lhs.value() == rhs.value();
}
```

```
template<class E>
 bool operator==(const Expression<E>& lhs, const double& rhs)
{
    return lhs.value() == rhs;
}
template<class E>
 bool operator==(const double& lhs, const Expression<E>& rhs)
{
    return lhs == rhs.value();
}

template<class E, class F>
 bool operator!=(const Expression<E>& lhs, const Expression<F>& rhs)
{
    return lhs.value() != rhs.value();
}
template<class E>
 bool operator!=(const Expression<E>& lhs, const double& rhs)
{
    return lhs.value() != rhs;
}
template<class E>
 bool operator!=(const double& lhs, const Expression<E>& rhs)
{
    return lhs != rhs.value();
}

template<class E, class F>
 bool operator<(const Expression<E>& lhs, const Expression<F>& rhs)
{
    return lhs.value() < rhs.value();
}
template<class E>
 bool operator<(const Expression<E>& lhs, const double& rhs)
{
    return lhs.value() < rhs;
}
template<class E>
 bool operator<(const double& lhs, const Expression<E>& rhs)
{
    return lhs < rhs.value();
}

//  And all the combinations of <, >, <=, >=, ==, != with either
//      two Numbers
//      a double and a Number
//      a Number and a double
```

and the unary +/- operators:

```
template <class RHS>
UnaryExpression<RHS, OPSubDL> operator-
(const Expression<RHS>& rhs)
{
    return 0.0 - rhs;
}

```

```
template <class RHS>
Expression<RHS> operator+
(const Expression<RHS>& rhs)
{
    return rhs;
}
```

This completes all binary and unary operators. Finally, we have the leaf expression, the custom number type:

```
//  The Number type, also an expression

//  CRTP
class Number : public Expression<Number>
{
    //  The value and node for this number, as normal
    double myValue;
    Node* myNode;

    //  Node creation on tape, as normal

    template <size_t N>
    Node* createMultiNode()
    {
        return tape->recordNode<N>();
    }

//  ...
```

The private helper createMultiNode(), templated on the number of arguments to an expression, creates the expression's node on tape. Another private helper, fromExpr(), flattens an expression assigned into a Number, computing derivatives through its DAG and storing them on the expression's multinomial node:

```
//  ...

    //  This is where, on assignment or construction from an expression,
    //      derivatives are pushed through the expression's DAG
    //      into the node
    template<class E>
    void fromExpr(const Expression<E>& e)
    {
        //  Build node
        auto* node = createMultiNode<E::numNumbers>();

        //  Push adjoints through expression DAG from 1 on top
        static_cast<const E&>(e)
            .pushAdjoint<E::numNumbers, 0>(*node, 1.0);

        //  Set my node
        myNode = node;
    }
//  ...
```

We create the node on tape with a call to *createMultiNode*(), and populate it with a call to *pushAdjoint*() on the top node of the expression's tree. Note the CRTP style call to *PushAdjoint*. It is given the template parameters $E :: numNumbers$, the count of the active inputs to the expression, known at compile time as explained earlier, and also given as a template parameter to *createMultiNode*(), and the count 0 of inputs already processed. It is also given as arguments the address of the expression's node on tape, where derivatives are stored once computed, along with the address of the adjoints of the expression's arguments, and 1.0, the adjoint (in the expression tree, not on tape) of the top node of the expression. Finally, we register the node as this number's node on tape.

We have seen the implementation of *pushAdjoint*() on the binary and unary expressions, but it is really on the leaves, that is, on the Number class, that *pushAdjoint*() populates the multinomial tape node. The execution of *pushAdjoint*() on the Number class occurs on the bottom of the expression tree. In particular, the argument *adjoint* accumulated the differential of the expression to this *Number*, so it may be stored on the expression's node:

```
//  ...

public:

    //  Expression template magic
    enum { numNumbers = 1 };

    //  Push adjoint
    //  This is where the derivatives and pointers are set on the node
    //  Push adjoint down the expression DAG

    //  N: count Numbers in the expression
    //  n: index of this Number in the expression
    template <size_t N, size_t n>
    void pushAdjoint(
        //  Node for the complete expression
        Node&           exprNode,
        //  Adjoint of this Number in the expression
        const double    adjoint)
        const
    {
        //  Register child adjoint pointer on the expression's node
        exprNode.pAdjPtrs[n] = Tape::multi
            ? myNode->pAdjoints
            : &myNode->mAdjoint;

        //  Register derivative on the expression's node
        exprNode.pDerivatives[n] = adjoint;
    }

//  ...
```

Next, we have (thread local) static tape access and constructors similar to the code of Chapter 10:

```
//  ...

public:

    //  Static access to tape, as normal
    static thread_local Tape* tape;

    //  Constructors

    Number() {}

    explicit Number(const double val) : myValue(val)
    {
        //  Create leaf
        myNode = createMultiNode<0>();
    }

    Number& operator=(const double val)
    {
        myValue = val;
        myNode = createMultiNode<0>();
        return *this;
    }

    //  No need for copy and assignment
    //  Default ones do the right thing:
    //      copy value and pointer to node

//  ...
```

The new thing is the construction or assignment from an expression, which calls *fromExpr*() above to flatten the assigned expression and create and populate the multinomial node by traversal of the expression's tree, whenever an expression is assigned to a Number:

```
//  ...

    //  Construct or assign from expression

    template <class E>
    Number(const Expression<E>& e) : myValue(e.value())
    {
        fromExpr<E>(static_cast<const E&>(e));
    }

    template <class E>
    Number& operator=
        (const Expression<E>& e)
    {
        myValue = e.value();
        fromExpr<E>(static_cast<const E&>(e));
        return *this;
    }

//  ...
```

Next, we have the same explicit conversion operators, accessors, and propagators as in traditional AAD code:

```
//  ...

    //  Explicit coversion to double
    explicit operator double& () { return myValue; }
    explicit operator double () const { return myValue; }

    //  All the normal accessors and propagators, as in normal AAD code

    //  Put on tape
    void putOnTape()
    {
        myNode = createMultiNode<0>();
    }

    //  Accessors: value and adjoint

    double& value()
    {
        return myValue;
    }
    double value() const
    {
        return myValue;
    }
    double& adjoint()
    {
        return myNode->adjoint();
    }
    double adjoint() const
    {
        return myNode->adjoint();
    }
    double& adjoint(const size_t n)
    {
        return myNode->adjoint(n);
    }
    double adjoint(const size_t n) const
    {
        return myNode->adjoint(n);
    }

    //  Reset all adjoints on the tape
    //      note we don't use this method
    void resetAdjoints()
    {
        tape->resetAdjoints();
    }

    //  Propagation

    //  Propagate adjoints
    //      from and to both INCLUSIVE
    static void propagateAdjoints(
        Tape::iterator propagateFrom,
        Tape::iterator propagateTo)
```

```
{
    auto it = propagateFrom;
    while (it != propagateTo)
    {
        it->propagateOne();
        --it;
    }
    it->propagateOne();
}

//  Convenient overloads

//  Set the adjoint on this node to 1,
//  Then propagate from the node
void propagateAdjoints(
    //  We start on this number's node
    Tape::iterator propagateTo)
{
    //  Set this adjoint to 1
    adjoint() = 1.0;
    //  Find node on tape
    auto it = tape->find(myNode);
    //  Reverse and propagate until we hit the stop
    while (it != propagateTo)
    {
        it->propagateOne();
        --it;
    }
    it->propagateOne();
}

//  These 2 set the adjoint to 1 on this node
void propagateToStart()
{
    propagateAdjoints(tape->begin());
}
void propagateToMark()
{
    propagateAdjoints(tape->markIt());
}

//  This one only propagates
//  Note: propagation starts at mark - 1
static void propagateMarkToStart()
{
    propagateAdjoints(prev(tape->markIt()), tape->begin());
}

//  Multi-adjoint propagation

//  Propagate adjoints
//      from and to both INCLUSIVE
static void propagateAdjointsMulti(
    Tape::iterator propagateFrom,
    Tape::iterator propagateTo)
```

```
    {
        auto it = propagateFrom;
        while (it != propagateTo)
        {
            it->propagateAll();
            --it;
        }
        it->propagateAll();
    }

//  ...
```

and, finally, the unary on-class operators:

```
//  ...

    //  Unary operators

    template <class E>
    Number& operator+=(const Expression<E>& e)
    {
        *this = *this + e;
        return *this;
    }

    template <class E>
    Number& operator*=(const Expression<E>& e)
    {
        *this = *this * e;
        return *this;
    }

    template <class E>
    Number& operator-=(const Expression<E>& e)
    {
        *this = *this - e;
        return *this;
    }

    template <class E>
    Number& operator/=(const Expression<E>& e)
    {
        *this = *this / e;
        return *this;
    }
};
```

which completes the expression-templated AAD code.

As mentioned earlier, we get the exact same results as previously, two to three times faster. Since a small portion of the instrumented code is templated, we actually accelerated AAD by a substantial factor.

Finally, we reiterate that this AAD framework encourages further improvements, both in the AAD library and instrumented code, with potential further acceleration. A systematic application of the *auto* type in the instrumented code delays the flattening of expressions, resulting in a smaller number of larger expressions on tape and a faster overall processing, expressions being processed faster than nodes on tape. It is also worth overloading more functions as expression building blocks, like we did for the Gaussian functions. The AADET framework is meant to evolve with client code and be extended to improve the speed of the entire application.

Debugging AAD Instrumentation

The AAD libraries of Chapters 10 and 15 were extensively tested for a long time in a vast number of contexts. There shouldn't be many bugs left there. In case readers do find bugs when experimenting with the libraries, we shall be most grateful if they kindly notify Wiley so we can correct the code in the online repository. Again, and following the many hours spent tracking and removing bugs and inefficiencies, we should expect the AAD libraries to be mostly clean.

This being said, the purpose of this publication is to let readers instrument their own code and differentiate it in constant time with the AAD libraries. It should be apparent from Chapters 12, 13, and 14 that a correct instrumentation is a long, difficult work, in our experience, more demanding than the development of the AAD libraries themselves. A lot may go wrong. This appendix offers general advice for debugging instrumented code, and reveals the one single tactic that saved us months of tedious debugging work while professionally developing AAD-powered systems, almost immediately identifying most AAD bugs in instrumented code. This debugging tactic is one of the key ingredients that allowed us to implement AAD throughout Danske Bank's production systems in a reasonably short time in the early 2010s.

First of all, it should be clear that the general advice of Section 3.20 for debugging and profiling serial and parallel code holds with AAD, and readers are encouraged to follow the steps articulated there in case their instrumented code doesn't produce the correct results, crashes, or takes too long. With AAD, however, a number of additional steps should be taken to improve debugging conditions:

- Switch to traditional AAD, without expression templates. Recall from Chapter 15 that this is done by changing the definition of the macro *AADET* to *false* in AAD.h. It is virtually impossible to debug expression templates, whereas traditional AAD provides human-readable information in the debugger's window. Incorrect instrumented code should be incorrect with both versions, and corrected code should remain correct in both cases. Traditional AAD is slower, but easier to debug, and it doesn't hurt switching off AADET for debugging as long as it is reestablished after the instrumented code is fixed.

- Switch off selective instrumentation and template the whole calculation when debugging. Reestablish selective instrumentation step by step after debugging, testing for correctness after de-templating every piece of code. Similarly, remove all explicit conversions while debugging and reestablish them afterwards, one by one.
- Wipe the tape instead of rewinding it in between calculations. To rewind and reuse the tape by overwriting is an optimization, but it makes it more difficult to assess and debug instrumented code. It is best always to wipe and start anew in between calculations for debugging. The easiest way to achieve this is to make *Tape* : : *rewind*() different for debug and release:

```
    //  Tape class, AADTape.h
    void rewind()
    {

#ifdef _DEBUG

        //  In debug mode, clear instead

        clear();

#else

        //  Normal code in rewind()

        if (multi)
        {
            myAdjointsMulti.rewind();
        }
        myDers.rewind();
        myArgPtrs.rewind();
        myNodes.rewind();

#endif

    }
```

Visual Studio automatically defines the macro _DEBUG for code compiled in the debug configuration. Release code is optimized for performance while debug code should maximize the clarity and relevance of the debugging information. For this reason, it is sometimes worth compiling different code for debug or release. With Visual Studio, this is easy to achieve with simple macros, as demonstrated above. When code is compiled in the debug configuration, every call to *rewind*() clears the tape instead. When compiled in release mode, the normal code for rewinding the tape is executed.

- Debug check-pointing with particular care. Inspect the results of partial differentiations separately. This may help identify which piece of the algorithm is incorrectly differentiated. When all pieces are correctly differentiated, the flaw may be in the check-pointing logic itself.

- Evidently, run algorithms over small data sets for debugging. AAD takes a long time in debug mode. Run no more than 100 simulations over big steps for Monte-Carlo simulations, or equivalently small data sets for other applications.

AAD is particularly hard to debug because errors typically manifest on adjoint propagation, whereas the cause always occurs at evaluation time, when operations are incorrectly recorded on tape. At back-propagation time when errors are revealed, however, all that remains is the flawed record on tape, and it is virtually impossible to identify the responsible operation in the code. For this reason, it may take hours and days, and a lot of guessing and logging to find the incorrect pieces of code that caused the flawed records. If only we could catch them at recording time.

In our many years of instrumenting AAD in professional software and academic presentations, the primary cause of bugs specific to AAD, in fact, almost every single AAD bug we ever encountered, was caused by operations with *arguments missing from the tape*. A line of code like:

```
Number z = x + y;
```

is executed, while x or y is not on tape. There are many reasons why it may be so. Maybe these variables are initial inputs, not the result of prior operations, and we forgot to put them on tape as discussed page 381? Maybe they were on tape, but the tape was wiped and they were not put back after that? Or maybe, in a multi-threaded environment, they are simply not on the right tape? Whatever the reason, it is very often the case that some arguments of a recorded operation are missing from the tape. This creates dangling pointers, in the record, to nonexistent argument adjoints, and invariably causes crashes or wrong results on back-propagation.

Once we realize that most AAD bugs are caused by arguments not on tape, it is very easy to catch those *at evaluation time*. Remember from Chapter 10 that the recording of every operation in the overloaded operators and functions calls the private method *node*() on the active arguments to acquire a pointer to their node on tape. It follows that we can insert code in this method, compiled only in a debug configuration, that checks that the argument node is on tape, at the time where the operation is recorded:

```
    // Number class, AADNumber.h
    Node& node() const
    {

#ifdef _DEBUG

```

```
7        //  Check that node is on tape,
8        //       only in debug
9
10       //  Find node on tape
11       auto it = tape->find(myNode);
12
13       //  Not found
14       if (it == tape->end())
15       {
16           throw runtime_error("Put a breakpoint here");
17       }
18
19   #endif
20
21       //  Normal code, executed in debug or release
22
23       return const_cast<Node&>(*myNode);
24
25   }
```

Run the code in debug, putting a breakpoint on line 16 . Execution will stop on the breakpoint whenever an operation is recorded with an argument not on tape, revealing the context of the operation in the debugger, which makes it easy to identify and fix the flaw.

For example, we put a bug in our Black and Scholes code of Chapter 6:

```
//  BlackScholes class, mcMdlBs.h
void setParamPointers()
{
    myParameters[0] = &mySpot;
    myParameters[1] = &myVol;
    myParameters[2] = &myRate;
    myParameters[3] = &myRate;
}
```

Can you see the bug? We wrongly set the pointer *myParameters*[3] to the address of *myRate*. It should be the dividend *myDiv*. As a result, the differential to the dividend is always 0 but what is returned to client code, under the wrong label "div," is the sensitivity to the interest rate. We have an incorrect differential, and it may be hard to identify the cause. There is a lot of code to go through, in the model, product, template algorithm, etc., and such typos are easily skipped. Running in debug mode with a breakpoint on line 16 in our *Number* : : *node*() code, execution stops immediately and the call stack points to a subtraction in *BlackScholes* : : *init*():

```
void init(
    const vector<Time>&         productTimeline,
    const vector<SampleDef>&    defline)
        override
{
    //  Pre-compute the standard devs and drifts over simulation timeline
            const T mu = myRate - myDiv;

    //  ...
```

which immediately tells us that *mu* does not record correctly, because either *myRate* or *myDiv* is not on tape. We know that parameters are put on tape altogether, so there must be a flaw in the registration of the parameters. It follows that we know to look into

$$BlackScholes :: setParamPointers()$$

at which point we find the bug immediately.

This is a trivial example, of course, where the bug could be identified by testing and reasoning without actual debugging. But the same tactic helped us identify a vast number of subtle hidden bugs in sophisticated code that would have been long and hard to find otherwise.

Conclusion

After an introduction to parallel computing, and the construction of a parallel simulation library, we explored in detail the theory and implementation of AAD. As a result, we could compute accurate market and model risk sensitivities over parallel simulations in fractions of a second.

We don't believe that such results could be achieved with a free or commercial AAD library applied as a black box over instrumented code. This kind of speed is only obtained with an intimate understanding of AAD and its implementation, combined with a deep expertise in the differentiated algorithms. Selective instrumentation, a clever path-wise back-propagation, and a systematic application of check-pointing are the key ingredients, along with an efficient AAD library, boosted with expression templates, and an efficient instrumented simulation code. That is why we covered both AAD and financial simulations in this publication. We also provided computationally efficient C++ code, one that is not meant to be used as a black box, but instead provides a flexible interface for expert users to accommodate many types of designs and applications. (Readers who made it to the conclusion are, by definition, expert users.)

While AAD over parallel simulations is central to effective, modern derivatives systems, it is not by itself sufficient for the efficient calculation and differentiation of complex derivatives books and xVA adjustments.

Other key technologies are at play there. In particular, we covered in detail the mathematical representation of financial products, including xVA, and their cash-flows, in Chapter 4, but we did not implement it in code, delivering instead a few sketchy examples for the production of numerical results. The representation and manipulation of transactions, down to cash-flows, in a consistent, modular, efficient, transparent, and flexible form, constitutes another keystone of modern computational finance, and the subject of a dedicated publication [11], co-authored by Jesper Andreasen.

Other key notions that we discussed but did not develop in code include regression proxies and model hierarchies. These are covered in a third (upcoming) volume, co-authored by Jesper Andreasen and Brian Huge. The third volume will also cover in detail the application of all those technologies to regulatory calculations like xVA, and the specific algorithms to calculate and differentiate of xVA *on an iPad mini.*

References

[1] P. Wolfe. Checking the calculation of gradients. *ACM TOMS*, 6(4):337–343, 1982.

[2] W. Baur and V. Strassen. The complexity of partial derivatives. *Theoretical Computer Science*, 22:317–300, 1983.

[3] A. Griewank. Who invented the reverse mode of differentiation? *Documenta Mathematica*, Extra Volume ISMP:389–400, 2012.

[4] N. Trefethen. Who invented the greatest numerical algorithms? www.comlab.ox.ac.uk/nick.trefethen, 2015.

[5] M. Giles and P. Glasserman. Smoking adjoints: Fast evaluation of greeks in Monte Carlo calculations. *Risk*, 2006.

[6] L. Andersen and V. Piterbarg. *Interest Rate Modeling, Volumes 1, 2, and 3.* Atlantic Financial Press, 2010.

[7] J. Andreasen. CVA on an iPad mini, part 2: The beast. Aarhus Kwant Factory PhD Course, 2014.

[8] J. Andreasen. Back to the future. *Risk*, 2005.

[9] J. Guyon and P. Henry-Labordere. The smile calibration problem solved. Available at SSRN: https://ssrn.com/abstract=1885032 or http://dx.doi.org/10.2139/ssrn.1885032, 2011.

[10] J. Andreasen. CVA on an iPad mini. *Global Derivatives*, 2014, Presentation.

[11] J. Andreasen and A. Savine. *Modern Computational Finance: Scripting for Derivatives and xVA*. Wiley Finance, 2018.

[12] B. Dupire. Pricing with a smile. *Risk*, 7(1):18–20, 1994.

[13] J. F. Carriere. Valuation of the early-exercise price for options using simulations and nonparametric regression. *Insurance: Mathematics and Economics*, 19(1):19–30, 1996.

[14] F. A. Longstaff and E. S. Schwartz. Valuing American options by simulation: A simple least-squares approach. *The Review of Financial Studies*, 14(1):113–147, 2001.

[15] A. Williams. *C++ Concurrency in Action*. Manning, 2012.

[16] G. Blacher and R. Smith. Leveraging GPU technology for the risk management of interest rates derivatives. *Global Derivatives*, 2015.

[17] N. Sherif. AAD vs GPUS: Banks turn to maths trick as chips lose appeal. *Risk Magazine*, 2015.

[18] S. Meyers. *Effective STL: 50 Specific Ways to Improve Your Use of the Standard Template Library*. Addison-Wesley Professional, 2001.

[19] R. Geva. The road to a "100× speed-up": Challenges and fiction in parallel computing. *Global Derivatives*, 2017.

[20] W. H. Press, S. A. Teukolsky, W. T. Vetterling, and B. P. Flannery. *Numerical Recipes: The Art of Scientific Computing*. Cambridge University Press, 2007.

[21] I. Cukic. *Functional Programming in C++*. Manning Publications, 2017.

[22] F. Black and M. Scholes. The pricing of options and corporate liabilities. *Journal of Political Economy*, 81(3):637–654, 1973.

[23] G. Hutton. *Programming in Haskell*. Cambridge University Press, 2016.

[24] S. Meyers. *Effective Modern C++: 42 Specific Ways to Improve Your Use of C++11 and C++14*. O'Reilly Media, 2014.

[25] E. Gamma, R. Helm, R. Johnson, and J. Vlissides. *Design Patterns: Elements of Reusable Object-Oriented Software*. Addison-Wesley Longman Publishing, 1995.

[26] J. C. Hull. *Options, Futures, and Other Derivatives*. Pearson, 2017.

[27] T. Bjork. *Arbitrage Theory in Continuous Time*. Oxford University Press, 2009.

[28] P. Carr and D. B. Madan. Option valuation using the fast Fourier transform. *Journal of Computational Finance*, 2(4):61–73, 1999.

[29] A. Brace, D. Gatarek, and M. Musiela. The market model of interest rate dynamics. *Mathematical Finance*, 7(2):127–154, 1997.

[30] J. Andreasen. CVA on an iPad mini, part 3: XVA algorithms. Aarhus Kwant Factory PhD Course, 2014.

[31] B. N. Huge and A. Savine. LSM reloaded: Differentiate XVA on your iPad mini. *SSRN*, 2017.

[32] J. M. Harrisson and S. R. Pliska. Martingales and stochastic integrals in the theory of continuous trading. *Stochastic Processes and Their Applications*, pp. 215–260, 1981.

[33] J. M. Harrisson and D. M. Kreps. Martingale and arbitrage in multiperiod securities markets. *Journal of Economic Theory*, pp. 381–408, 1979.

[34] B. Dupire. Functional to calculus. Portfolio Research Paper, Bloomberg, 2009.

[35] H. Geman, N. El Karoui, and J. C. Rochet. Changes of numeraire, changes of probability measure and option pricing. *Journal of Applied Probability*, 32, 1995.

[36] P. S. Hagan, D. Kumar, A. S. Lesniewski, and D. E. Woodward. Managing smile risk. *Wilmott Magazine*, 1:84–108, 2002.

[37] L. Bachelier. Theorie de la speculation. *Annales Scientifiques de l'Ecole Normale Superieure*, 3(17):21–86, 1900.

[38] G. Amblard and J. Lebuchoux. Models for CMS caps. *Risk*, 2000.

[39] A. Savine. A brief history of discounting. Lecture Notes.

[40] J. Gatheral. A parsimonious arbitrage-free implied volatility parameterization with application to the valuation of volatility derivatives. *Global Derivatives*, 2004.

[41] J. Andreasen and B. Huge. Volatility interpolation. *Risk*, 2011.

[42] S. L. Heston. A closed-form solution for options with stochastic volatility with applications to bond and currency options. *The Review of Financial Studies*, 6(2):327–343, 1993.

[43] J. Gatheral. *The Volatility Surface: A Practitioner's Guide*. Wiley Finance, 2006.

[44] A. Savine. A theory of volatility. *Proceedings of the International Conference on Mathematical Finance*, pp. 151–167, 2001. Recent Developments in Mathematical Finance, Shanghai, China, 10–13 May 2001; also available on SSRN.

[45] B. Dupire. Arbitrage pricing with stochastic volatility. Working Paper, 1992.

[46] A. Savine. Volatility. Lecture Notes from University of Copenhagen. tinyUrl.com/ant1savinePub, password: 54v1n3 folder: Vol.

[47] P. Carr, K. Ellis, and V. Gupta. Static hedging of exotic options. *Journal of Finance*, 53(3):1165–1190, 1998.

[48] S. Nielsen, M. Jonsson, and R. Poulsen. The fundamental theorem of derivative trading: Exposition, extensions and experiments. *Quantitative Finance*, 2017.

[49] D. Heath, R. Jarrow, and A. Morton. Bond pricing and the term structure of interest rates: A new methodology for contingent claim valuation. *Ecnonometrica*, 60(1):77–105, 1992.

[50] B. Dupire. A unified theory of volatility. Working Paper, 1996.

[51] L. Bergomi. *Stochastic Volatility Modeling*. Chapman & Hall, 2016.

[52] J. Hull and A. White. The pricing of options on assets with stochastic volatilities. *Journal of Finance*, 42:281–300, 1987.

[53] J. Andreasen and B. N. Huge. Zabr: Expansions for the masses. Available at SSRN: https://ssrn.com/abstract=1980726 or http://dx.doi.org/10.2139/ssrn.1980726, 2011.

[54] A. Lipton. *Mathematical Methods for Foreign Exchange*. World Scientific, 2001.

[55] O. Cheyette. Markov representation of the Heath-Jarrow-Morton model. BARRA, 1992.

[56] O. Vasicek. An equilibrium characterization of the term structure. *Journal of Financial Economics*, 5(2):177–188, 1977.

[57] J. Hull and A. White. One-factor interest-rate models and the valuation of interest-rate derivative securities. *Journal of Financial and Quantitative Analysis*, 28(2):235–254, 1993.

[58] N. El Karoui and J. C. Rochet. A pricing formula for options on coupon bonds. Working Paper, 1989.

[59] P. Ritchken and L. Sankarasubramanian. Volatility structures of forward rates and the dynamics of the term structure. *Mathematical Finance*, 5:55–72, 1995.

[60] J. Andreasen. Finite difference methods. *Lecture Notes from Universities of Copenhagen and Aarhus*.

[61] A. Lipton. The vol smile problem. *Risk*, 2002.

[62] P. Glasserman. *Monte Carlo Methods in Financial Engineering*. Springer, 2003.

[63] P. Jaeckel. *Monte Carlo Methods in Finance*. Wiley Finance, 2002.

[64] M. Broadie, P. Glasserman, and S. Kou. A continuity correction for discrete barrier options. *Mathematical Finance*, 7(4):325–348, 1997.

[65] C. Bishop. *Pattern Recognition and Machine Learning*. Springer Verlag, 2006.

[66] B. Moro. The full Monte. *Risk*, 1995.

[67] L. Andersen and J. Andreasen. Volatility skews and extensions of the Libor market model. *Applied Mathematical Finance*, 7:1–32, 2000.

[68] J. Andreasen and B. Huge. Random grids. *Risk*, 2011.
[69] B. Dupire and A. Savine. *Dimension Reduction and Other Ways of Speeding Monte Carlo Simulation. Risk Publications*, 1998.
[70] P. L'Ecuyer. Combined multiple recursive number generators. *Operations Research*, 1996.
[71] I. M. Sobol. Sobol's original publication. *USSR Computational Mathematics and Mathematical Physics*, 7(4):86–112, 1967.
[72] S. Joe and F. Y. Kuo. Remark on algorithm 659: Implementing Sobol's quasirandom sequence generator. *ACM Transactions in Mathematical Software*, 29:49–57, 2003. Also available on web.maths.unsw.edu.au/~fkuo/sobol.
[73] S. Joe and F. Y. Kuo. Constructing Sobol sequences with better two-dimensional projections. *SIAM Journal of Scientific Computing*, 30:2635–2654, 2008. Also available on web.maths.unsw.edu.au/~fkuo/sobol.
[74] M. S. Joshi. *C++ Design Patterns and Derivatives Pricing*. Cambridge University Press, 2008.
[75] E. Schlogl. *Quantitative Finance: An Object-Oriented Approach in C++*. Chapman & Hall, 2013.
[76] D. Vandevoorde, N. M. Josuttis, and D. Gregor. *C++ Templates: The Complete Guide, 2nd Edition*. Addison-Wesley, 2017.
[77] A. Savine. Stabilize risks of discontinuous payoffs with fuzzy logic. *Global Derivatives*, 2016.
[78] L. Bergomi. Theta (and other greeks): Smooth barriers. *QuantMinds*, Lisbon, 2018.
[79] S. Dalton. *Financial Applications Using Excel Add-in Development in C / C++*. Wiley Finance, 2007.
[80] J. Andreasen. CVA on an iPad mini, part 1: Intro. Aarhus Kwant Factory PhD Course, 2014.
[81] J. Andreasen. CVA on an iPad mini, part 4: Cutting the IT edge. Aarhus Kwant Factory PhD Course, 2014.
[82] J. Andreasen. Tricks and tactics for FRTB. *Global Derivatives*, 2018.
[83] R. E. Wengert. A simple automatic derivative evaluation program. *Communications of the ACM*, 7(8):463–464, 1964.
[84] L. Capriotti. No more bumping: The promises and challenges of adjoint algorithmic differentiation. Invited talk at the Seventh Euro Algorithmic Differentiation Workshop, Oxford University, 2008.
[85] L. Capriotti. Fast greeks by algorithmic differentiation. *The Journal of Computational Finance*, 14(3), 2011.
[86] L. Capriotti and M. Giles. Fast correlation greeks by adjoint algorithmic differentiation. *Risk*, 2010.
[87] L. Capriotti and M. Giles. Adjoint greeks made easy. *Risk*, 2012.
[88] U. Naumann. *The Art of Differentiating Computer Programs: An Introduction to Algorithmic Differentiation. SIAM*, 2012.
[89] R. J. Hogan. Fast reverse-mode automatic differentiation using expression templates in C++. *ACM Transactions in Mathematical Software*, 40(4), 2014.
[90] H.-J. Flyger, B. Huge, and A. Savine. Practical implementation of AAD for derivatives risk management, XVA and RWA. *Global Derivatives*, 2015.

[91] R. J. Barnes. Matrix differentiation. https://atmos.washington.edu/~dennis/MatrixCalculus.pdf.

[92] M. Giles. An extended collection of matrix derivative results for forward and reverse mode algorithmic differentiation. An extended version of a paper that appeared in the proceedings of AD2008, the 5th International Conference on Automatic Differentiation, 2008.

[93] B. Huge. AD of matrix calculations: Regression and Cholesky. Danske Bank, Working Paper.

[94] L. Capriotti, Y. Jiang, and A. Macrina. Real-time risk management: An AAD-PDE approach. Available at SSRN: https://ssrn.com/abstract=2630328 or http://dx.doi.org/10.2139/ssrn.2630328, 2015.

[95] O. Scavenius. Rate surfaces calibration and risk using auto-differentiation. WBS Interest Rate Conference, London, 2012.

[96] A. Savine. Automatic differentiation for financial derivatives. *Global Derivatives*, 2014.

[97] A. Reghai, O. Kettani, and M. Messaoud. CVA with greeks and AAD. *Risk*, 2015.

[98] R. Merton. Option pricing when underlying stock returns are discontinuous. *Journal of Financial Economics*, pp. 125–144, 1976.

[99] J. Andreasen and B. Huge. Wots me delta? *Danske Markets*, 2014.

[100] A. Alexandrescu. *Modern C++ Design: Generic Programming and Design Patterns Applied*. Addison-Wesley, 2001.

Index